Vibrations: Analysis and Control Mechanism

Vibrations: Analysis and Control Mechanism

Edited by **Rene Sava**

Clanrye International

New Jersey

Published by Clanrye International,
55 Van Reypen Street,
Jersey City, NJ 07306, USA
www.clanryeinternational.com

Vibrations: Analysis and Control Mechanism
Edited by Rene Sava

International Standard Book Number: 978-1-63240-515-9 (Hardback)

Printed in the United States of America.

Contents

Preface

This book has been a concerted effort by a group of academicians, researchers and scientists, who have contributed their research works for the realization of the book. This book has materialized in the wake of emerging advancements and innovations in this field. Therefore, the need of the hour was to compile all the required researches and disseminate the knowledge to a broad spectrum of people comprising of students, researchers and specialists of the field.

The analysis and control mechanism of vibrations are elucidated in this extensive book. Vibration is a phenomenon that we can notice in various systems since its impact can vary vastly from the personal discomfort generated from the unevenness of a road to the downfall of a building or a bridge at the time of earthquake. This book is an abridgement of research works on vibration examination and control. It reflects on new techniques that help us understand and mitigate this phenomenon. This book deals with novel developments on vibration analysis and comprises of a series of case studies that illustrate new approaches on vibration control. The applications differ and include areas like vehicle suspension systems, wind turbines and civil engineering structures.

At the end of the preface, I would like to thank the authors for their brilliant chapters and the publisher for guiding us all-through the making of the book till its final stage. Also, I would like to thank my family for providing the support and encouragement throughout my academic career and research projects.

Editor

Section 1

New Theoretical Developments on Vibration Analysis and Control

Chapter 1

Active Vibration Control Using a Kautz Filter

Samuel da Silva, Vicente Lopes Junior and Michael J. Brennan

Additional information is available at the end of the chapter

1. Introduction

Impulse response functions (IRFs) have been largely used in experimental modal analysis in order to extract the modal parameters (natural frequencies, damping factors and modal forms) in different areas. IRFs occupy a prominent place in applications of aeronautical, machinery and automobile industries, mainly when the system has coupled modes. Additionally, IRFs have practical advantages for use in control theory for many reasons, e.g.:

- For very complex systems, they can be determined by experimental tests, or using data of input and output measured by load cells or accelerometers, or directly with an impact hammer.
- The identified model is essentially nonparametric.
- Normally a finite impulse response (FIR) model of the structure is employed. Thus, the stability can be warranted a priori. Additionally, the many adaptive controlers are based on an FIR structure and it is easy to perform a recursive estimation.

In general, IRFs can be identified by impact tests with an instrumented hammer or by using numerical algorithms implemented in commercial software. IRFs can be determined with those algorithms through different methods, e. g., the covariance method based on the sum of convolutions of the measured input forces. However, there is an over parametrization that is a drawback when the lag memory is high. Fortunately, an expansion of the IRFs into orthonormal basis functions can enhance the procedure of reducing the number of parameters [15]. For describing mechanical vibrating systems, Kautz filters are interesting orthogonal functions set in Hilbert space [21] that include a priori knowledge about the dominant poles. The eigenvalues associated to vibrating mechanical systems are conjugated complex poles, so, the IRFs can be expanded in orthonormal basis functions with those conditions. Kautz filters are orthogonal funcions that can be used for this purpose. These filters can decrease the computational cost and accelerate the convergence rate providing a good estimate of the IRFs [14].

Kautz filters have found several applications, e.g., acoustic and audio [20], circuit theory [17], experimental modal analysis in mechanical systems [2, 12–14], vibration control [6], model reduction [4], robust control [18], predictive control [19], general system identification [5, 16, 22, 25], non-linear system identification with Volterra models [7–11], etc. Although it may seem that the mathematical and theoretical aspects of Kautz filters are more interesting for academic purposed, some practical applications can be found in the literature. For example, the flight testing certification of aircrafts for aeroelastic stability was completely charecterized through a series connection of Kautz filters in [1]. The application used a simulated nonlinear prototypical two-dimensional wing section and F/A-18 active aeroelastic wing ground vibration test data.

In specific control applications with Kautz filters, the strategies are, normally, based on active noise control using feedforward compensation, e. g. as performed in [26]. It is well-known that Wiener theory can be used to describe internal model control to change the control architecture from feedforward to feedback [3]. However, feedback compensation can also be directly implemented. Thus, the goal of the present chapter is to apply Kautz filters for active vibration control. The main steps and characteristics involved in this procedure are described. Specifically, this chapter emphasises the following:

- Feedback control, considering dynamic canceling.
- It is not necessary to have a complete mathematical model and the controller is designed directly in the digital domain for fast practical implementation.
- The control method is based on experimental IRFs (nonparametric) and in orthonormal basis functions. Thus, the method is grey-box because prior knowledge of the mechanical vibrating system treated is assumed (poles of Kautz filter to represent the system). Additionally, complex vibration system can be controlled.
- An example of a single-degree of freedom mechanical model is used to illustrate the main steps.
- Additionally, an experimental example by using a clamped beam with PZT actuator and PVDF sensor is presented.

The chapter is organized as follows. First, the IRF identification and covariance method is reviewed briefly, followed by the Kautz filter with multiple poles for expansion of impulse response. After, a vibration control strategy is described and example applications involving single-input-single-output vibrating systems are used to illustrate the approach. Finally, the results are discussed and suggestions for a non-linear identification procedure are proposed.

2. Impulse response function

The output $\tilde{y}(k)$ of a linear discrete-time and invariant system can be written as:

$$\tilde{y}(k) = \sum_{i=0}^{\infty} h(i)u(k-i) \tag{1}$$

where the sequences $\{u(k), k = 0, 1, \ldots, T_f\}$ and $\{\tilde{y}(k), k = 0, 1 \ldots, T_f\}$ are the sampled input and output signals, respectively; T_f is the final time, and $h(k)$ is the impulse response function

(IRF). The measured output signal is given by $y(k) = \tilde{y}(k) + w(k)$, where $w(k)$ is a white or colored noise. The eq. (1) represents a sum of convolution between the input signal $u(k)$ and the IRF $h(k)$. In mechanical and vibrating systems applications, the IRF can be obtained by impact tests with a hammer or by using numerical algorithms based on time or frequency measured signals.

Normally, to obtain the IRF, eq. (1), is truncated in N terms by considering $|h(k)| < \epsilon, \forall j > N$, where ϵ is a residue. In this case, eq. (1) can be given as:

$$y(k) \approx \sum_{i=0}^{N} h(i)u(k-i) \tag{2}$$

The approach in eq. (2) changes an infinite impulse response model (IIR) into a finite impulse response model (FIR). The most common method to identify the $h(k)$ is by using the correlation functions due to the robustness to noise issues yielding to the classical Wiener-Hopf equation:

$$R_{uy}(k) \approx \sum_{i=0}^{N} h(i)R_{uu}(k-i) \tag{3}$$

where the correlation function $R_{uu}(k)$ and cross-correlation function $R_{uy}(k)$ can be estimated experimentally. Based on eq. (3), a least-square (LS) identification method can be performed to estimate the expansion coefficients in the time-series that describes the FIR model $h(k)$. This approach for estimating an IRF has some advantages over other estimators, for instance:

- the stability of the identified model is guaranteed a priori, since the model is FIR.
- the model is assumed to be described only for arbitrary zeros and poles at the origin of the complex plane.
- the model is linear in the parameters, hence the LS approach can be performed.

However, this identification technique often leads to conservative results because a common vibration system is hardly ever represented by a FIR model. Thus, the practical drawback is that a large number of parameters $h(k)$ must be considered in order to obtain a good approach in eq. (3). In order to overcome this drawback, a set of orthonormal basis functions can be employed to expand the covariance method and reduces the number of parameters. Next section provides some considerations in this sense.

3. Covariance method expanded in orthonormal basis functions

The IRF $h(k)$ can alternatively be written using α_j, $j = 0, 1, \ldots, J$, as expansion coefficients described by z-function $\Psi_j(z)$:

$$h(k) = \sum_{j=0}^{J} \alpha_j \psi_j(k), \qquad k = 0, 1, \ldots, N \tag{4}$$

where $\psi_j(k)$ is the IRF of the transfer function $\Psi_j(z)$. The z transform of eq. (4) is given by a linear combination of the functions $\Psi_j(z)$:

$$H(z) \approx \alpha_0\Psi_0(z) + \alpha_1\Psi_1(z) + \cdots + \alpha_J\Psi_J(z) = \sum_{j=0}^{J} \alpha_j\Psi_j(z) \tag{5}$$

The convergence of $\Psi_j(z)$ is related to the completeness properties of these subsets of functions. If the functions $\Psi_j(z)$ are properly chosen (poles placement), the order $J << N$. Thus, it is easier to identify the coefficient α_j using eq. (4) [2, 12, 14, 15, 22, 24], which can be written in a matrix form:

$$\begin{Bmatrix} h(0) \\ h(1) \\ \vdots \\ h(N) \end{Bmatrix} = \begin{bmatrix} \psi_0(0) & \psi_1(0) & \cdots & \psi_J(0) \\ \psi_0(1) & \psi_1(1) & \cdots & \psi_J(1) \\ \vdots & \vdots & \ddots & \vdots \\ \psi_0(N) & \psi_1(N) & \cdots & \psi_J(N) \end{bmatrix} \begin{Bmatrix} \alpha_0 \\ \alpha_1 \\ \vdots \\ \alpha_J \end{Bmatrix} \tag{6}$$

By incorporating the eq. (4) into Wiener-Hopf equation, eq. (3), one can obtain:

$$\begin{aligned} R_{uy}(k) &\approx \sum_{i=0}^{N} h(i)R_{uu}(k-i) \equiv \sum_{i=0}^{N}\sum_{j=0}^{J} \alpha_j\psi_j(i)R_{uu}(k-i) \\ &= \sum_{j=0}^{J} \alpha_j \sum_{i=0}^{N} \psi_j(i)R_{uu}(k-i) = \sum_{j=0}^{J} \alpha_j v_j(k) \end{aligned} \tag{7}$$

where $v_j(k)$, $k = 0, \cdots, N$ is the input signal $R_{uu}(k)$ processed by each element of the discrete-time function $\psi_j(k)$, $j = 0, 1, \ldots, J$, which forms the approximation base and is the IRF of the orthogonal function:

$$v_j(k) = \sum_{i=0}^{N} \psi_j(i)R_{uu}(k-i) \tag{8}$$

Eq. (8) is basically a filtering of the input signal $R_{uu}(k)$ by a set of filter $\psi_j(k)$. Finally, the eq. (7) is used to describe:

$$\begin{Bmatrix} R_{uy}(0) \\ R_{uy}(1) \\ \vdots \\ R_{uy}(N) \end{Bmatrix} = \begin{bmatrix} v_0(0) & v_1(0) & \cdots & v_J(0) \\ v_0(1) & v_1(1) & \cdots & v_J(1) \\ \vdots & \vdots & \ddots & \vdots \\ v_0(N) & v_1(N) & \cdots & v_J(N) \end{bmatrix} \begin{Bmatrix} \alpha_0 \\ \alpha_1 \\ \vdots \\ \alpha_J \end{Bmatrix} \tag{9}$$

The effectiveness of the model is limited by the choice of the filters $\Psi_j(z)$. Thus, the choice of the basis functions is very important. For describing mechanical vibration and flexible systems, the Kautz functions have been demonstrated to provide a good generalization by including complex poles in the z-domain [2, 14].

4. Kautz filter

The Kautz filters can be given by [16, 22, 24]:

$$\Psi_{2n}(z) = \frac{\sqrt{(1-c^2)(1-b^2)}z}{z^2+b(c-1)z-c}\left[\frac{-cz^2+b(c-1)z+1}{z^2+b(c-1)z-c}\right]^{n-1} \tag{10}$$

$$\Psi_{2n-1}(z) = \frac{\sqrt{1-c^2}z(z-b)}{z^2+b(c-1)z-c}\left[\frac{-cz^2+b(c-1)z+1}{z^2+b(c-1)z-c}\right]^{n-1} \tag{11}$$

where the constants b and c are relative to the poles $\beta = \sigma + j\omega$ and $\beta^* = \sigma - j\omega$ in the j-th filter through the relations:

$$b = \frac{(\beta_j + \beta_j^*)}{(1+\beta_j\beta_j^*)}, \tag{12}$$

$$c = -\beta\beta_j^* \tag{13}$$

A sequence of filters is utilized with different poles in each section describing the modal behavior in the frequency range of interest. A question is relative for choosing the poles and the IRFs iteratively based on application of eq. (2) and output experimental signal $y_e(k)$. An error signal can be written by:

$$e(k) = \hat{y}(k) - y_e(k) \tag{14}$$

where $\hat{y}(k)$ is the predicted output signal by the IRF $\hat{h}(k)$ estimated considering Kautz basis defined by the poles β_j and β_j^* in the z-domain:

$$\hat{y}(k) = \sum_{i=0}^{N} \hat{h}(i)u(k-i) \tag{15}$$

The optimization problem can be described by objective function that employs an Euclidean norm and the Kautz poles are functions of the frequencies and damping factors that are the optimization parameters. These parameters can be restricted in a range searching. This optimization problem can be solved by several classical approaches. A detailed explanation in this point can be found in [12].

5. Active vibration control strategy

If an IRF is well identified through covariance method expanded with Kautz filters, a model in z-domain can be described by applying the $z-$ transform in the IRF $h(k)$[1]:

$$H(z) = \sum_{n=0}^{+\infty} h(n)z^{-n} \approx \alpha_0\Psi_0(z) + \alpha_1\Psi_1(z) + \cdots + \alpha_J\Psi_J(z) \tag{16}$$

[1] Considering $h(k)$ is a causal sequence.

A controller can be inserted in the direct branch of the control loop to try to reject the disturbance. This controller $G(z)$ has a digital structure given by:

$$G(z) = L(z)H^{-1}(z) \tag{17}$$

where $H^{-1}(z)$ is the inverse of the identified transfer function of the system and $L(z)$ has the desirable dynamic. The compensator $L(z)$ can have a second order structure or any format with a damping ratio ξ_c bigger than the uncontrolled damping ratio. The control project is to find a gain and the $G(z)$ formed to reduce the damping of the system. For practical implementation, these equations can be programmed directly in the discrete-time domain by using the mathematical convolution operator.

It is worth to point out that one consider only the control of stable systems described by $H(z)$ experimentally identified. Consequently, the transfer function $H(z)$ has all poles within the unitary circle because $H(z)$ is identified using the Kautz poles that are set to be stable. The auxiliary function $L(z)$ is proposed to warrant stability and the required performance in the closed-loop system

Two examples are used to show the approach proposed. The first one is a single-degree-of-freedom model that is a simple and easy example for the interested reader reproduce it. The second one is based on active vibration control in a smart structure with PZT actuator and PVDF sensor for presenting its use employing experimental data.

The results are illustrated in a single-degree-of-freedom model given by:

$$\ddot{x}(t) + 2\xi\omega_n\dot{x}(t) + \omega_n^2 x(t) = f(t) \tag{18}$$

where $x(t)$ is the displacement vector, the over dot is the time derivative, ξ is the damping factor, ω_n is the natural frequency in rad/s and $f(t)$ is the excitation force. To simulate the uncontrolled responses, it were used the values of $\xi = 0.01$ and $\omega_n = 62.83$ rad/s that correspond to 10 Hz. The motion equation from eq. (18) is solved numerically through the Runge-Kutta method with a sampling rate of 100 Hz, that corresponds to a time sample of $dt = 0.01$ s, with 2048 samples. The force used was a white noise random with level of amplitude of the 3 N. The fig. (1) shows the input and output signal simulated for uncontrolled condition.

An important step to identify the IRFs is the choice of Kautz poles that need to reflect adequately the dominant dynamics of the vibrating systems. In real-world application the choice of the poles is a complicated problem. However, a simple power spectral density of the output signal (in our example the displacement) can give an orientation to help in the selection. If the system is more complicated, an optimization procedure could be used [12]. Figure (2) shows the power spectral density of the displacement. Clearly, it seems a peak value close to 10 Hz that is a possible candidate of natural frequency. The frequency response function (FRF) experimental is also estimated through spectral analysis only to compare the values of the natural frequency and damping factor, fig. (3).

Based on the frequency of 10 Hz, a continuous pole in $s-$domain given by $s_{1,2} = -0.6283 \pm 62.82j$, where j is the imaginary unit, is set. Kautz filter is described in discrete-time domain, so, it is necessary to convert the pole to $z-$domain. The relationship $\beta = e^{sdt}$ can be used to

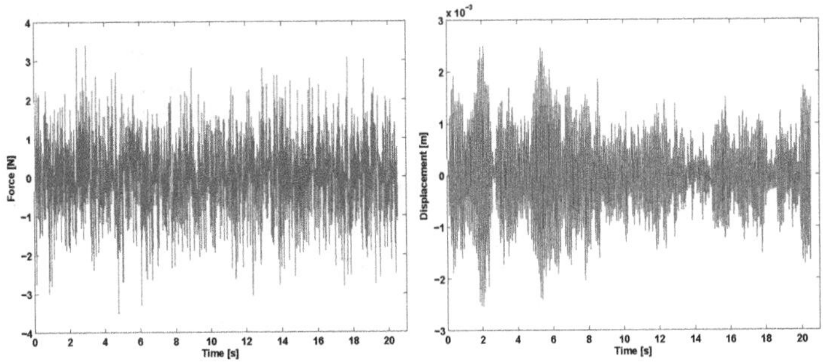

(a) Input force (disturbance). (b) Output displacement.

Figure 1. Response of the system for the uncontrolled case.

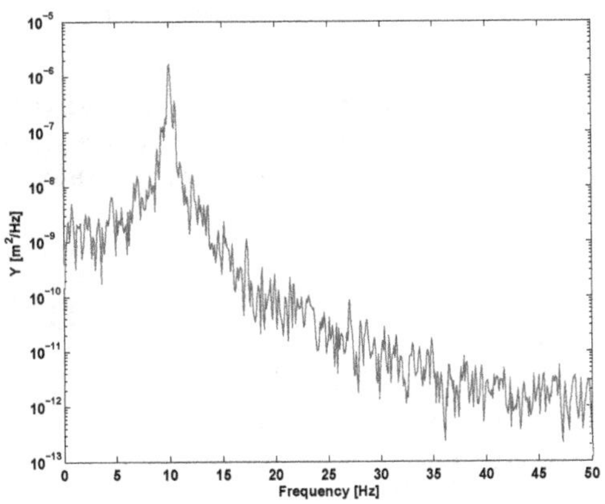

Figure 2. Power spectral density of the output signal (displacement) estimated using Welch method with Hanning window, 25 % of overlap and two sections.

obtain the discrete Kautz poles given by $\beta = 0.8040 + 0.5841j$ and $\beta^* = 0.8040 + 0.5841j$. Once the system is a SISO and with only one degree of freedom, only one section of Kautz filter is employed, $J = 1$ (2 terms), and $N = 600$ samples are considered to be enough to complete description of the memory lag. The constants b and c are computed through eq. (12) and (13) and the eqs. (10) and (11) are utilized to construct the Kautz filter given by:

$$\Psi_0(z) = \frac{0.0926}{z^2 - 1.608z + 0.9875} \tag{19}$$

$$\Psi_1(z) = \frac{0.1575z - 0.1275}{z^2 - 1.608z + 0.9875} \tag{20}$$

The impulse response of the two sections of the Kautz filter are used to process the correlation function of the input signal $f(t)$, through eq. (8). Equation (9) is solved by LS approach in

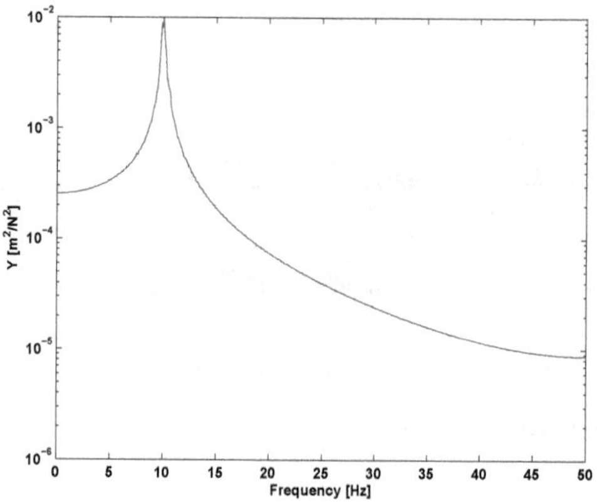

Figure 3. Frequency response function identified using spectral estimate H_1 through Welch method with Hanning window, 25 % of overlap and two sections.

order to identify the expansion coefficients α_0 and α_1. With these values, eq. (6) is used to identify the IRF. Figure (4) presents the result of the identification process and compare with the analytical IRF. It is observed a good concordance between the experimental identified and the theoretical IRF.

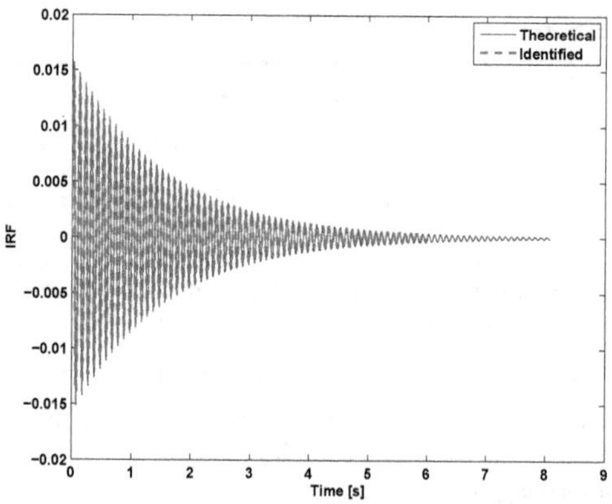

Figure 4. Impulse response function comparison between analytical and identified by Kautz filters.

Once the IRF is identified, an experimental FIR model representative of the system is now known. This $H(z)$ model is used to represent a controller $G(z)$ inserted in the direct branch of the control loop with unitary feedback, by using the following expression:

$$G(z) = L(z)H^{-1}(z) \tag{21}$$

where $H^{-1}(z)$ is the inverse of the transfer function of the system identified experimentally, $H(z)$, described by:

$$H(z) \approx \alpha_0 \Psi_0(z) + \alpha_1 \Psi_1(z) \tag{22}$$

and $L(z)$ is a desirable dynamic to the system. The controller used has the following structure of a second order system:

$$L(s) = K \frac{\omega_n^2}{s^2 + 2\xi_c \omega_n s + \omega_n^2} \tag{23}$$

where ξ_c is the damping factor of the controlled system and K is a control gain. The structure in eq. (23) is continuous in the $s-$domain, and for application in a digital format is necessary to use a bilinear transform (Tustin's method). It is chosen a gain of $K = 3 \times 10^{-4}$ and $\xi_c = 0.08$. These values are chosen based on the adequate behavior for the controlled system in the closed loop and with a low level of control actuator force required. The natural frequency in closed loop is maintained the same of the uncontrolled system. Thus, the digital compensator $L(z)$ is given by:

$$L(z) = 10^{-5} \frac{5.543z + 5.358}{z^2 - 1.541z + 0.9044} \tag{24}$$

Finally, the feedback transfer function $M(z)$ is given by:

$$M(z) = \frac{L(z)H^{-1}(z)H(z)}{1 + L(z)H^{-1}(z)H(z)} \tag{25}$$

that corresponds to:

$$M(z) = 10^{-5} \frac{5.543z^3 - 3.184z^2 - 3.244z + 4.846}{z^4 - 3.082z^3 + 4.183z^2 - 2.787z + 0.8179} \tag{26}$$

Clearly the effectiveness of the controller depends on the correct identification of the $H(z)$ to allow a perfect cancelation. Figure (5) shows the frequency response function comparison between uncontrolled and controlled system where it is seen that the peak decrease by increase actively the damping with the digital compensator. Figure (6) shows the output displacement without and with control. The disturbance force is considered with the same level and type of the tests used in the uncontrolled condition.

A cantilever aluminium beam with a PZT actuator patch and a piezoelectric sensor (PVDF) symmetrically bonded to both sides of the beam is used to illustrate the process of IRF identification and design of a digital controller for active vibration reduction. The PZT and PVDF are bonded attached collocated near to the clamped end of the beam, as seen in fig. (7). The PZT patch is the model QP10N from ACX with size of $50 \times 20 \times 0.254$ mm of length, width and thickness, respectively. The PVDF has dimensions of $30 \times 10 \times 0.205$ mm of length, width and thickness, respectively, and it is bonded with a distance of 5 mm of the clamped end. The complete experimental setup is shown in figs. (7) and (8).

A white noise signal is generated in the computer, converted to analogic domain with a D/A converter and pre-processed by a voltage amplifier with gain of 20 V/V before application in the PZT actuator. The output signal is measured with the PVDF and linked directly with the

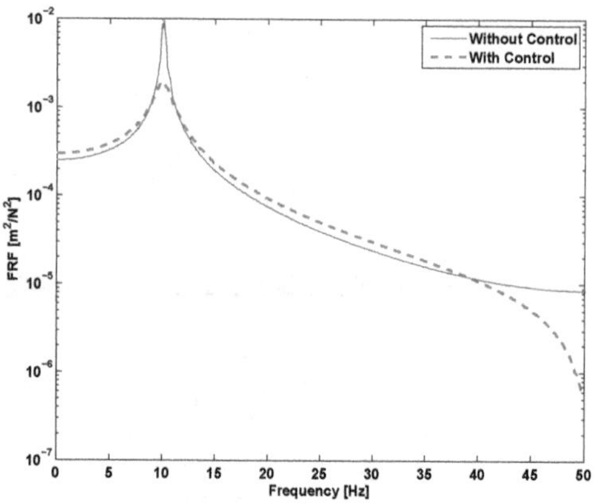

Figure 5. Frequency response function comparison between uncontrolled and controlled condition.

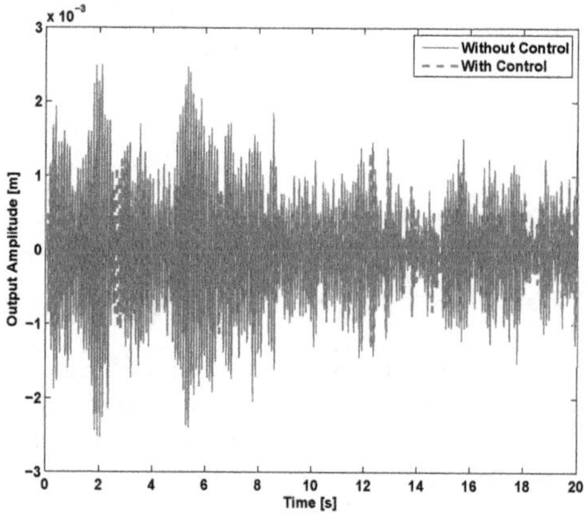

Figure 6. Output displacement comparison between uncontrolled and controlled condition.

charge amplifier and pre-processed with a A/D converter. All experimental signals are saved and processed with a *dSPACE* 1104 acquisition board with a sample rate of 1 kHz and with 5 seconds of test duration. Figure (9) shows the time series signals of PZT actuator (input) and PVDF sensor (output) for uncontrolled system.

The first step in this approach is the choosing an adequate set of poles for the Kautz Filters. As the mathematical model is unknown, one needs to start by availing the power spectral density of the PVDF sensor (output) as suggested in the first example. Figure (10) presents the power spectral density of the output signal (PVDF) estimated using Welch method with Hanning window, 50 % of overlap and 5 sections. The peaks in frequencies of 13, 78, 211, 355 and 434 Hz can be considered candidates for natural frequencies. For comparison purposes,

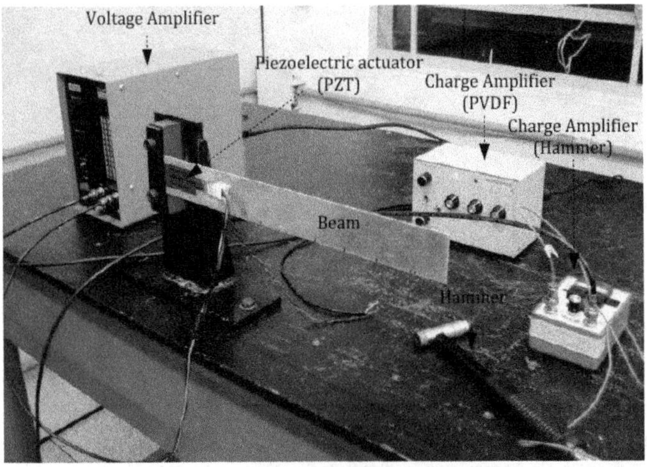

(a) Overall experimental setup.

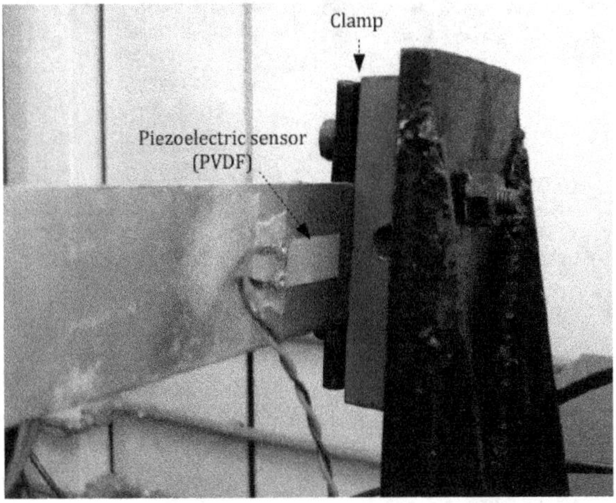

(b) Detail of the PVDF Sensor.

Figure 7. View of the experimental setup.

the frequency response function (FRF) experimental is estimated through spectral analysis to observe the values of the natural frequencies and damping factors, fig. (11).

Based on the spectral analysis one must choose the continuous poles candidates given by $s_i = -\xi_i\omega_{ni} \pm j\omega_{ni}\sqrt{1-\xi_i^2}$, $i = 1, 2, 3, 4, 5$. The most difficult parameters to be identified are the damping factors. Several trial and error tests were performed until to reach an adequate

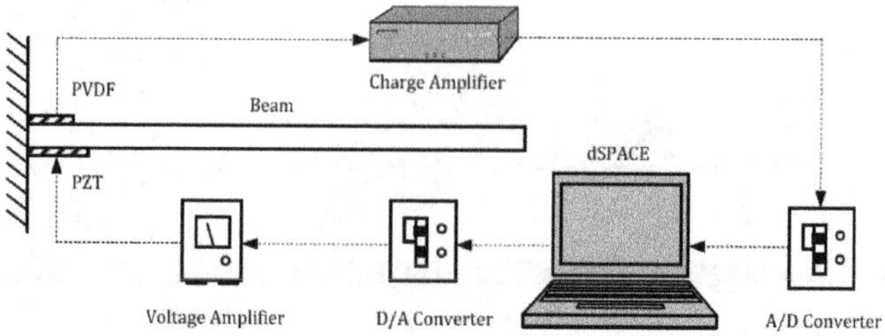

Figure 8. Experimental setup.

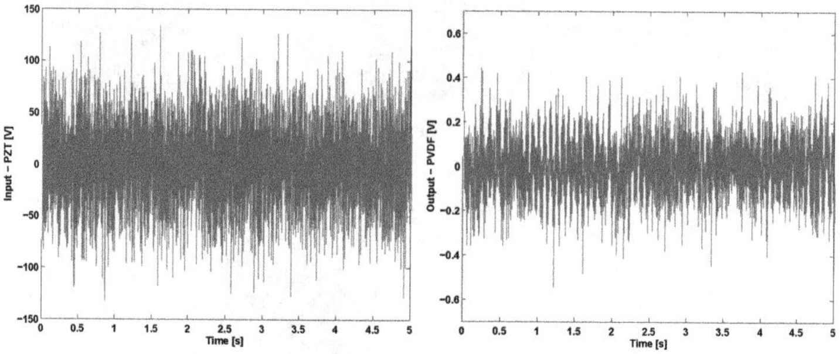

(a) PZT Actuator. (b) PVDF Sensor.

Figure 9. Response of the experimental tests in the time domain for the uncontrolled case.

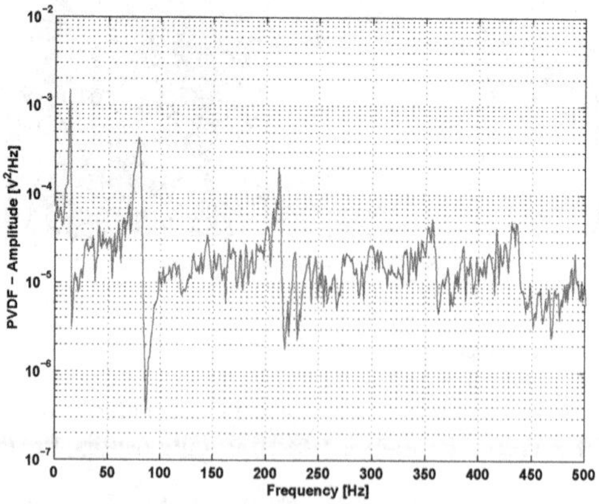

Figure 10. Power spectral density of the output signal (PVDF) estimated using Welch method with Hanning window, 50 % of overlap and 5 sections.

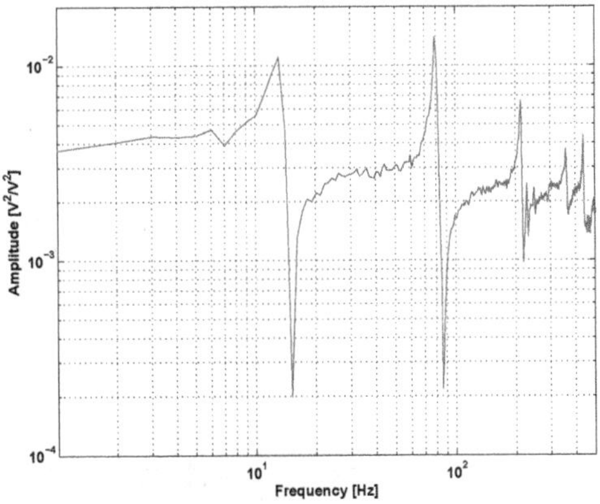

Figure 11. Frequency response function identified using spectral estimate H_1 through Welch method with Hanning window, 50 % of overlap and 5 sections.

result. A reasonable identification were reached based on the parameters given by:

$$\omega_{n1} = 81.68 \quad \text{rad/s} \qquad \xi_1 = 0.04 \qquad s_1 = -3.2673 \pm 81.6160j \tag{27}$$

$$\omega_{n2} = 490.08 \quad \text{rad/s} \qquad \xi_2 = 0.019 \qquad s_2 = -9.3117 \pm 490j \tag{28}$$

$$\omega_{n3} = 1.3258 \times 10^3 \quad \text{rad/s} \qquad \xi_3 = 0.02 \qquad s_3 = -26.515 \pm 1325.5j \tag{29}$$

$$\omega_{n4} = 2.23 \times 10^3 \quad \text{rad/s} \qquad \xi_4 = 0.1 \qquad s_4 = -223.05 \pm 2219.4j \tag{30}$$

$$\omega_{n5} = 2.72 \times 10^3 \quad \text{rad/s} \qquad \xi_5 = 0.1 \qquad s_5 = -276.7 \pm 2713.2j \tag{31}$$

Once the fourth and fifth modes are apparently well damped by analysing the frequency response the correspond poles are also considered well damped (not dominants). The Kautz filter is described in the discrete-time domain. So, it is necessary to convert to $z-$domain. The relationship $\beta_i = e^{s_i dt}$ can be used to obtain the five pair of complex discrete Kautz poles given by:

$$\beta_1 = 0.9934 \pm 0.0813j \tag{32}$$

$$\beta_2 = 0.8742 \pm 0.4663j \tag{33}$$

$$\beta_3 = 0.2365 \pm 0.9447j \tag{34}$$

$$\beta_4 = -0.4833 \pm 0.6376j \tag{35}$$

$$\beta_5 = -0.6925 \pm 0.3163j \tag{36}$$

The cantilever beam is a SISO system, but with apparent five modes in the frequency range computed of interest. So, they are used 5 sections of Kautz filters, $J = 4$ and $N = 1200$ samples that are considered to be enough to complete the view of the memory lag. The constants b and c are computed and the eqs. (10) and (11) are utilized to construct the Kautz filters.

Figure (12) shows the comparison between the IFFT of the FRF from H_1 estimated and the IRF identified through Kautz filter.

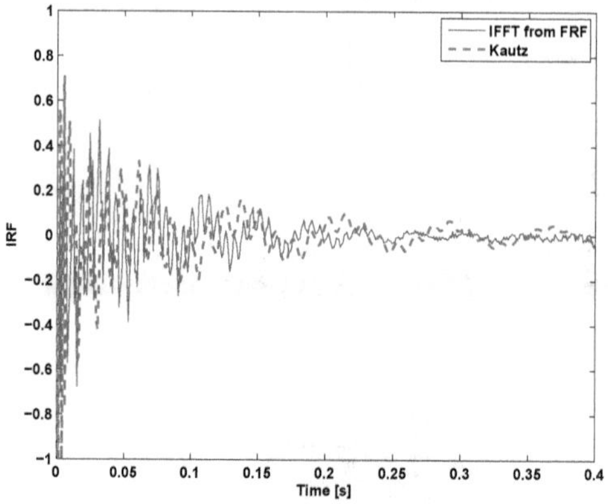

Figure 12. Impulse response function comparison between IFFT of the estimated FRF and identified by Kautz filters.

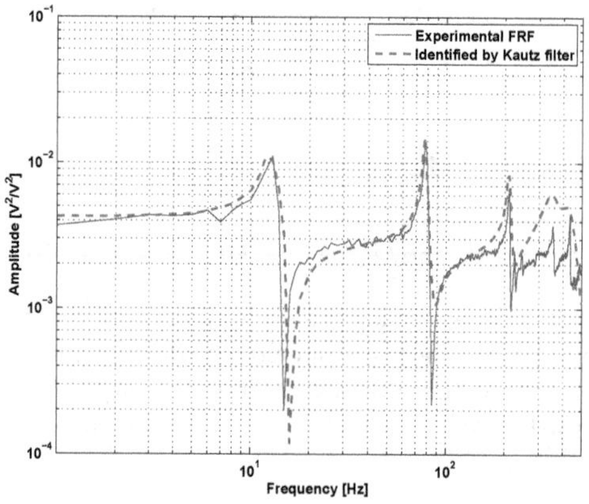

Figure 13. FRF comparison between estimated FRF through H_1 spectral estimate and identified by Kautz filters.

Although, it seems that are not a complete visual agreement between the curves, the FRF seen in figure (13) presents a good agreement. It is worth to comments that with the same experimental data, [23] identified a state-space model through Eigensystem Realization Algorithms (ERA) combined with Observer/Kalman filter Identification (OKID). The results presented with Kautz filter allowed a better identification in this frequency range comparing than with ERA/OKID.

Figure (14) shows the output response of th PVDF estimated by a convolution between the IRF identified by Kautz filter with the input excitation from PZT actuator. The estimated output can be compared with the experimental measured response (see fig. 9(b)).

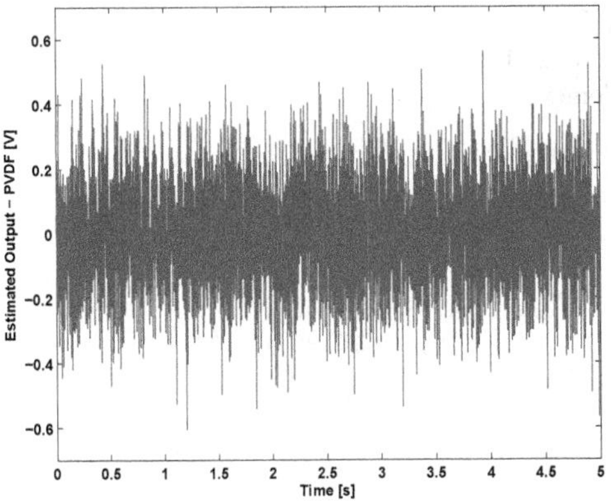

Figure 14. PVDF output estimated by IRF identified with Kautz filters.

The controller is designed based on the inverse of the identified system described by eq. (16), called by $H^{-1}(z)$, in series with a compensator $L(z)$. The $L(z)$ is chosen by combination of 3 second-order system realized in parallel structure:

$$L(z) = K\left(L_1(z) + L_2(z) + L_3(z)\right) \tag{37}$$

where $K = 1.5 \times 10^{-3}$ is a controller gain and the transfer functions are defined by:

$$L_1(z) = \frac{0.003316z + 0.003298}{z^2 - 1.977z + 0.9838} \tag{38}$$

$$L_2(z) = \frac{0.1122z + 0.1068}{z^2 - 1.644z + 0.8633} \tag{39}$$

$$L_3(z) = \frac{0.6691z + 0.5813}{z^2 - 0.4215z + 0.6718} \tag{40}$$

It is important to observe that the three compensators, $L_1(z)$, $L_2(z)$ and $L_3(z)$ have the natural frequencies corresponding to the first three modes of the systems, but with an increase in the level of damping factor for reducing the vibration level in the closed-loop system. The compensator $L(z)$ in its final form is given by:

$$L(z) = \frac{0.001177z^5 - 0.003011z^4 + 0.001994z^3 + 0.001218z^2 - 0.002219z + 0.0008492}{z^6 - 4.043z^5 + 7.297z^4 - 7.907z^3 + 5.676z^2 - 2.592z + 0.5706} \tag{41}$$

It was decided to control only the 3 first modes for two main reasons:

- The fourth and fifth modes are not dominant.
- Additionally, these modes are not well identified by the Kautz filter. One included the damping factor in these modes with these values shown above in order to correct identification the anti-ressonance region.

Figure (15) shows the FRF comparison between the uncontrolled (estimated by Kautz filter) and controlled system where is possible to observe the reduction in the resonance peak caused by the controller implemented.

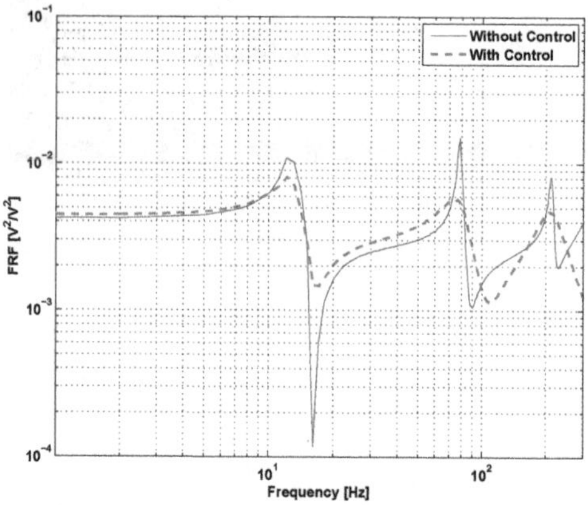

Figure 15. FRF comparison between uncontrolled and controlled condition. Input: PZT actuator - Output: PVDF sensor.

Another advantage of this procedure face to state-feedback approaches is relative to the controlability and observability conditions. If one use procedures identification for obtaining a state-space realization, e. g. ERA/OKID as made by [23], is necessary to verify a prior the observability and controlability conditions. In some situations some modes are not controllable and observable adequately with a specific realization. Once the technique used in this chapter is not described in state-space variables and it is based on input/output variables with non-parametric IRF model, these kinds of drawbacks are avoided.

This chapter has described a procedure for non-parametric system identification of an impulse response function (IRF) based on input and output experimental data. Orthogonal functions are used to reduce the number of samples to be identified. A simple active vibration control procedure with a digital compensator that seeks to cancel the plant dynamic is also described. Once the IRF in the uncontrolled condition is well estimated by Kautz filters, the control strategy presented can increase the damping in a satisfactory level with low actuator requirements. Single-input-single-output vibrating systems have been used to illustrate the performance and the main aspects for practical implementation. This procedure can also be extended for nonlinear systems using Hammerstein or Wiener block-oriented models.

Acknowledgements

The authors are thankful for the financial support provided by National Council for Scientific and Technological Development (CNPq/Brasil), INCT and São Paulo Research Foundation (FAPESP). The authors acknowledge the helpful suggestions of the Editor. The authors also are thankful the help of Prof. Dr. Gustavo Luiz Chagas ManhŃes de Abreu and Sanderson Manoel da Conceição for providing the experimental data in the clamped beam.

Author details

Samuel da Silva, Vicente Lopes Junior and Michael J. Brennan
UNESP - Univ Estadual Paulista, Faculdade de Engenharia de Ilha Solteira, Departamento de Engenharia Mecânica, Av. Brasil 56, Centro, Ilha Solteira, SP, Brasil

6. References

[1] Baldelli, D. H., Lind, R. & Brenner, M. [2005]. Nonlinear aeroelastic/aeroservoelastic modeling by block-oriented identification, *Journal of Guidance, Control and Dynamics* 28(5): 1056–1064.

[2] Baldelli, D. H., Mazzaro, M. C. & Peña, R. S. S. [2001]. Robust identification of lightly damped flexible structures by means of orthonormal bases, *IEEE Transactions on Control Systems Technology* 9(5): 696–707.

[3] Brennan, M. J. & Kim, S. M. [2001]. Feedforward and feedback control of sound and vibration - a Wiener filter approach, *Journal of Sound and Vibration* 246(2): 281–296.

[4] Brinker, A. C. & Belt, H. J. W. [1998]. Using kautz models in model reduction, *in* A. Prochazka, J. Uhlír, P. J. W. Rayner & N. Kingsbury (eds), *Signal Analysis and prediction*, 1st edn, Birkhauser Boston.

[5] Campello, R. J. G. B., Oliveira, G. H. C. & Amaral, W. C. [2007]. Identificação e controle de processos via desenvolvimento em séries ortonormais. parte a: Identificação, *Revista Controle & Automação* 18(3): 301–321.

[6] D. Mayer, S. H. H. H. [2001]. Application of Kautz models for adaptive vibration control, *in* IMECE (ed.), *American Society of Mechanical Engineers (Veranst.)*, ASME International Mechanical Engineering Congress and Exposition, NewYork.

[7] da Rosa, A., Campello, R. J. G. B. & do Amaral, W. C. [2006]. Desenvolvimento de modelos de Volterra usando funções de Kautz e sua aplicação à modelagem de um sistema de levitaçãoo magnética, *XVI Congresso Brasileiro de Automática.*

[8] da Rosa, A., Campello, R. J. G. B. & do Amaral, W. C. [2007]. Choice of free parameters of discrete-time Volterra models using Kautz functions, *Automatica* 43(6): 1084–1091.

[9] da Rosa, A., Campello, R. J. G. B. & do Amaral, W. C. [2008]. An optimal expansion of Volterra models using independent Kautz bases for each kernel dimension, *International Journal of Control* 81(6): 962–975.

[10] da Rosa, A., Campello, R. J. G. B. & do Amaral, W. C. [2009]. Exact search directions for optimization of linear and nonlinear models based on generalized orthonormal functions, *IEEE Transactions os Automatic Control* 54(12): 2757–2772.

[11] da Silva, S. [2011a]. Non-linear model updating of a three-dimensional portal frame based on Wiener series, *International Journal of Non-linear Mechanics* 46: 312–320.

[12] da Silva, S. [2011b]. Non-parametric identification of mechanical systems by Kautz filter with multiple poles, *Mechanical Systems and Signal Processing* 25(4): 1103–1111.

[13] da Silva, S., Cogan, S. & Foltête, E. [2010]. Nonlinear identification in estructural dynamics based on Wiener series and Kautz filter., *Mechanical Systems and Signal Processing* 24(1): 52–58.

[14] da Silva, S., Dias Júnior, M. & Lopes Junior, V. [2009]. Identification of mechanical systems through Kautz filter, *Journal of Vibration and Control* 15(6): 849–865.

[15] Heuberger, P. S. C., Van Den Hof, P. M. J. & Wahlberg, B. [2005]. *Modelling and Identification with Rational Orthogonal Basis Functions*, 1st edn, Springer.

[16] Heuberger, P., Van Den Hof, P. & Bosgra, O. H. [1995]. A generalized orthonormal basis of linear dynamical systems, *IEE Transactions on Automatic Control* 40(3): 451–465.

[17] Kautz, W. H. [1954]. Transient synthesis in the time domain, *IRE Transactions on Circuit Theory* 1(1): 29 – 39.

[18] Oliveira, G. H. C., Amaral, W. C., Favier, G. & Dumont, G. A. [2000]. Constrained robust predictive controller for uncertain processes modeled by orthonormal series functions, *Automatica* 36(4): 563–571.

[19] Oliveira, G. H. C., Campello, R. J. G. B. & Amaral, W. C. [2007]. Identificação e controle de processor via desenvolvimento em séries ortonormais. parte b: Controle preditivo, *Revista Controle & Automática* 18(3): 322–336.

[20] Paetero, T. & Karjalainen, M. [2003]. Kautz filters and generalized frequency resolution: Theory and audio aplications, *Audio Engineering Society* 51(1/2): 27–44.

[21] Sansone, G. [1958]. *Orthogonal Functions*, Vol. 9, Dover Publications.

[22] Van Den Hof, P. M. J., Heuberger, P. S. & Bokor, J. [1995]. System identification with generalized orthonormal basis functions, *Automatica* 31(12): 1821–1834.

[23] Vasques, C. H., Conceição, S. M., Abreu, G. L. C. M., Lopes Jr., V. & Brennan, M. J. [2011]. Identification and control of systems submitted to mechanical vibration, *21st International Congress of Mechanical Engineering - COBEM 2011*, Natal, RN, Brasil.

[24] Wahlberg, B. [1994]. System identification using Kautz models, *IEEE Transactions on Automatic Control* 39(6): 1276 – 1282.

[25] Wahlberg, B. & Makila, P. M. [1996]. On approximation of stable linear dynamical systems using Laguerre and Kautz functions, *Automatica* 32(5): 693–708.

[26] Zeng, J. & de Callafon, R. [2005]. Filters parametrized by orthonormal basis functions for active noise control, *ASME International Mechanical Engineering Congress and Exposition - IMECE 2005*.

Chapter 2

Vibration Control by Exploiting Nonlinear Influence in the Frequency Domain

Xingjian Jing

Additional information is available at the end of the chapter

1. Introduction

In the control theory of linear systems, system transfer function provides a coordinate-free and equivalent description for system dynamic characteristics, by which it is convenient to conduct analysis and design. Therefore, frequency domain methods are commonly used by engineers and widely applied in engineering practice. However, although the analysis and design of linear systems in the frequency domain have been well established, the frequency domain analysis for nonlinear systems is not straightforward. Nonlinear systems usually have very complicated output frequency characteristics and dynamic behaviour such as harmonics, inter-modulation, chaos and bifurcation. Investigation and understanding of these nonlinear phenomena in the frequency domain are far from full development. Frequency domain methods for nonlinear analysis have been investigated for many years. There are several different approaches to the analysis and design for nonlinear systems, such as describing functions [5, 13], harmonic balance [18], and frequency domain methods developed from the absolute stability theory [10], for example the well-known Popov circle theorem [12, 21] etc. Investigation of nonlinear systems in the frequency domain can also be done based on the Volterra series expansion theory [11, 15, 16, 19, 20]. There are a large class of nonlinear systems which have a convergent Volterra series expansion [2, 17]. For this class of nonlinear systems, referred to as Volterra systems, the generalized frequency response function (GFRF) was defined in [4], which is similar to the transfer function of linear systems. To obtain the GFRFs for Volterra systems described by nonlinear differential equations, the probing method can be used [16]. Once the GRFRs are obtained for a practical system, system output spectrum can then be evaluated [9]. These form a fundamental basis for the analysis of nonlinear Volterra systems in the frequency domain and provide an elegant and useful method for the frequency domain analysis of a class of nonlinear systems. Many techniques developed (e.g. the GFRFs) can be regarded as an important extension of frequency domain theories for linear systems to nonlinear cases.

In this study, understanding of nonlinearity in the frequency domain is investigated from a novel viewpoint for Volterra systems. The system output spectrum is shown to be an alternating series with respect to some model parameters under certain conditions. This property has great significance in that the system output spectrum can therefore be easily suppressed by tuning the corresponding parameters. This provides a novel insight into the nonlinear influence in a system. The sufficient (and necessary) conditions in which the output spectrum can be transformed into an alternating series are studied. These results are illustrated by two example studies which investigated a single degree of freedom (SDOF) spring-damping system with a cubic nonlinear damping. The results established in this study demonstrate a novel characteristic of the nonlinear influence in the frequency domain, and provide a novel insight into the analysis and design of nonlinear vibration control systems.

The chapter is organised as follows. Section 2 provides a detailed background of this study. The novel nonlinear characteristic and its influence are discussed in Section 3. Section 4 gives a sufficient and necessary condition under which system output spectrum can be transformed into an alternating series. A conclusion is given in Section 5. A nomenclature section which explains the main notations used in this paper is given in Appendix A.

2. Frequency response functions of nonlinear systems

There are a class of nonlinear systems for which the input-output relationship can be sufficiently approximated by a Volterra series (of a maximum order N) around the zero equilibrium as [2, 17]

$$y(t)=\sum_{n=1}^{N}\int_{-\infty}^{\infty}\cdots\int_{-\infty}^{\infty}h_n(\tau_1,\cdots,\tau_n)\prod_{i=1}^{n}u(t-\tau_i)d\tau_i \tag{1}$$

where $h_n(\tau_1,\cdots,\tau_n)$ is the nth-order Volterra kernel which is a real valued function of $\tau_1,\cdots,\tau_n$. For the same class of nonlinear systems, it can also be modelled by the following nonlinear differential equation (NDE)

$$\sum_{m=1}^{M}\sum_{p=0}^{m}\sum_{k_1,k_m=0}^{K}c_{p,m-p}(k_1,\cdots,k_m)\prod_{i=1}^{p}\frac{d^{k_i}y(t)}{dt^{k_i}}\prod_{i=p+1}^{m}\frac{d^{k_i}u(t)}{dt^{k_i}}=0 \tag{2}$$

where $\left.\frac{d^k x(t)}{dt^k}\right|_{k=0}=x(t)$, $\sum_{k_1,k_{p+q}=0}^{K}(\cdot)=\sum_{k_1=0}^{K}(\cdot)\cdots\sum_{k_{p+q}=0}^{K}(\cdot)$, M is the maximum degree of nonlinearity in terms of y(t) and u(t), and K is the maximum order of the derivative. In this model, the parameters such as $c_{0,1}(.)$ and $c_{1,0}(.)$ are referred to as linear parameters corresponding to coefficients of linear terms in the model, *i.e.*, $\frac{d^k y(t)}{dt^k}$ and $\frac{d^k u(t)}{dt^k}$ for k=0,1,…,K; and $c_{p,q}(\cdot)$ for $p+q>1$ are referred to as nonlinear parameters corresponding to nonlinear terms in the model of the form $\prod_{i=1}^{p}\frac{d^{k_i}y(t)}{dt^{k_i}}\prod_{i=p+1}^{p+q}\frac{d^{k_i}u(t)}{dt^{k_i}}$, *e.g.*, $y(t)^p u(t)^q$. The value $p+q$ is referred to as the nonlinear degree of parameter $c_{p,q}(\cdot)$.

By using the probing method [16], a recursive algorithm for the computation of the nth-order generalized frequency response function (GFRF) for the NDE model (2) is provided in [1]. Therefore, the output spectrum of model (2) can be evaluated as [9]

$$Y(j\omega)=\sum_{n=1}^{N}\frac{1}{\sqrt{n}(2\pi)^{n-1}}\int_{\omega_1+\cdots+\omega_n=\omega} H_n(j\omega_1,\cdots,j\omega_n)\prod_{i=1}^{n}U(j\omega_i)d\sigma_\omega \tag{3}$$

which is truncated at the largest order N and where,

$$H_n(j\omega_1,\cdots,j\omega_n)=\int_{-\infty}^{\infty}\cdots\int_{-\infty}^{\infty}h_n(\tau_1,\cdots,\tau_n)\exp(-j(\omega_1\tau_1+\cdots+\omega_n\tau_n))d\tau_1\cdots d\tau_n \tag{4}$$

is known as the nth-order GFRF defined in [4], and $h_n(\tau_1,\cdots,\tau_n)$ is the nth-order Volterra kernel introduced in (1). When the system input is a multi-tone function described by

$$u(t)=\sum_{i=1}^{\overline{K}}|F_i|\cos(\omega_i t+\angle F_i) \tag{5}$$

(where F_i is a complex number, $\angle F_i$ is the argument, $|F_i|$ is the modulus, and $\overline{K}$ is a positive integer), the system output frequency response can be evaluated as [9]:

$$Y(j\omega)=\sum_{n=1}^{N}\frac{1}{2^n}\sum_{\omega_{k_1}+\cdots+\omega_{k_n}=\omega} H_n(j\omega_{k_1},\cdots,j\omega_{k_n})F(\omega_{k_1})\cdots F(\omega_{k_n}) \tag{6}$$

where $F(\omega_{k_i})$ can be explicitly written as $F(\omega_{k_i})=\left|F_{|k_i|}\right|e^{j\angle F_{|k_i|}\cdot \mathrm{sig}(k_i)}$ for $k_i\in\{\pm1,\cdots,\pm\overline{K}\}$ in stead of the form in [9], $\mathrm{sgn}(a)=\begin{cases}1 & a\ge 0\\ -1 & a<0\end{cases}$, and $\omega_{k_i}\in\{\pm\omega_1,\cdots,\pm\omega_{\overline{K}}\}$.

In order to explicitly reveal the relationship between model parameters and the frequency response functions above, the parametric characteristics of the GFRFs and output spectrum are studied in [6]. The nth-order GFRF can then be expressed into a more straightforward polynomial form as

$$H_n(j\omega_1,\cdots,j\omega_n)=CE\left(H_n(j\omega_1,\cdots,j\omega_n)\right)\cdot f_n(j\omega_1,\cdots,j\omega_n) \tag{7}$$

where $CE\left(H_n(j\omega_1,\cdots,j\omega_n)\right)$ is referred to as the parametric characteristic of the nth-order GFRF $H_n(j\omega_1,\cdots,j\omega_n)$, which can be recursively determined as

$$CE(H_n(j\omega_1,\cdots,j\omega_n))=C_{0,n}\oplus\left(\mathop{\oplus}_{q=1}^{n-1}\mathop{\oplus}_{p=1}^{n-q}C_{p,q}\otimes CE(H_{n-q-p+1}(\cdot))\right)\oplus\left(\mathop{\oplus}_{p=2}^{n}C_{p,0}\otimes CE(H_{n-p+1}(\cdot))\right) \tag{8}$$

with terminating condition $CE\left(H_1(j\omega_i)\right)=1$. Note that CE is a new operator with two operations "$\otimes$" and "$\oplus$" defined in [6,7] (the definition of CE can be referred to Appendix

B and more detailed discussions in [22]), and $C_{p,q}$ is a vector consisting of all the (p+q)th degree nonlinear parameters, i.e.,

$$C_{p,q} = [c_{p,q}(0,\cdots,0), c_{p,q}(0,\cdots,1),\cdots,c_{p,q}(\underbrace{K,\cdots,K}_{p+q=m})]$$

In Equation (8), $f_n(j\omega_1,\cdots,j\omega_n)$ is a complex valued vector with the same dimension as $CE(H_n(j\omega_1,\cdots,j\omega_n))$. In [7], a mapping function $\varphi_n(CE(H_n(\cdot));\omega_1,\cdots,\omega_n)$ from the parametric characteristic $CE(H_n(j\omega_1,\cdots,j\omega_n))$ to its corresponding correlative function $f_n(j\omega_1,\cdots,j\omega_n)$ is established as

$$\begin{aligned}&\varphi_{n(\overline{s})}(c_{p_0,q_0}(\cdot)c_{p_1,q_1}(\cdot)\cdots c_{p_k,q_k}(\cdot);\omega_{l(1)}\cdots\omega_{l(n(\overline{s}))})\\ &= \sum_{\substack{\text{all the 2-partitions}\\ \text{for } \overline{s} \text{ satisfying}\\ s_1(\overline{s})=c_{p,q}(\cdot) \text{ and } p>0}} \Bigg\{ f_1(c_{p,q}(\cdot),n(\overline{s});\omega_{l(1)}\cdots\omega_{l(n(\overline{s}))})\cdot \sum_{\substack{\text{all the p-partitions}\\ \text{for } \overline{s}/c_{p,q}(\cdot)}} \sum_{\substack{\text{all the different}\\ \text{permutations}\\ \text{of } \{s_{x_1},\cdots,s_{x_p}\}}} \Bigg[f_{2a}(s_{\overline{x}_1}\cdots s_{\overline{x}_p}(\overline{s}/c_{p,q}(\cdot));\omega_{l(1)}\cdots\omega_{l(n(\overline{s})-q)}) \\ &\qquad \cdot\prod_{i=1}^{p}\varphi_{n(s_{\overline{x}_i}(\overline{s}/c_{p,q}(\cdot)))}(s_{\overline{x}_i}(\overline{s}/c_{p,q}(\cdot));\omega_{l(\overline{X}(i)+1)}\cdots\omega_{l(\overline{X}(i)+n(s_{\overline{x}_i}(\overline{s}/c_{p,q}(\cdot))))})\Bigg]\Bigg\}\end{aligned} \quad (9a)$$

where the terminating condition is k=0 and $\varphi_1(1;\omega_i) = H_1(j\omega_i)$ (which is the transfer function when all nonlinear parameters are zero), $\{s_{\overline{x}_1},\cdots s_{\overline{x}_p}\}$ is a permutation of $\{s_{x_1},\cdots s_{x_p}\}$, $\omega_{l(1)}\cdots\omega_{l(n(\overline{s}))}$ represents the frequency variables involved in the corresponding functions, $l(i)$ for i=1… $n(\overline{s})$ is a positive integer representing the index of the frequency variables, $\overline{s} = c_{p_0,q_0}(\cdot)c_{p_1,q_1}(\cdot)\cdots c_{p_k,q_k}(\cdot)$, $n(s_x(\overline{s})) = \sum_{i=1}^{x}(p_i+q_i) - x + 1$, x is the number of the parameters in s_x, $\sum_{i=1}^{x}(p_i+q_i)$ is the sum of the subscripts of all the parameters in s_x. Moreover,

$$\overline{X}(i) = \sum_{j=1}^{i-1} n(s_{\overline{x}_j}(\overline{s}/c_{pq}(\cdot))) \quad (9b)$$

$$L_n(j\varpi) = -\sum_{k_1=0}^{K} c_{1,0}(k_1)(j\varpi)^{k_1} \qquad \forall \varpi \in R \quad (9c)$$

$$f_1(c_{p,q}(\cdot),n(\overline{s});\omega_{l(1)}\cdots\omega_{l(n(\overline{s}))}) = (\prod_{i=1}^{q}(j\omega_{l(n(\overline{s})-q+i)})^{k_{p+i}} \Big/ L_{n(\overline{s})}(j\sum_{i=1}^{n(\overline{s})}\omega_{l(i)}) \quad (9d)$$

$$f_{2a}(s_{\overline{x}_1}\cdots s_{\overline{x}_p}(\overline{s}/c_{p,q}(\cdot));\omega_{l(1)}\cdots\omega_{l(n(\overline{s})-q)}) = \prod_{i=1}^{p}(j\omega_{l(\overline{X}(i)+1)} + \cdots + j\omega_{l(\overline{X}(i)+n(s_{\overline{x}_i}(\overline{s}/c_{pq}(\cdot))))})^{k_i} \quad (9e)$$

The mapping function $\varphi_n(CE(H_n(\cdot));\omega_1,\cdots,\omega_n)$ enables the complex valued function $f_n(j\omega_1,\cdots,j\omega_n)$ to be analytically and directly determined in terms of the first order GFRF and nonlinear parameters. Therefore, the nth-order GFRF can directly be written into a more straightforward and meaningful polynomial function in terms of the first order GFRF and model parameters by using the mapping function $\varphi_n(CE(H_n(\cdot));\omega_1,\cdots,\omega_n)$ as

$$H_n(j\omega_1,\cdots,j\omega_n) = CE\big(H_n(j\omega_1,\cdots,j\omega_n)\big)\cdot\varphi_n(CE\big(H_n(\cdot)\big);\omega_1,\cdots,\omega_n) \tag{10}$$

Using (10), Equation (3) can be written as

$$Y(j\omega) = \sum_{n=1}^{N} CE\big(H_n(j\omega_1,\cdots,j\omega_n)\big)\cdot\overline{F}_n(j\omega) \tag{11a}$$

where $\overline{F}_n(j\omega) = \frac{1}{\sqrt{n}(2\pi)^{n-1}}\int_{\omega_1+\cdots+\omega_n=\omega}\varphi_n(CE(H_n(\cdot);\omega_1,\cdots,\omega_n)\cdot\prod_{i=1}^{n}U(j\omega_i)d\sigma_\omega$. Similarly, Equation (6) can be written as

$$Y(j\omega) = \sum_{n=1}^{N} CE\Big(H_n(j\omega_{k_1},\cdots,j\omega_{k_n})\Big)\cdot\tilde{F}_n(\omega) \tag{11b}$$

where $\tilde{F}_n(j\omega) = \frac{1}{2^n}\sum_{\omega_{k_1}+\cdots+\omega_{k_n}=\omega}\varphi_n(CE(H_n(\cdot));\omega_{k_1},\cdots,\omega_{k_n})\cdot F(\omega_{k_1})\cdots F(\omega_{k_n})$. Note that the expressions for output spectrum above are all truncated at the largest order N. The significance of the expressions in (10-11) is that, the explicit relationship between any model parameters and the frequency response functions can be demonstrated clearly and thus it is convenient to be used for system analysis and design.

Example 1. Consider a simple example to demonstrate the results above. Suppose all the other nonlinear parameters in (2) are zero except $c_{1,1}(1,1)$, $c_{0,2}(1,1)$, $c_{2,0}(1,1)$. For convenience, $c_{1,1}(1,1)$ is written as $c_{1,1}$ and so on. Consider the parametric characteristic of $H_3(.)$, which can easily be derived from (8),

$$\begin{aligned} &CE\big(H_3(j\omega_1,\cdots,j\omega_3)\big)\\ &= C_{0,3}\oplus C_{1,1}\otimes C_{0,2}\oplus C_{1,1}^2\oplus C_{1,1}\otimes C_{2,0}\oplus C_{2,1}\oplus C_{1,2}\oplus C_{2,0}\otimes C_{0,2}\oplus C_{2,0}^2\oplus C_{3,0}\\ &= C_{1,1}\otimes C_{0,2}\oplus C_{1,1}^2\oplus C_{1,1}\otimes C_{2,0}\oplus C_{2,0}\otimes C_{0,2}\oplus C_{2,0}^2 \end{aligned}$$

Note that $C_{1,1}= c_{1,1}$, $C_{0,2}=c_{0,2}$, $C_{2,0}=c_{2,0}$. Thus,

$$CE(H_3(j\omega_1,\cdots,j\omega_3)) = [c_{1,1}c_{0,2}, c_{1,1}^2, c_{1,1}c_{2,0}, c_{2,0}c_{0,2}, c_{2,0}c_{1,1}, c_{2,0}^2]$$

Using (9abc), the correlative functions of each term in $CE\big(H_3(j\omega_1,\cdots,j\omega_3)\big)$ can all be obtained. For example, for the term $c_{1,1}c_{0,2}$, it can be derived directly from (9abc) that

$$\begin{aligned} &\varphi_{n(\overline{s})}(c_{1,1}(\cdot)c_{0,2}(\cdot);\omega_{l(1)}\cdots\omega_{l(n(\overline{s}))}) = \varphi_3(c_{1,1}(\cdot)c_{0,2}(\cdot);\omega_1\cdots\omega_3)\\ &= f_1(c_{1,1}(\cdot),3;\omega_1\cdots\omega_3)\cdot f_{2a}(s_1(c_{1,1}(\cdot)c_{0,2}(\cdot)/c_{1,1}(\cdot));\omega_1,\omega_2)\cdot\varphi_2(s_1(c_{0,2}(\cdot));\omega_1,\omega_2)\\ &= f_1(c_{1,1}(\cdot),3;\omega_1\cdots\omega_3)\cdot f_{2a}(c_{0,2}(\cdot);\omega_1,\omega_2)\cdot\varphi_2(c_{0,2}(\cdot);\omega_1,\omega_2)\\ &= \frac{j\omega_3}{L_3(j\omega_1+\cdots+j\omega_3)}\cdot(j\omega_1+j\omega_2)\cdot\frac{j\omega_1 j\omega_2}{L_2(j\omega_1+j\omega_2)} = \frac{j\omega_1 j\omega_2 j\omega_3(j\omega_1+j\omega_2)}{L_3(j\omega_1+\cdots+j\omega_3)L_2(j\omega_1+j\omega_2)} \end{aligned}$$

Proceed with the process above, the whole correlative function of $CE\left(H_3(j\omega_1,\cdots,j\omega_3)\right)$ can be obtained, and then (10-11ab) can be determined. This demonstrates a new way to analytically compute the high order GFRFs, and the final results can directly be written into a polynomial form as (10-11ab), for example in this case

$$\begin{aligned} H_3(j\omega_1,\cdots,j\omega_3) &= [c_{1,1}c_{0,2}, c_{1,1}^2, c_{1,1}c_{2,0}, c_{2,0}c_{0,2}, c_{2,0}c_{1,1}, c_{2,0}^2] \cdot \varphi_3(CE(H_3(j\omega_1,\cdots,j\omega_3)); \omega_1,\cdots,\omega_3) \\ &= c_{1,1}c_{0,2} \cdot \varphi_3(c_{1,1}c_{0,2};\omega_1,\cdots,\omega_3) + c_{1,1}^2 \cdot \varphi_3(c_{1,1}^2;\omega_1,\cdots,\omega_3) + \ldots + c_{2,0}^2 \cdot \varphi_3(c_{2,0}^2;\omega_1,\cdots,\omega_3) \end{aligned}$$

As discussed in [7], it can be seen from Equations (10-11ab) and Example 1 that the mapping function $\varphi_n(CE(H_n(\cdot));\omega_1,\cdots,\omega_n)$ can facilitate the frequency domain analysis of nonlinear systems such that the relationship between the frequency response functions and model parameters, and the relationship between the frequency response functions and $H_1(j\omega_{l(1)})$ can be demonstrated explicitly, and some new properties of the GFRFs and output spectrum can be revealed. In practice, the output spectrum of a nonlinear system can be expanded as a power series with respect to a specific model parameter of interest by using (11ab) for N→∞. The nonlinear effect on system output spectrum incurred by this model parameter which may represents the physical characteristic of a structural unit in the system can then be analysed and designed by studying this power series in the frequency domain. Note that the fundamental properties of this power series (e.g. convergence) are to a large extent dominated by the properties of its coefficients, which are explicitly determined by the mapping function $\varphi_n(CE(H_n(\cdot));\omega_1,\cdots,\omega_n)$. Thus studying the properties of this power series is now equivalent to studying the properties of the mapping function $\varphi_n(CE(H_n(\cdot));\omega_1,\cdots,\omega_n)$. Therefore, the mapping function $\varphi_n(CE(H_n(\cdot));\omega_1,\cdots,\omega_n)$ introduced above provides an important and significant technique for this frequency domain analysis to study the nonlinear influence on system output spectrum.

In this study, a novel property of the nonlinear influence on system output spectrum is revealed by using the new mapping function $\varphi_n(CE(H_n(\cdot));\omega_1,\cdots,\omega_n)$ and frequency response functions defined in Equations (10-11). It is shown that the nonlinear terms in a system can drive the system output spectrum to be an alternating series under certain conditions when the system subjects to a sinusoidal input, and the system output spectrum is shown to have some interesting properties in engineering practice when it can be expanded into an alternating series with respect to a specific model parameter of interest. This provides a novel insight into the nonlinear effect incurred by nonlinear terms in a nonlinear system to the system output spectrum.

3. Alternating phenomenon in the output spectrum and its influence

The alternating phenomena and its influence are discussed in this section to point out the significance of this novel property, and then the conditions under which system output spectrum can be expressed into an alternating series are studied in the following section.

For any nonlinear parameter (simply denoted by c) in model (2), the output spectrum (11ab) can be expanded with respect to this parameter into a power series as

$$Y(j\omega) = F_0(j\omega) + cF_1(j\omega) + c^2F_2(j\omega) + \cdots + c^\rho F_\rho(j\omega) + \cdots \tag{12}$$

Note that when c represents a nonlinearity from input terms, Equation (12) may be a finite series; in other cases, it is definitely an infinite series, and if only the first ρ terms in the series (12) are considered, there is a truncation error denoted by $o(\rho)$. As demonstrated in Example 1, $F_i(j\omega)$ for i=0,1,2,... are some scalar frequency functions and can be obtained from $\bar{F}_i(j\omega)$ or $\tilde{F}_i(j\omega)$ in (11a,b) by using the mapping function $\varphi_n(CE(H_n(\cdot));\omega_1,\cdots,\omega_n)$. Clearly, $F_i(j\omega)$ dominates the fundamental properties of this power series such as convergence. Thus these properties of this power series can be revealed by studying the property of $\varphi_n(CE(H_n(\cdot));\omega_1,\cdots,\omega_n)$. This will be discussed more in the next section. In this section, the alternating phenomenon of this power series and its influence are discussed.

For any $\upsilon \in \mathbb{C}$, define an operator as

$$\text{sgn}_c(\upsilon) = \left[\text{sgn}_r(\text{Re}(\upsilon)) \quad \text{sgn}_r(\text{Im}(\upsilon))\right]$$

where $\text{sgn}_r(x) = \begin{cases} +1 & x > 0 \\ 0 & x = 0 \\ -1 & x < 0 \end{cases}$ for $x \in \mathbb{R}$.

Definition 1 (Alternating series). Consider a power series of form (12) with c>0. If $\text{sgn}_c(F_i(j\omega)) = -\text{sgn}_c(F_{i+1}(j\omega))$ for i=0,1,2,3,..., then the series is an alternating series.

The series (12) can be written into two series as

$$\begin{aligned} Y(j\omega) &= \text{Re}(Y(j\omega)) + j(\text{Im}(Y(j\omega))) \\ &= \text{Re}(F_0(j\omega)) + c\,\text{Re}(F_1(j\omega)) + c^2\,\text{Re}(F_2(j\omega)) + \cdots + c^\rho\,\text{Re}(F_\rho(j\omega)) + \cdots \\ &\quad + j(\text{Im}(F_0(j\omega)) + c\,\text{Im}(F_1(j\omega)) + c^2\,\text{Im}(F_2(j\omega)) + \cdots + c^\rho\,\text{Im}(F_\rho(j\omega)) + \cdots) \end{aligned} \tag{13}$$

From definition 1, if $Y(j\omega)$ is an alternating series, then $\text{Re}(Y(j\omega))$ and $\text{Im}(Y(j\omega))$ are both alternating. When (12) is an alternating series, there are some interesting properties summarized in Theorem 1. Denote

$$Y(j\omega)_{1\to\rho} = F_0(j\omega) + cF_1(j\omega) + c^2F_2(j\omega) + \cdots + c^\rho F_\rho(j\omega) \tag{14}$$

Theorem 1. Suppose (12) is an alternating series at a ω ($\in \mathbb{R}_+$) for c>0, then:

(1) if there exist T>0 and R>0 such that for i>T

$$\min\left\{-\frac{\text{Re}(F_i(j\omega))}{\text{Re}(F_{i+1}(j\omega))}, -\frac{\text{Im}(F_i(j\omega))}{\text{Im}(F_{i+1}(j\omega))}\right\} > R$$

then (12) has a radius of convergence R, the truncation error for a finite order $\rho > T$ is $|o(\rho)| \le c^{\rho+1} |F_{\rho+1}(j\omega)|$, and for all $n \ge 0$,

$$|Y(j\omega)| \in \Pi_n = [|Y(j\omega)_{1\to T+2n+1}|, |Y(j\omega)_{1\to T+2n}|] \text{ and } \Pi_{n+1} \subset \Pi_n;$$

(2) $|Y(j\omega)|^2 = Y(j\omega)Y(-j\omega)$ is also an alternating series with respect to parameter c; Furthermore, $|Y(j\omega)|^2 = Y(j\omega)Y(-j\omega)$ is alternating only if $\mathrm{Re}(Y(j\omega))$ is alternating;

(3) there exists a constant $\bar{c} > 0$ such that $\frac{\partial |Y(j\omega)|}{\partial c} < 0$ for $0 < c < \bar{c}$.

Proof. See Appendix C.□

The first point in Theorem 1 shows that only if there exists a positive constant $R>0$, the series must be convergent under $0<c<R$, its truncation error and limit value can therefore be easily evaluated. The other two points of Theorem 1 imply that the magnitude of an alternating series can be suppressed by choosing a proper value for the parameter c. Therefore, once the system output spectrum can be expressed into an alternating series with respect to a model parameter (say c), it is easier to find a proper value for c such that the output spectrum is convergent, and the magnitude can be suppressed. Moreover, it is also shown that the lowest limit of the magnitude of the output spectrum that can be reached is larger than $|Y(j\omega)_{1\to T+1}|$ and the truncation error of the output spectrum is less than the absolute value of the term of the largest order at the truncated point.

Example 2. Consider a single degree of freedom (SDOF) spring-damping system with a cubic nonlinear damping which can be described by the following differential equation

$$m\ddot{y} = -k_0 y - B\dot{y} - c\dot{y}^3 + u(t) \tag{15}$$

Note that k_0 represents the spring characteristic, B the damping characteristic and c is the cubic nonlinear damping characteristic. This system is a simple case of NDE model (2) and can be written into the form of NDE model with M=3, K=2, $c_{1,0}(2) = m$, $c_{1,0}(1) = B$, $c_{1,0}(0) = k_0$, $c_{3,0}(111) = c$, $c_{0,1}(0) = -1$ and all the other parameters are zero.

Note that there is only one nonlinear term in the output in this case, the nth-order GFRF for system (15) can be derived according to the algorithm in [1], which can be recursively determined as

$$H_n(j\omega_1,\cdots,j\omega_n) = \frac{c_{3,0}(1,1,1)H_{n,3}(j\omega_1,\cdots,j\omega_n)}{L_n(j\omega_1+\cdots+j\omega_n)}$$

$$H_{n,3}(\cdot) = \sum_{i=1}^{n-2} H_i(j\omega_1,\cdots,j\omega_i)H_{n-i,2}(j\omega_{i+1},\cdots,j\omega_n)(j\omega_1+\cdots+j\omega_i)$$

$$H_{n,1}(j\omega_1,\cdots,j\omega_n) = H_n(j\omega_1,\cdots,j\omega_n)(j\omega_1+\cdots+j\omega_n)$$

Proceeding with the recursive computation above, it can be seen that $H_n(j\omega_1,\cdots,j\omega_n)$ is a polynomial of $c_{3,0}(111)$, and substituting these equations above into (11) gives another polynomial for the output spectrum. By using the relationship (10) and the mapping function $\varphi_n(CE(H_n(\cdot));\omega_1,\cdots,\omega_n)$, these results can be obtained directly as follows.

For simplicity, let $u(t)=F_d\sin(\Omega t)$ $(F_d>0)$. Then $F(\omega_{k_l})=-jk_lF_d$, for $k_l=\pm1$, $\omega_{k_l}=k_l\Omega$, and $l=1,\cdots,n$ in (11b). By using (8) or Proposition 5 in [6], it can be obtained that

$$CE(H_{2n+1}(j\omega_1,\cdots,j\omega_{2n+1}))=(c_{3,0}(1,1,1))^n \text{ and } CE(H_{2n}(j\omega_1,\cdots,j\omega_{2n}))=0 \text{ for n=0,1,2,3,...} \quad (16)$$

Therefore, for n=0,1,2,3,...

$$H_{2n+1}(j\omega_1,\cdots,j\omega_{2n+1})=c^n\cdot\varphi_{2n+1}(CE(H_{2n+1}(\cdot));\omega_1,\cdots,\omega_{2n+1}) \text{ and } H_{2n}(j\omega_1,\cdots,j\omega_{2n})=0 \quad (17)$$

Then the output spectrum at frequency Ω can be computed as (N is the largest order after truncated)

$$Y(j\Omega)=\sum_{n=0}^{\lfloor N-1/2\rfloor} c^n\cdot\widetilde{F}_{2n+1}(\Omega)+\cdots \quad (18)$$

where $\widetilde{F}_{2n+1}(j\Omega)$ can be computed as

$$\begin{aligned}\tilde{F}_{2n+1}(j\Omega)&=\frac{1}{2^{2n+1}}\sum_{\omega_{k_1}+\cdots+\omega_{k_{2n+1}}=\Omega}\varphi_{2n+1}(CE(H_{2n+1}(\cdot));\omega_{k_1},\cdots,\omega_{k_{2n+1}})\cdot(-jF_d)^{2n+1}\cdot k_1k_2\cdots k_{2n+1}\\&=\frac{1}{2^{2n+1}}\sum_{\omega_{k_1}+\cdots+\omega_{k_{2n+1}}=\Omega}\varphi_{2n+1}(CE(H_{2n+1}(\cdot));\omega_{k_1},\cdots,\omega_{k_{2n+1}})\cdot(-1)^{n+1}j(F_d)^{2n+1}\cdot(-1)^n\\&=-j(\frac{F_d}{2})^{2n+1}\sum_{\omega_{k_1}+\cdots+\omega_{k_{2n+1}}=\Omega}\varphi_{2n+1}(CE(H_{2n+1}(\cdot));\omega_{k_1},\cdots,\omega_{k_{2n+1}})\end{aligned} \quad (19)$$

and $\varphi_{2n+1}(CE(H_{2n+1}(\cdot));\omega_1,\cdots,\omega_{2n+1})=\varphi_{2n+1}(c_{3,0}(1,1,1)^n;\omega_1,\cdots,\omega_{2n+1})$ can be obtained according to equations (9a-c). For example,

$$\varphi_3(c_{3,0}(111);\omega_1,\omega_2,\omega_3)=\frac{1}{L_3(j\sum_{i=1}^{3}\omega_i)}\cdot\prod_{i=1}^{3}(j\omega_i)\cdot\prod_{i=1}^{3}H_1(j\omega_i)=\frac{\prod_{i=1}^{3}(j\omega_i)}{L_3(j\sum_{i=1}^{3}\omega_i)}\cdot\prod_{i=1}^{3}H_1(j\omega_i)$$

$$\varphi_5(c_{3,0}(111)c_{3,0}(111);\omega_1,\cdots,\omega_5)$$

$$= f_1(c_{3,0}(111),5;\omega_1,\cdots,\omega_5)\cdot \sum_{\substack{\text{all the 3-partitions}\\ \text{for } c_{3,0}(111)}} \sum_{\substack{\text{all the different}\\ \text{permutations of } \{0,0,1\}}} \Big[f_{2a}(s_{\bar{X}_1}\cdots s_{\bar{X}_p}(c_{3,0}(111));\omega_1\cdots\omega_5)$$

$$\cdot\prod_{i=1}^{3}\varphi_{n(s_{\bar{X}_i}(\bar{s}/c_{p,q}(\cdot)))}(s_{\bar{X}_i}(c_{3,0}(111));\omega_{l(\bar{X}(i)+1)}\cdots\omega_{l(\bar{X}(i)+n(s_{\bar{X}_i}(\bar{s}/c_{p,q}(\cdot))))})\Big]$$

$$= f_1(c_{3,0}(111),5;\omega_1,\cdots,\omega_5)\cdot\begin{pmatrix} f_{2a}(s_0s_0s_1(c_{3,0}(111));\omega_1\cdots\omega_5)\varphi_1(1;\omega_1)\varphi_1(1;\omega_2)\varphi_3(c_{3,0}(111);\omega_3\cdots\omega_5) \\ +f_{2a}(s_0s_1s_0(c_{3,0}(111));\omega_1\cdots\omega_5)\varphi_1(1;\omega_1)\varphi_3(c_{3,0}(111);\omega_2\cdots\omega_4)\varphi_1(1;\omega_5) \\ +f_{2a}(s_1s_0s_0(c_{3,0}(111));\omega_1\cdots\omega_5)\varphi_3(c_{3,0}(111);\omega_1\cdots\omega_3)\varphi_1(1;\omega_4)\varphi_1(1;\omega_5) \end{pmatrix}$$

$$= \frac{1}{L_5(j\sum_{i=1}^{5}\omega_i)}\cdot\left(\frac{(j\sum_{i=3}^{5}\omega_i)\prod_{i=1}^{5}(j\omega_i)}{L_3(j\sum_{i=3}^{5}\omega_i)}+\frac{(j\sum_{i=2}^{4}\omega_i)\prod_{i=1}^{5}(j\omega_i)}{L_3(j\sum_{i=2}^{4}\omega_i)}+\frac{(j\sum_{i=1}^{3}\omega_i)\prod_{i=1}^{5}(j\omega_i)}{L_3(j\sum_{i=1}^{3}\omega_i)}\right)\cdot\prod_{i=1}^{5}H_1(j\omega_i)$$

where $\omega_i \in \{\Omega,-\Omega\}$, and so on. Substituting these results into Equations (18-19), the output spectrum is clearly a power series with respect to the parameter c. When there are more nonlinear terms, it is obvious that the computation process above can directly result in a straightforward multivariate power series with respect to these nonlinear parameters. To check the alternating phenomenon of the output spectrum, consider the following values for each linear parameter: m=240, k_0=16000, B=296, F_d=100, and $\Omega = 8.165$. Then it is obtained that

$$Y(j\Omega)=\tilde{F}_1(\Omega)+c\tilde{F}_3(\Omega)+c^2\tilde{F}_5(\Omega)+\cdots$$

$$=-j(\frac{F_d}{2})H_1(j\Omega)+3(\frac{F_d}{2})^3\frac{\Omega^3|H_1(j\Omega)|^2H_1(j\Omega)}{L_1(j\Omega)}$$

$$+3(\frac{F_d}{2})^5\frac{\Omega^5|H_1(j\Omega)|^4H_1(j\Omega)}{L_1(j\Omega)}(\frac{j6\Omega}{L_1(j\Omega)}+\frac{j3\Omega}{L_1(j3\Omega)}+\frac{-j3\Omega}{L_1(-j\Omega)})+\cdots$$

=(-0.02068817126756 + 0.00000114704116i)

+(5.982851578532449e-006 -6.634300276113922e-010i)c

+(-5.192417616715994e-009 +3.323565122085705e-011i)c^2+… (20a)

The series is alternating. In order to check the series further, computation of $\varphi_{2n+1}(c_{3,0}(1,1,1)^n;\omega_1,\cdots,\omega_{2n+1})$ can be carried out for higher orders. It can also be verified that the magnitude square of the output spectrum (20a) is still an alternating series, *i.e.*,

$$|Y(j\Omega)|^2 = (4.280004317115985\text{e-}004)-(2.475485177721052\text{e-}007)c$$

$$+(2.506378395908398e\text{-}010)c^2 - \ldots \qquad (20b)$$

As pointed in Theorem 1, it is easy to find a c such that (20a-b) are convergent and their limits are decreased. From (20b) and according to Theorem 1, it can be computed that $0.01671739 < |Y(j\Omega)| < 0.0192276 < 0.0206882$ for c=600. This can be verified by Figure 1. Figure 1 is a result from simulation tests, and shows that the magnitude of the output spectrum is decreasing when c is increasing. This property is of great significance in practical engineering systems for output suppression through structural characteristic design or feedback control.

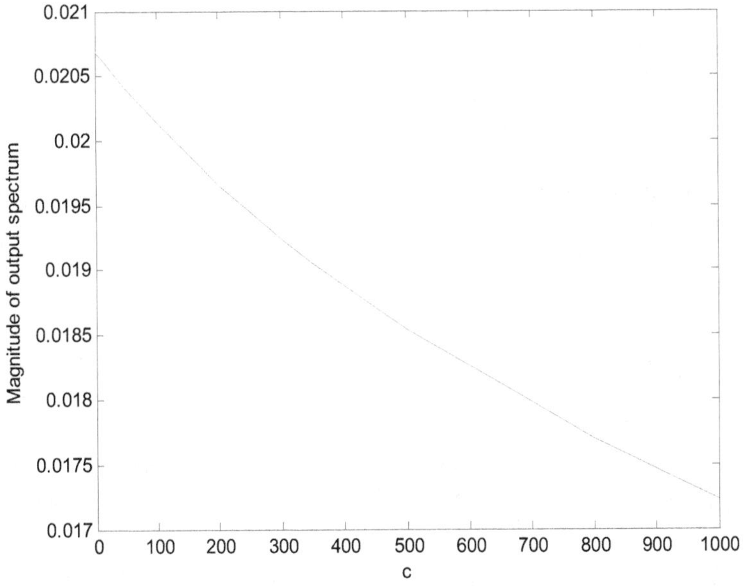

Figure 1. Magnitude of output spectrum

4. Alternating conditions

In this section, the conditions under which the output spectrum described by Equation (12) can be expressed into an alternating series with respect to any nonlinear parameter are studied. Suppose the system subjects to a harmonic input $u(t) = F_d \sin(\Omega t)$ $(F_d > 0)$ and only the output nonlinearities (*i.e.*, $c_{p,0}(.)$ with $p \geq 2$) are considered. For convenience, assume that there is only one nonlinear parameter $c_{p,0}(.)$ in model (2) and all the other nonlinear parameters are zero. The results for this case can be extended to the general one.

Under the assumptions above, it can be obtained from the parametric characteristic analysis in [6] as demonstrated in Example 2 and Equation (11b) that

$$\begin{aligned} Y(j\Omega) &= Y_1(j\Omega) + Y_p(j\Omega) + \cdots + Y_{(p-1)n+1}(j\Omega) + \cdots \\ &= \tilde{F}_1(\Omega) + c_{p,0}(\cdot)\tilde{F}_p(\Omega) + \cdots + c_{p,0}(\cdot)^n \tilde{F}_{(p-1)n+1}(\Omega) + \cdots \end{aligned} \tag{21a}$$

where $\omega_{k_i} \in \{\pm\Omega\}$, $\tilde{F}_{(p-1)n+1}(j\Omega)$ can be computed from (11b), and n is a positive integer. Noting that $F(\omega_{k_l}) = -jk_l F_d$, $k_l = \pm 1$, $\omega_{k_l} = k_l\Omega$, and $l = 1,\cdots,n$ in (11b),

$$\tilde{F}_{(p-1)n+1}(j\Omega) = \frac{1}{2^{(p-1)n+1}} \sum_{\omega_{k_1}+\cdots+\omega_{k_{(p-1)n+1}}=\Omega} \varphi_{(p-1)n+1}(c_{p,0}(\cdot)^n;\omega_{k_1},\cdots,\omega_{k_{(p-1)n+1}})\cdot(-jF_d)^{(p-1)n+1}\cdot k_1k_2\cdots k_{(p-1)n+1} \tag{21b}$$

If p is an odd integer, then $(p-1)n+1$ is also an odd integer. Thus there should be $(p-1)n/2$ frequency variables being $-\Omega$ and $(p-1)n/2+1$ frequency variables being Ω such that $\omega_{k_1} + \cdots + \omega_{k_{(p-1)n+1}} = \Omega$. In this case,

$$(-jF_d)^{(p-1)n+1}\cdot k_1k_2\cdots k_{(p-1)n+1} = (-1)\cdot j\cdot\left(j^2\right)^{(p-1)n/2}\cdot(F_d)^{(p-1)n+1}\cdot(-1)^{(p-1)n/2} = -j(F_d)^{(p-1)n+1}$$

If p is an even integer, then $(p-1)n+1$ is an odd integer for $n=2k$ ($k=1,2,3,\ldots$) and an even integer for $n=2k-1$ ($k=1,2,3,\ldots$). When n is an odd integer, $\omega_{k_1} + \cdots + \omega_{k_{(p-1)n+1}} \neq \Omega$ for $\omega_{k_l} \in \{\pm\Omega\}$. This gives that $\tilde{F}_{(p-1)n+1}(j\Omega)$ =0. When n is an even integer, $(p-1)n+1$ is an odd integer. In this case, it is similar to that p is an odd integer. Therefore, for $n>0$

$$\tilde{F}_{(p-1)n+1}(j\Omega) = \begin{cases} -j\left(\dfrac{F_d}{2}\right)^{(p-1)n+1} \displaystyle\sum_{\omega_{k_1}+\cdots+\omega_{k_{(p-1)n+1}}=\Omega} \varphi_{(p-1)n+1}(c_{p,0}(\cdot)^n;\omega_{k_1},\cdots,\omega_{k_{(p-1)n+1}}) & \text{if p is odd or n is even} \\ 0 & \text{else} \end{cases} \tag{21c}$$

From Equations (21a-c) it is obvious that the property of the new mapping $\varphi_{(p-1)n+1}(c_{p,0}(\cdot)^n;\omega_{k_1},\cdots,\omega_{k_{(p-1)n+1}})$ plays a key role in the series. To develop the alternating conditions for series (21a), the following results can be obtained.

Lemma 1. That $\varphi_{(p-1)n+1}(c_{p,0}(\cdot)^n;\omega_{k_1},\cdots,\omega_{k_{(p-1)n+1}})$ is symmetric or asymmetric has no influence on $\tilde{F}_{(p-1)n+1}(j\Omega)$.

Lemma 1 is obvious since $\sum_{\omega_{k_1}+\cdots+\omega_{k_{(p-1)n+1}}=\Omega}(\cdot)$ includes all the possible permutations of $(\omega_{k_1},\cdots,\omega_{k_{2n+1}})$. Although there are many choices to obtain the asymmetric $\varphi_{(p-1)n+1}(c_{p,0}(\cdot)^n;\omega_{k_1},\cdots,\omega_{k_{(p-1)n+1}})$ which may be different at different permutation $(\omega_{k_1},\cdots,\omega_{k_{(p-1)n+1}})$, they have no effect on the analysis of $\tilde{F}_{(p-1)n+1}(j\Omega)$.

The following lemma is straightforward.

Lemma 2. For $\upsilon_1,\upsilon_2,\nu \in \mathbb{C}$, suppose $\mathrm{sgn}_c(\upsilon_1) = -\mathrm{sgn}_c(\upsilon_2)$. If $\mathrm{Re}(\nu)\mathrm{Im}(\nu) = 0$, then $\mathrm{sgn}_c(\upsilon_1\nu) = -\mathrm{sgn}_c(\upsilon_2\nu)$. If $\mathrm{Re}(\nu)\mathrm{Im}(\nu) = 0$ and $\nu \neq 0$, then $\mathrm{sgn}_c(\upsilon_1/\nu) = -\mathrm{sgn}_c(\upsilon_2/\nu)$ □.

Theorem 2. The output spectrum in (21a-c) is an alternating series with respect to parameter $c_{p,0}(k_1,k_2,\ldots,k_p)$ satisfying $c_{p,0}(.)>0$ and $p=2r+1$ for r=1,2,3,...

(1) if and only if

$$\mathrm{sgn}_c\left(\sum_{\omega_{k_1}+\cdots+\omega_{k_{(p-1)n+1}}=\Omega}(-1)^{n-1}\varphi_{(p-1)n+1}(c_{p,0}(\cdot)^n;\omega_{l(1)}\cdots\omega_{l((p-1)n+1)})\right)=const\text{ , i.e.,}$$

$$\mathrm{sgn}_c\left(\begin{array}{l}\dfrac{H_1(j\Omega)}{L_{(p-1)n+1}(j\Omega)}\displaystyle\sum_{\omega_{k_1}+\cdots+\omega_{k_{(p-1)n+1}}=\Omega}\ \sum_{\substack{\text{all the different combinations}\\ \text{of }\{\bar{x}_1,\bar{x}_2,\ldots,\bar{x}_p\}\text{ satisfying}\\ \bar{x}_1+\cdots+\bar{x}_p=n-1,\ 0\le\bar{x}_i\le n-1}}\left[\prod_{i=1}^{p}\varphi'_{(p-1)\bar{x}_i+1}(c_{p,0}(\cdot)^{\bar{x}_i};\omega_{l(\bar{X}(i)+1)}\cdots\omega_{l(\bar{X}(i)+(p-1)\bar{x}_i+1)})\right.\\ \left.\cdot\dfrac{n_x^*(\bar{x}_1,\cdots,\bar{x}_p)}{n_k^*(k_1,\cdots,k_p)}\cdot\displaystyle\sum_{\substack{\text{all the different}\\ \text{permutations of}\\ \{k_1,\cdots,k_p\}}}\prod_{i=1}^{p}(j\omega_{l(\bar{X}(i)+1)}+\cdots+j\omega_{l(\bar{X}(i)+(p-1)\bar{x}_i+1)})^{k_i}\right]\end{array}\right)=const \tag{22}$$

where $const$ is a two-dimensional constant vector whose elements are +1, 0 or -1;

$$\begin{array}{l}\varphi'_{(p-1)n+1}(c_{p,0}(\cdot)^n;\omega_{l(1)}\cdots\omega_{l((p-1)n+1)})\\ =\dfrac{-1}{L_{(p-1)n+1}(j\omega_{l(1)}+\cdots+j\omega_{l((p-1)n+1)})}\cdot\displaystyle\sum_{\substack{\text{all the different combinations}\\ \text{of }\{\bar{x}_1,\bar{x}_2,\ldots,\bar{x}_p\}\text{ satisfying}\\ \bar{x}_1+\cdots+\bar{x}_p=n-1,\ 0\le\bar{x}_i\le n-1}}\left[\prod_{i=1}^{p}\varphi'_{(p-1)\bar{x}_i+1}(c_{p,0}(\cdot)^{\bar{x}_i};\omega_{l(\bar{X}(i)+1)}\cdots\omega_{l(\bar{X}(i)+(p-1)\bar{x}_i+1)})\right.\\ \left.\cdot\dfrac{n_x^*(\bar{x}_1,\cdots,\bar{x}_p)}{n_k^*(k_1,\cdots,k_p)}\cdot\displaystyle\sum_{\substack{\text{all the different}\\ \text{permutations of}\\ \{k_1,\cdots,k_p\}}}\prod_{i=1}^{p}(j\omega_{l(\bar{X}(i)+1)}+\cdots+j\omega_{l(\bar{X}(i)+(p-1)\bar{x}_i+1)})^{k_i}\right]\end{array}$$

the termination is $\varphi'_1(1;\omega_i)=1$; $n_k^*(k_1,\cdots,k_p)=\dfrac{p!}{n_1!n_2!\cdots n_e!}$, $n_1+\ldots+n_e=p$, e is the number of distinct differentials k_i appearing in the combination, n_i is the number of repetitions of k_i, and a similar definition holds for $n_x^*(\bar{x}_1,\cdots,\bar{x}_p)$.

(2) or if $k_1=k_2=\ldots=k_p=k$ in $c_{p,0}(.)$, $\mathrm{Re}(\dfrac{H_1(j\Omega)}{L_{(p-1)n+1}(j\Omega)})\mathrm{Im}(\dfrac{H_1(j\Omega)}{L_{(p-1)n+1}(j\Omega)})=0$, and

$$\mathrm{sgn}_c\left(\sum_{\omega_{k_1}+\cdots+\omega_{k_{(p-1)n+1}}=\Omega}\ \sum_{\substack{\text{all the different combinations}\\ \text{of }\{\bar{x}_1,\bar{x}_2,\ldots,\bar{x}_p\}\text{ satisfying}\\ \bar{x}_1+\cdots+\bar{x}_p=n-1,\ 0\le\bar{x}_i\le n-1}}\left[n_x^*(\bar{x}_1,\cdots,\bar{x}_p)\cdot\prod_{i=1}^{p}\varphi''_{(p-1)\bar{x}_i+1}(c_{p,0}(\cdot)^{\bar{x}_i};\omega_{l(\bar{X}(i)+1)}\cdots\omega_{l(\bar{X}(i)+(p-1)\bar{x}_i+1)})\right]\right)=const \tag{23}$$

where if $\bar{x}_i = 0$, $\varphi''_{(p-1)\bar{x}_i+1}(c_{p,0}(\cdot)^{\bar{x}_i};\omega_{l(\bar{X}(i)+1)}\cdots\omega_{l(\bar{X}(i)+(p-1)\bar{x}_i+1)})=1$, otherwise,

$$\varphi''_{(p-1)\bar{x}_i+1}(c_{p,0}(\cdot)^{\bar{x}_i};\omega_{l(\bar{X}(i)+1)}\cdots\omega_{l(\bar{X}(i)+(p-1)\bar{x}_i+1)})$$

$$=\frac{(j\omega_{l(\bar{X}(i)+1)}+\cdots+j\omega_{l(\bar{X}(i)+(p-1)\bar{x}_i+1)})^k}{-L_{(p-1)\bar{x}_i+1}(j\omega_{l(\bar{X}(i)+1)}+\cdots+j\omega_{l(\bar{X}(i)+(p-1)\bar{x}_i+1)})}\cdot$$

$$\sum_{\substack{\text{all the different combinations}\\ \text{of } \{x_1,x_2,\ldots,x_p\} \text{ satisfying}\\ x_1+\cdots+x_p=\bar{x}_i-1,\ 0\le x_j\le \bar{x}_i-1}} n_x^*(x_1,\cdots,x_p)\cdot\prod_{j=1}^{p}\varphi''_{(p-1)x_j+1}(c_{p,0}(\cdot)^{x_j};\omega_{l(\bar{X}'(j)+1)}\cdots\omega_{l(\bar{X}'(j)+(p-1)x_j+1)})$$

The recursive terminal of $\varphi''_{(p-1)\bar{x}_i+1}(c_{p,0}(\cdot)^{\bar{x}_i};\omega_{l(\bar{X}(i)+1)}\cdots\omega_{l(\bar{X}(i)+(p-1)\bar{x}_i+1)})$ is $\bar{x}_i=1$.

Proof. See Appendix D. □

Theorem 2 provides a sufficient and necessary condition for the output spectrum series (21a-c) to be an alternating series with respect to a nonlinear parameter $c_{p,0}(k_1,k_2,\ldots,k_p)$ satisfying $c_{p,0}(.)>0$ and $p=2r+1$ for r=1,2,3,.... Similar results can also be established for any other nonlinear parameters. Regarding nonlinear parameter $c_{p,0}(k_1,k_2,\ldots,k_p)$ satisfying $c_{p,0}(.)>0$ and $p=2r$ for r=1,2,3,...., it can be obtained from (21a-c) that

$$Y(j\Omega)=\tilde{F}_1(\Omega)+c_{p,0}(\cdot)^2\tilde{F}_{2(p-1)+1}(\Omega)+\cdots+c_{p,0}(\cdot)^{2n}\tilde{F}_{2(p-1)n+1}(\Omega)+\cdots$$

$\tilde{F}_{2(p-1)n+1}(\Omega)$ for n=1,2,3,... should be alternating so that $Y(j\Omega)$ is alternating. This yields that

$$\operatorname{sgn}_c\left(\sum_{\omega_{k_1}+\cdots+\omega_{k_{2(p-1)n+1}}=\Omega}\varphi_{2(p-1)n+1}(c_{p,0}(\cdot)^{2n};\omega_{l(1)}\cdots\omega_{l(2(p-1)n+1)})\right)$$

$$=-\operatorname{sgn}_c\left(\sum_{\omega_{k_1}+\cdots+\omega_{k_{2(p-1)(n+1)+1}}=\Omega}\varphi_{2(p-1)(n+1)+1}(c_{p,0}(\cdot)^{2(n+1)};\omega_{l(1)}\cdots\omega_{l(2(p-1)(n+1)+1)})\right)$$

Clearly, this is different from the conditions in Theorem 2. It may be more difficult for the output spectrum to be alternating with respect to $c_{p,0}(.)>0$ with $p=2r$ (even degree) than with respect to $c_{p,0}(.)>0$ with $p=2r+1$ (odd degree).

Note that Equation (21a) is based on the assumption that there is only nonlinear parameter $c_{p,0}(.)$ and all the other nonlinear parameters are zero. If the effects from the other nonlinear parameters are considered, Equation (21a) can be written as

$$Y(j\Omega)=\tilde{F}'_1(\Omega)+c_{p,0}(\cdot)\tilde{F}'_p(\Omega)+\cdots+c_{p,0}(\cdot)^n\tilde{F}'_{(p-1)n+1}(\Omega)+\cdots \tag{24a}$$

where

$$\tilde{F}'_{(p-1)n+1}(\Omega) = \tilde{F}_{(p-1)n+1}(\Omega) + \delta_{(p-1)n+1}(\Omega; C_{p',q'} \setminus c_{p,0}(.)) \tag{24b}$$

$C_{p',q'}$ includes all the nonlinear parameters in the system. Based on the parametric characteristic analysis in [6] and the new mapping function $\varphi_n(CE(H_n(\cdot)); \omega_1, \cdots, \omega_n)$ defined in [7], (24b) can be determined easily. For example, suppose p is an odd integer larger than 1, then $\tilde{F}_{(p-1)n+1}(j\Omega)$ is given in (21c), and $\delta_{(p-1)n+1}(\Omega; C_{p',q'} \setminus c_{p,0}(.))$ can be computed as

$$\delta_{(p-1)n+1}(\Omega; C_{p',q'} \setminus c_{p,0}(.)) = \sum_{\substack{\text{all the monomails consisting of the parameters in } C_{p',q'} \setminus c_{p,0}(\cdot) \\ \text{satisfying } np + \sum(p'_i + q'_i) \text{ is odd and less than N}}} \left[-j \left(\frac{F_d}{2} \right)^{n(c_{p,0}{}^n s(\cdot))} \cdot \sum_{\omega_{k_1} + \cdots + \omega_{k_{n(c_{p,0}{}^n s(\cdot))}} = \Omega} \varphi_{n(c_{p,0}{}^n s(\cdot))}(c_{p,0}{}^n s(C_{p',q'} \setminus c_{p,0}(\cdot)); \omega_{k_1} \cdots \omega_{k_{n(c_{p,0}{}^n s(\cdot))}}) \right]$$

where $s(C_{p',q'} \setminus c_{p,0}(\cdot))$ denotes a monomial consisting of some parameters in $C_{p',q'} \setminus c_{p,0}(\cdot)$.

It is obvious that if (21a) is an alternating series, then (24a) can still be alternating under a proper design of the other nonlinear parameters (for example the other parameters are sufficiently small). Moreover, from the discussions above, it can be seen that whether the system output spectrum is an alternating series or not with respect to a specific nonlinear parameter is greatly dependent on the system linear parameters.

Example 3. To demonstrate the theoretical results above, consider again model (15) in Example 2. Let $u(t) = F_d \sin(\Omega t)$ $(F_d > 0)$. The output spectrum at frequency Ω is given in (18-19). From Lemma 3 in Appendix D, it can be derived for this case that

$$\varphi_{2n+1}(c_{3,0}(\cdot)^n; \omega_{l(1)} \cdots \omega_{l(2n+1)}) = \frac{(-1)^{n-1} \prod_{i=1}^{2n+1} [(j\omega_{l(i)})^k H_1(j\omega_{l(i)})]}{L_{2n+1}(j\omega_{l(1)} + \cdots + j\omega_{l(2n+1)})} \cdot \sum_{\substack{\text{all the different combinations} \\ \text{of } \{\bar{x}_1, \bar{x}_2, \bar{x}_3\} \text{ satisfying} \\ \bar{x}_1 + \bar{x}_2 + \bar{x}_3 = n-1,\ 0 \le \bar{x}_i \le n-1}} n_x^*(\bar{x}_1, \bar{x}_2, \bar{x}_3) \cdot \prod_{i=1}^{3} \varphi''_{2\bar{x}_i+1}(c_{3,0}(\cdot)^{\bar{x}_i}; \omega_{l(\bar{X}(i)+1)} \cdots \omega_{l(\bar{X}(i)+2\bar{x}_i+1)}) \tag{25a}$$

where, if $\bar{x}_i = 0$, $\varphi''_{(p-1)\bar{x}_i+1}(c_{p,0}(\cdot)^{\bar{x}_i}; \omega_{l(\bar{X}(i)+1)} \cdots \omega_{l(\bar{X}(i)+(p-1)\bar{x}_i+1)}) = 1$, otherwise,

$$\varphi''_{2\bar{x}_i+1}(c_{3,0}(\cdot)^{\bar{x}_i}; \omega_{l(\bar{X}(i)+1)} \cdots \omega_{l(\bar{X}(i)+2\bar{x}_i+1)}) = \frac{(j\omega_{l(\bar{X}(i)+1)} + \cdots + j\omega_{l(\bar{X}(i)+2\bar{x}_i+1)})^k}{-L_{2\bar{x}_i+1}(j\omega_{l(\bar{X}(i)+1)} + \cdots + j\omega_{l(\bar{X}(i)+2\bar{x}_i+1)})} \cdot \sum_{\substack{\text{all the different combinations} \\ \text{of } \{x_1, x_2, x_3\} \text{ satisfying} \\ x_1 + x_2 + x_3 = \bar{x}_i - 1,\ 0 \le x_j \le \bar{x}_i - 1}} n_x^*(x_1, x_2, x_3) \cdot \prod_{j=1}^{3} \varphi''_{2x_j+1}(c_{3,0}(\cdot)^{x_j}; \omega_{l(\bar{X}'(j)+1)} \cdots \omega_{l(\bar{X}'(j)+2x_j+1)}) \tag{25b}$$

Note that the terminal condition for (25b) is at $\overline{x}_i = 1$, i.e.,

$$\varphi''_{2\overline{x}_i+1}(c_{3,0}(\cdot)^{\overline{x}_i};\omega_{l(\overline{X}(i)+1)}\cdots\omega_{l(\overline{X}(i)+2\overline{x}_i+1)})\Big|_{\overline{x}_i=1} = \varphi''_3(c_{3,0}(\cdot);\omega_{l(1)}\cdots\omega_{l(3)}) = \frac{(j\omega_{l(1)}+\cdots+j\omega_{l(3)})^k}{-L_3(j\omega_{l(1)}+\cdots+j\omega_{l(3)})} \quad (25c)$$

Therefore, from (25a-c) it can be easily shown that $\varphi_{2n+1}(c_{3,0}(\cdot)^n;\omega_1\cdots\omega_{2n+1})$ can be written as

$$\varphi_{2n+1}(c_{3,0}(\cdot)^n;\omega_1\cdots\omega_{2n+1})$$
$$= \frac{(-1)^{n-1}\prod_{i=1}^{2n+1} j\omega_i H_1(j\omega_i)}{L_{2n+1}(j\omega_1+\cdots+j\omega_{2n+1})} \cdot \sum_{\substack{\text{all the combination } (x_1,x_2,\ldots,x_{n-1}) \\ \text{satisfying } x_i\in\{2j+1|1\le j\le n-1\} \\ x_1\ge x_2\ge\ldots\ge x_{n-1},\text{ and} \\ \text{"=" happens only if } x_i+x_{i+1}\le 2n-2}} r_X(x_1,x_2,\ldots,x_{n-1}) \prod_{i=1}^{n-1} \frac{j\omega_{l(1)}+\cdots+j\omega_{l(x_i)}}{-L_{x_i}(j\omega_{l(1)}+\cdots+j\omega_{l(x_i)})} \quad (26)$$

where $r_X(x_1,x_2,\ldots,x_{n-1})$ is a positive integer which can be explicitly determined by (25ab) and represents the number of all the involved combinations which have the same $\prod_{i=1}^{n-1} \frac{j\omega_{l(1)}+\cdots+j\omega_{l(x_i)}}{-L_{x_i}(j\omega_{l(1)}+\cdots+j\omega_{l(x_i)})}$. Therefore, according to the sufficient condition in Theorem 2, it can be seen from (26) that the output spectrum (18) is an alternating series only if the following two conditions hold:

(a1) $\mathrm{Re}(\frac{H_1(j\Omega)}{L_{2n+1}(j\Omega)})\mathrm{Im}(\frac{H_1(j\Omega)}{L_{2n+1}(j\Omega)}) = 0$

(a2) $$\mathrm{sgn}_c\left(\sum_{\omega_{k_1}+\cdots+\omega_{k_{2n+1}}=\Omega} \sum_{\substack{\text{all the combination } (x_1,x_2,\ldots,x_{n-1}) \\ \text{satisfying } x_i\in\{2j+1|1\le j\le n-1\} \\ x_1\ge x_2\ge\ldots\ge x_{n-1},\text{ and} \\ \text{"=" happens only if } x_i+x_{i+1}\le 2n-2}} r_X(x_1,x_2,\ldots,x_{n-1}) \prod_{i=1}^{n-1} \frac{j\omega_{l(1)}+\cdots+j\omega_{l(x_i)}}{-L_{x_i}(j\omega_{l(1)}+\cdots+j\omega_{l(x_i)})}\right) = const$$

Suppose $\Omega = \sqrt{\frac{k_0}{m}}$ which is a natural resonance frequency of model (15). It can be derived that

$$L_{2n+1}(j\Omega) = -\sum_{k_1=0}^{K} c_{1,0}(k_1)(j\Omega)^{k_1} = -(m(j\Omega)^2 + B(j\Omega) + k_0) = -jB\Omega$$

$$H_1(j\Omega) = \frac{-1}{L_1(j\Omega)} = \frac{1}{jB\Omega}$$

It is obvious that condition (a1) above is satisfied if $\Omega = \sqrt{\frac{k_0}{m}}$. Considering condition (a2), it can be derived that

$$\frac{j\omega_{l(1)}+\cdots+j\omega_{l(x_i)}}{-L_{x_i}(j\omega_{l(1)}+\cdots+j\omega_{l(x_i)})}=\frac{j\varepsilon(x_i)\Omega}{-L_{x_i}(j\varepsilon(x_i)\Omega)} \tag{27a}$$

where $\varepsilon(x_i)\in\{\pm(2j+1)\,|\,0\le j\le\lceil n+1\rceil\}$, and $\lceil n+1\rceil$ denotes the odd integer not larger than n+1. Especially, when $\varepsilon(x_i)=\pm1$, it yields that

$$\frac{j\omega_{l(1)}+\cdots+j\omega_{l(x_i)}}{-L_{x_i}(j\omega_{l(1)}+\cdots+j\omega_{l(x_i)})}=\frac{\pm j\Omega}{-L_{x_i}(\pm j\Omega)}=\frac{\pm j\Omega}{\pm jB\Omega}=\frac{1}{B} \tag{27b}$$

when $|\varepsilon(x_i)|>1$,

$$\begin{aligned}\frac{j\omega_{l(1)}+\cdots+j\omega_{l(x_i)}}{-L_{x_i}(j\omega_{l(1)}+\cdots+j\omega_{l(x_i)})}&=\frac{j\varepsilon(x_i)\Omega}{-L_{x_i}(j\varepsilon(x_i)\Omega)}=\frac{j\varepsilon(x_i)\Omega}{m(j\varepsilon(x_i)\Omega)^2+B(j\varepsilon(x_i)\Omega)+k_0}\\&=\frac{j\varepsilon(x_i)\Omega}{(1-\varepsilon(x_i)^2)k_0+j\varepsilon(x_i)\Omega B}=\frac{1}{B+j(\varepsilon(x_i)-\frac{1}{\varepsilon(x_i)})\sqrt{k_0m}}\end{aligned} \tag{27c}$$

If $B<<\sqrt{k_0m}$, then it gives

$$\frac{j\omega_{l(1)}+\cdots+j\omega_{l(x_i)}}{-L_{x_i}(j\omega_{l(1)}+\cdots+j\omega_{l(x_i)})}\approx\frac{1}{j(\varepsilon(x_i)-\frac{1}{\varepsilon(x_i)})\sqrt{k_0m}} \tag{27d}$$

Note that in all the combinations involved in the summation operator in (26) or condition (a2), *i.e.*,

$$\sum_{\omega_{k_1}+\cdots+\omega_{k_{2n+1}}=\Omega}\ \sum_{\substack{\text{all the combination }(x_1,x_2,\ldots,x_{n-1})\\ \text{satisfying } x_i\in\{2j+1|1\le j\le n-1\}\\ x_1\ge x_2\ge\ldots\ge x_{n-1},\text{ and}\\ \text{"=" happens only if } x_i+x_{i+1}\le 2n-2}}(\cdot)$$

There always exists a combination such that

$$\prod_{i=1}^{n-1}\frac{j\omega_{l(1)}+\cdots+j\omega_{l(x_i)}}{-L_{x_i}(j\omega_{l(1)}+\cdots+j\omega_{l(x_i)})}=\frac{1}{B^{n-1}} \tag{28}$$

Note that (27b) holds for both $\varepsilon(x_i)=\pm1$, thus there is no combination such that

$$\prod_{i=1}^{n-1}\frac{j\omega_{l(1)}+\cdots+j\omega_{l(x_i)}}{-L_{x_i}(j\omega_{l(1)}+\cdots+j\omega_{l(x_i)})}=-\frac{1}{B^{n-1}}$$

Also noting that $B<<\sqrt{k_0m}$, all these show that

$$\max_{\substack{\text{all the involved}\\ \text{combinations}}}\left(\left|\prod_{i=1}^{n-1}\frac{j\omega_{l(1)}+\cdots+j\omega_{l(x_i)}}{-L_{x_i}(j\omega_{l(1)}+\cdots+j\omega_{l(x_i)})}\right|\right)=\frac{1}{B^{n-1}}$$

which happens in the combination where (28) holds.

Because there are n+1 frequency variables to be +Ω and n frequency variables to be −Ω such that $\omega_1 + \cdots + \omega_{2n+1} = \Omega$ in (18-19), there are more combinations where $\varepsilon(x_i) > 0$ that is $(\varepsilon(x_i) - \frac{1}{\varepsilon(x_i)})\sqrt{k_0 m} > 0$ in (27c-d). Thus there are more combinations where Im($\frac{j\omega_{l(1)} + \cdots + j\omega_{l(x_i)}}{-L_{x_i}(j\omega_{l(1)} + \cdots + j\omega_{l(x_i)})}$) is negative. Using (27b) and (27d), it can be shown under the condition that $B << \sqrt{k_0 m}$,

$$\max_{\substack{\text{all the involved}\\\text{combinations}}} \left(\left| \mathrm{Im}\left(\prod_{i=1}^{n-1} \frac{j\omega_{l(1)} + \cdots + j\omega_{l(x_i)}}{-L_{x_i}(j\omega_{l(1)} + \cdots + j\omega_{l(x_i)})} \right) \right| \right) \approx \frac{1}{B^{n-2}(\varepsilon(x_i) - \frac{1}{\varepsilon(x_i)})\sqrt{k_0 m}} \Big|_{\varepsilon(x_i)=3} = \frac{1}{2.7B^{n-2}\sqrt{k_0 m}}$$

This happens in the combinations where the argument of $\prod_{i=1}^{n-1} \frac{j\omega_{l(1)} + \cdots + j\omega_{l(x_i)}}{-L_{x_i}(j\omega_{l(1)} + \cdots + j\omega_{l(x_i)})}$ is either -90^0 or +90^0. Note that there are more cases in which the arguments are -90^0. If the argument is -180^0, the absolute value of the corresponding imaginary part will be not more than

$$\max_{\substack{\text{the combination}\\\text{whose argument is}\\-180^0}} \left(\left| \mathrm{Im}\left(\prod_{i=1}^{n-1} \frac{j\omega_{l(1)} + \cdots + j\omega_{l(x_i)}}{-L_{x_i}(j\omega_{l(1)} + \cdots + j\omega_{l(x_i)})} \right) \right| \right) \approx \frac{1}{B^{n-4}(\varepsilon(x_i) - \frac{1}{\varepsilon(x_i)})^3 \sqrt{k_0 m}^{\,3}} \Big|_{\varepsilon(x_i)=3} = \frac{1}{2.7^3 B^{n-4}\sqrt{k_0 m}^{\,3}}$$

which is much less than $\frac{1}{2.7B^{n-2}\sqrt{k_0 m}}$.

Therefore, if B is sufficiently smaller than $\sqrt{k_0 m}$, the following two inequalities can hold for n>1

$$\mathrm{Re}\left(\sum_{\substack{\text{all the combination } (x_1, x_2, \ldots, x_{n-1})\\ \text{satisfying } x_i \in \{2j+1 | 1 \le j \le n-1\}\\ x_1 \ge x_2 \ge .. \ge x_{n-1}, \text{ and}\\ \text{"=" happens only if } x_i + x_{i+1} \le 2n-2}} r_X(x_1, x_2, \ldots, x_{n-1}) \prod_{i=1}^{n-1} \frac{j\omega_{l(1)} + \cdots + j\omega_{l(x_i)}}{-L_{x_i}(j\omega_{l(1)} + \cdots + j\omega_{l(x_i)})} \right) > 0$$

$$\mathrm{Im}\left(\sum_{\substack{\text{all the combination } (x_1, x_2, \ldots, x_{n-1})\\ \text{satisfying } x_i \in \{2j+1 | 1 \le j \le n-1\}\\ x_1 \ge x_2 \ge .. \ge x_{n-1}, \text{ and}\\ \text{"=" happens only if } x_i + x_{i+1} \le 2n-2}} r_X(x_1, x_2, \ldots, x_{n-1}) \prod_{i=1}^{n-1} \frac{j\omega_{l(1)} + \cdots + j\omega_{l(x_i)}}{-L_{x_i}(j\omega_{l(1)} + \cdots + j\omega_{l(x_i)})} \right) < 0$$

That is, condition (a2) holds for n>1 under $B << \sqrt{k_0 m}$ and $\Omega = \sqrt{\frac{k_0}{m}}$. Hence, (18) is an alternating series if the following two conditions hold:

(b1) B is sufficiently smaller than $\sqrt{k_0 m}$,

(b2) The input frequency is $\Omega = \sqrt{\frac{k_0}{m}}$.

In Example 2, note that $\Omega = \sqrt{\frac{k_0}{m}} \approx 8.165$, B=296<< $\sqrt{k_0 m}$ =1959.592. These are consistent with the theoretical results established above. As it has been checked numerically in Example 2 that (18) is an alternating series, the theoretical results above are well verified by the real system.

Therefore, it can be seen that, at the driving frequency the system output spectrum (subject to a cubic nonlinear damping) can be designed to be an alternating series by properly designing system parameters (see conditions (b1-b2) above) and therefore can be suppressed as shown in Example 2 by properly choosing a value for the nonlinear parameter c. This result clearly demonstrate the mechanism for the nonlinear effect of the cubic nonlinear damping in the frequency domain.

More simulation studies about the properties of the cubic nonlinear damping can be referred to the simulation results in [8], where the effects of the cubic nonlinear damping are studied in details and compared with a linear damping. The case study here theoretically shows for the first time why and when these nonlinear effects happen and what the underlying mechanism is.

Based on the discussions in Examples 2-3, it can be concluded that, the results of this study provide a new systematic method for the analysis and design of the nonlinear effect for a class of nonlinearities in the frequency domain.

5. Conclusions

Nonlinear influence on system output spectrum is investigated in this study from a novel perspective based on Volterra series expansion in the frequency domain. For a class of system nonlinearities, it is shown that system output spectrum can be expanded into an alternating series with respect to nonlinear parameters of the model under certain conditions and this alternating series has some interesting properties for engineering practices. Although there may be several existing methods such as perturbation analysis that can achieve similar objectives for some simple cases in practice, this study proposes a novel viewpoint on the nonlinear effect (i.e., alternating series) and on the analysis of nonlinear effect (i.e., the GFRFs-based) for a class of nonlinearities in the frequency domain. As some important properties of a linear system (e.g. stability) are determined by the positions of the poles of its transfer function, the fact of alternating series should be a natural characteristic of some important nonlinear effects for nonlinear systems in the frequency domain. This study provides some fundamental results for characterizing and understanding of nonlinear effects in the frequency domain from this novel viewpoint. The GFRFs-based analysis provides a useful technique for the analysis of nonlinear systems which is just similar to the transfer function based analysis for linear systems. The method demonstrated in this paper has been used for the analysis and design of nonlinear damping systems. Further study will focus on more detailed design and analysis methods based on these results for practical systems.

Author details

Xingiian Jing
Department of Mechanical Engineering, Hong Kong Polytechnic University, Hung Hom, Kowloon, Hong Kong

Acknowledgement

The author gratefully acknowledges the support of a GRF project of Hong Kong RGC (Ref 517810), the Department General Research Funds and Internal Competitive Research Grants of Hong Kong Polytechnic University for this work.

Appendix

Appendix A: Nomenclature

$c_{p,q}(k_1,\cdots,k_{p+q})$ - A model parameter in the NDE model, k_i is the order of the derivative, p represents the order of the involved output nonlinearity, q is the order of the involved input nonlinearity, and $p+q$ is the nonlinear degree of the parameter.

$H_n(j\omega_1,\cdots,j\omega_n)$ - The nth-order GFRF

$C_{p,q}=[c_{p,q}(0,\cdots,0),c_{p,q}(0,\cdots,1),\cdots,c_{p,q}(\underbrace{K,\cdots,K}_{p+q=m})]$ - A parameter vector consisting of all the nonlinear parameters of the form ..

$CE(.)$ - The coefficient extraction operator

$CE(H_n(j\omega_1,\cdots,j\omega_n))$ - The parametric characteristics of the nth-order GFRF

$f_n(j\omega_1,\cdots,j\omega_n)$ - The correlative function of $CE(H_n(j\omega_1,\cdots,j\omega_n))$

$\otimes$ - The reduced Kronecker product defined in the CE operator

$\oplus$ - The reduced vectorized summation defined in the CE operator

$\underset{(*)}{\otimes}(\cdot)$ and $\underset{(*)}{\oplus}(\cdot)$ - The multiplication and addition by the reduced Kronecker product "$\otimes$" and vectorized sum "$\oplus$" of the terms in (.) satisfying (*), respectively

$\overset{k}{\underset{i=1}{\otimes}} C_{p,q}=C_{p,q}\otimes\cdots\otimes C_{p,q}$ - can be simply written as $C_{p,q}^k$.

$c_{p_0,q_0}(\cdot)c_{p_1,q_1}(\cdot)\cdots c_{p_k,q_k}(\cdot)$ - A monomial consisting of nonlinear parameters

$s_{x_1}s_{x_2}\cdots s_{x_p}$ - A *p*-partition of a monomial $c_{p_0,q_0}(\cdot)c_{p_1,q_1}(\cdot)\cdots c_{p_k,q_k}(\cdot)$

s_{x_i} - A monomial of x_i parameters of $\{c_{p_0,q_0}(\cdot),\cdots,c_{p_k,q_k}(\cdot)\}$ of the involved monomial, $0 \le x_i \le k$, and $s_0=1$

$\varphi_n : S_C(n) \to S_f(n)$ - A new mapping function from the parametric characteristics to the correlative functions, $S_C(n)$ is the set of all the monomials in the parametric characteristics and $S_f(n)$ is the set of all the involved correlative functions in the nth order GFRF.

$n(s_x(\overline{s}))$ - The order of the GFRF where the monomial $s_x(\overline{s})$ is generated

Appendix B: The Coefficient Extraction (CE) operator [6,7,22,23]

Consider a series

$$H_{CF} = c_1 f_1 + c_2 f_2 + \cdots + c_\sigma f_\sigma \in \Xi$$

where the coefficients c_i (i=1,..., σ) are different monomial functions in a set P_c of some parameters in a set C_s which takes values in $\mathbb{C}$, f_i for i=1,...,n are some complex valued scalar functions in a set P_f which are independent of the parameters in C_s, Ξ denotes all the finite order series with coefficients in P_c timing some functions in P_f, C=[c_1,c_2,..., c_σ], and F=[f_1,f_2,..., f_σ]^T. Define a **Coefficient Extraction** operator $CE : \Xi \to \mathbb{C}^\sigma$ for this series such that

$$CE(H_{CF}) = [c_1, c_2, \cdots, c_\sigma] = C \in \mathbb{C}^\sigma$$

where $\mathbb{C}^\sigma$ is the σ-dimensional complex valued vector space. This operator has the following properties, also acting as operation rules:

1. Reduced vectorized sum "$\oplus$".

   $CE(H_{C_1F_1} + H_{C_2F_2}) = CE(H_{C_1F_1}) \oplus CE(H_{C_2F_2}) = C_1 \oplus C_2 = [C_1, C_2']$, $\quad C_2' = VEC(\overline{C}_2 - \overline{C}_1 \cap \overline{C}_2)$, where $\overline{C}_1 = \{C_1(i) \mid 1 \le i \le |C_1|\}$, $\overline{C}_2 = \{C_2(i) \mid 1 \le i \le |C_2|\}$, $VEC(.)$ is a vector consisting of all the elements in set (.). C_2' is a vector including all the elements in C_2 except the same elements as those in C_1.

2. Reduced Kronecker product "$\otimes$".

$$CE(H_{C_1F_1} \cdot H_{C_2F_2}) = CE(H_{C_1F_1}) \otimes CE(H_{C_2F_2}) = C_1 \otimes C_2 = VEC\left\{ C_3(i) \left| \begin{matrix} C_3 = [C_1(1)C_2, \cdots, C_1(|C_1|)C_2] \\ 1 \le i \le |C_3| \end{matrix} \right. \right\}$$

which implies that there are no repetitive elements in $C_1 \otimes C_2$.

3. Invariant. (a) $CE(\alpha \cdot H_{CF}) = CE(H_{CF})$ $\quad \forall \alpha \in \mathbb{C}$ but is not a parameter of interest; (b) $CE(H_{CF_1} + H_{CF_2}) = CE(H_{C(F_1+F_2)}) = C$

4. Unitary. If H_{CF} is not a function of c_i for i=1…n, $CE(H_{CF})=1$.

   When there is a unitary 1 in $CE(H_{CF})$, there is a nonzero constant term in the corresponding series H_{CF} which has no relation with the coefficients c_i (for i=1…n). In addition, if H_{CF}=0, then CE(H_{CF})=0.

5. Inverse. $CE^{-1}(C)=H_{CF}$.
6. $CE(H_{C_1F_1})\approx CE(H_{C_2F_2})$ if the elements of C_1 are the same as those of C_2, where "$\approx$" means equivalence, *i.e.*, both series are in fact the same result considering the order of c_if_i in the series has no effect on the value of a function series H_{CF}. This further implies that the CE operator is also commutative and associative, for instance, $CE(H_{C_1F_1}+H_{C_2F_2})=C_1\oplus C_2\approx CE(H_{C_2F_2}+H_{C_1F_1})=C_2\oplus C_1$. Hence, the results by the CE operator with respect to the same purpose may be different but all correspond to the same function series and are thus equivalent.
7. Separable and interested parameters only. A parameter in a series can only be extracted if the parameter is interested and the series is separable with respect to this interested parameter. Thus the operation result is different for different purposes. □

Appendix C: Proof of theorem 1

(1) $Y(j\omega)$ is convergent if and only if $\mathrm{Re}(Y(j\omega))$ and $\mathrm{Im}(Y(j\omega))$ are both convergent. Since $Y(j\omega)$ is an alternating series, $\mathrm{Re}(Y(j\omega))$ and $\mathrm{Im}(Y(j\omega))$ are both alternating from Definition 1. Then according to [3], $\mathrm{Re}(Y(j\omega))$ is convergent if $\left|\mathrm{Re}(c^iF_i(j\omega))\right|>\left|\mathrm{Re}(c^{i+1}F_{i+1}(j\omega))\right|$ and $\lim_{i\to\infty}\left|\mathrm{Re}(c^iF_i(j\omega))\right|=0$. Therefore, if there exists T>0 such that $\left|\mathrm{Re}(c^iF_i(j\omega))\right|>\left|\mathrm{Re}(c^{i+1}F_{i+1}(j\omega))\right|$ for i>T and $\lim_{i\to\infty}\left|\mathrm{Re}(c^iF_i(j\omega))\right|=0$, the alternating series $\mathrm{Re}(Y(j\omega))$ is also convergent. Now since there exist T>0 and R>0 such that $-\dfrac{\mathrm{Re}(F_i(j\omega))}{\mathrm{Re}(F_{i+1}(j\omega))}>R$ for i>T and note c<R, it can be obtained that for i>T

$$-\frac{\mathrm{Re}(c^{i+1}F_{i+1}(j\omega))}{\mathrm{Re}(c^iF_i(j\omega))}=-\frac{\mathrm{Re}(cF_{i+1}(j\omega))}{\mathrm{Re}(F_i(j\omega))}=\left|\frac{\mathrm{Re}(cF_{i+1}(j\omega))}{\mathrm{Re}(F_i(j\omega))}\right|<\frac{c}{R}<1$$

i.e., $\left|\mathrm{Re}(c^iF_i(j\omega))\right|>\left|\mathrm{Re}(c^{i+1}F_{i+1}(j\omega))\right|$ for i>T and c<R. Moreover, it can also be obtained that for n>0

$$\left|\mathrm{Re}(F_{T+n}(j\omega))\right|<\frac{1}{R^n}\left|\mathrm{Re}(F_T(j\omega))\right|$$

It further yields that

$$\left|\mathrm{Re}(c^{T+n}F_{T+n}(j\omega))\right|<(\frac{c}{R})^n c^T\left|\mathrm{Re}(F_T(j\omega))\right|$$

That is, $\lim_{n\to\infty}\left|\mathrm{Re}(c^{T+n}F_{T+n}(j\omega))\right|=0$. Therefore, $\mathrm{Re}(Y(j\omega))$ is convergent. Similarly, it can be proved that $\mathrm{Im}(Y(j\omega))$ is convergent. This proves that $Y(j\omega)$ is convergent. According to [3], the truncation errors for the convergent alternating series $\mathrm{Re}(Y(j\omega))$ and $\mathrm{Im}(Y(j\omega))$ are bounded by

$$\left|o_R(\rho)\right|\le c^{\rho+1}\left|\mathrm{Re}(F_{\rho+1}(j\omega))\right| \text{ and } \left|o_I(\rho)\right|\le c^{\rho+1}\left|\mathrm{Im}(F_{\rho+1}(j\omega))\right|$$

Therefore, the truncation error for the series $Y(j\omega)$ is

$$\left|o(\rho)\right|=\sqrt{o_R(\rho)^2+o_I(\rho)^2}\le c^{\rho+1}\left|F_{\rho+1}(j\omega)\right|$$

Since $\mathrm{Re}(Y(j\omega))$ and $\mathrm{Im}(Y(j\omega))$ are both alternating series and the absolute value of each term in $\mathrm{Re}(Y(j\omega))$ and $\mathrm{Im}(Y(j\omega))$ are monotone decreasing, i.e., $\left|\mathrm{Re}(c^iF_i(j\omega))\right|>\left|\mathrm{Re}(c^{i+1}F_{i+1}(j\omega))\right|$ and $\left|\mathrm{Im}(c^iF_i(j\omega))\right|>\left|\mathrm{Im}(c^{i+1}F_{i+1}(j\omega))\right|$ for $i>T$, then it can be shown for $\mathrm{Re}(Y(j\omega))$ and $\mathrm{Im}(Y(j\omega))$ that for $n\ge 0$

$$\left|\mathrm{Re}(Y(j\omega)_{1\to T+1})\right|<\cdots<\left|\mathrm{Re}(Y(j\omega)_{1\to T+2n+1})\right|<\left|\mathrm{Re}(Y(j\omega))\right|<\left|\mathrm{Re}(Y(j\omega)_{1\to T+2n})\right|<\cdots<\left|\mathrm{Re}(Y(j\omega)_{1\to T})\right|$$

$$\left|\mathrm{Im}(Y(j\omega)_{1\to T+1})\right|<\cdots<\left|\mathrm{Im}(Y(j\omega)_{1\to T+2n+1})\right|<\left|\mathrm{Im}(Y(j\omega))\right|<\left|\mathrm{Im}(Y(j\omega)_{1\to T+2n})\right|<\cdots<\left|\mathrm{Im}(Y(j\omega)_{1\to T})\right|$$

Therefore, $\left|Y(j\omega)_{1\to T+1}\right|<\cdots<\left|Y(j\omega)_{1\to T+2n+1}\right|<\left|Y(j\omega)\right|<\left|Y(j\omega)_{1\to T+2n}\right|<\cdots<\left|Y(j\omega)_{1\to T}\right|$.

(2)

$$\begin{aligned}\left|Y(j\omega)\right|^2&=Y(j\omega)Y(-j\omega)\\&=(F_0(j\omega)+cF_1(j\omega)+c^2F_2(j\omega)+\cdots)(F_0(-j\omega)+cF_1(-j\omega)+c^2F_2(-j\omega)+\cdots)\\&=\sum_{n=0,1,2,\ldots}c^n\sum_{i=0}^{n}F_i(j\omega)F_{n-i}(-j\omega)\end{aligned}$$

It can be verified that the (2k)th terms in the series are positive and the (2k+1)th terms are negative for k=0,1,2,…. Moreover, it is not difficult to obtain that it needs only the real parts of the terms in $Y(j\omega)$ to be alternating for $\left|Y(j\omega)\right|^2=Y(j\omega)Y(-j\omega)$ to be alternating.

(3)

$$\begin{aligned}\frac{\partial\left|Y(j\omega)\right|}{\partial c}&=\frac{1}{2\left|Y(j\omega)\right|}\frac{\partial\left|Y(j\omega)\right|^2}{\partial c}\\&=\frac{1}{2\left|Y(j\omega)\right|}\left\{\mathrm{Re}(F_0(j\omega)F_1(-j\omega))+c\sum_{n=1,2,\ldots}nc^{n-1}\sum_{i=0}^{n}F_i(j\omega)F_{n-i}(-j\omega)\right\}\end{aligned}$$

Since $\mathrm{Re}(F_0(j\omega)F_1(-j\omega))<0$, there must exist $\bar{c}>0$ such that $\frac{\partial\left|Y(j\omega)\right|}{\partial c}<0$ for $0<c<\bar{c}$. This completes the proof. □

Appendix D: Proof of theorem 2

In order to prove Theorem 2, the following lemma is needed, which provides a fundamental technique for the derivation of the main results in Theorem 2 by exploiting the recursive nature of $\varphi_{n(\bar{s})}(c_{p,0}(\cdot)^n;\omega_{l(1)}\cdots\omega_{l(n(\bar{s}))})$.

Lemma 3. Consider a nonlinear parameter denoted by $c_{p,q}(k_1,k_2,\ldots,k_{p+q})$.

(1) If $p\geq 2$ and q=0, then

$$\varphi_{n(\bar{s})}(c_{p,0}(\cdot)^n;\omega_{l(1)}\cdots\omega_{l(n(\bar{s}))})=\varphi_{(p-1)n+1}(c_{p,0}(\cdot)^n;\omega_{l(1)}\cdots\omega_{l((p-1)n+1)})$$

$$=\frac{(-1)^{n-1}\prod_{i=1}^{(p-1)n+1}H_1(j\omega_{l(i)})}{L_{(p-1)n+1}(j\omega_{l(1)}+\cdots+j\omega_{l((p-1)n+1)})}\cdot\sum_{\substack{\text{all the different combinations}\\ \text{of }\{\bar{x}_1,\bar{x}_2,\ldots,\bar{x}_p\}\text{ satisfying}\\ \bar{x}_1+\cdots+\bar{x}_p=n-1,\ 0\leq\bar{x}_i\leq n-1}}\left[\prod_{i=1}^{p}\varphi'_{(p-1)\bar{x}_i+1}(c_{p,0}(\cdot)^{\bar{x}_i};\omega_{l(\bar{X}(i)+1)}\cdots\omega_{l(\bar{X}(i)+(p-1)\bar{x}_i+1)})\right.$$
$$\left.\cdot\frac{n_x^*(\bar{x}_1,\cdots,\bar{x}_p)}{n_k^*(k_1,\cdots,k_p)}\cdot\sum_{\substack{\text{all the different}\\ \text{permutations of}\\ \{k_1,\cdots,k_p\}}}\prod_{i=1}^{p}(j\omega_{l(\bar{X}(i)+1)}+\cdots+j\omega_{l(\bar{X}(i)+(p-1)\bar{x}_i+1)})^{k_i}\right]$$

where,

$$\varphi'_{(p-1)n+1}(c_{p,0}(\cdot)^n;\omega_{l(1)}\cdots\omega_{l((p-1)n+1)})$$

$$=\frac{-1}{L_{(p-1)n+1}(j\omega_{l(1)}+\cdots+j\omega_{l((p-1)n+1)})}\cdot\sum_{\substack{\text{all the different combinations}\\ \text{of }\{\bar{x}_1,\bar{x}_2,\ldots,\bar{x}_p\}\text{ satisfying}\\ \bar{x}_1+\cdots+\bar{x}_p=n-1,\ 0\leq\bar{x}_i\leq n-1}}\left[\prod_{i=1}^{p}\varphi'_{(p-1)\bar{x}_i+1}(c_{p,0}(\cdot)^{\bar{x}_i};\omega_{l(\bar{X}(i)+1)}\cdots\omega_{l(\bar{X}(i)+(p-1)\bar{x}_i+1)})\right.$$
$$\left.\cdot\frac{n_x^*(\bar{x}_1,\cdots,\bar{x}_p)}{n_k^*(k_1,\cdots,k_p)}\cdot\sum_{\substack{\text{all the different}\\ \text{permutations of}\\ \{k_1,\cdots,k_p\}}}\prod_{i=1}^{p}(j\omega_{l(\bar{X}(i)+1)}+\cdots+j\omega_{l(\bar{X}(i)+(p-1)\bar{x}_i+1)})^{k_i}\right]$$

the termination is $\varphi'_1(1;\omega_i)=1$. $n_k^*(k_1,\cdots,k_p)=\dfrac{p!}{n_1!n_2!\cdots n_e!}$, $n_1+\ldots+n_e=p$, e is the number of distinct differentials k_i appearing in the combination, n_i is the number of repetitions of k_i, and a similar definition holds for $n_x^*(\bar{x}_1,\cdots,\bar{x}_p)$.

(2) If $p\geq 2$, q=0 and $k_1=k_2=\ldots=k_p=k$, then

$$\varphi_{(p-1)n+1}(c_{p,0}(\cdot)^n;\omega_{l(1)}\cdots\omega_{l((p-1)n+1)})$$

$$=\frac{(-1)^{n-1}\prod_{i=1}^{(p-1)n+1}[(j\omega_{l(i)})^kH_1(j\omega_{l(i)})]}{L_{(p-1)n+1}(j\omega_{l(1)}+\cdots+j\omega_{l((p-1)n+1)})}$$

$$\cdot\sum_{\substack{\text{all the different combinations}\\ \text{of }\{\bar{x}_1,\bar{x}_2,\ldots,\bar{x}_p\}\text{ satisfying}\\ \bar{x}_1+\cdots+\bar{x}_p=n-1,\ 0\leq\bar{x}_i\leq n-1}} n_x^*(\bar{x}_1,\cdots,\bar{x}_p)\cdot\prod_{i=1}^{p}\varphi''_{(p-1)\bar{x}_i+1}(c_{p,0}(\cdot)^{\bar{x}_i};\omega_{l(\bar{X}(i)+1)}\cdots\omega_{l(\bar{X}(i)+(p-1)\bar{x}_i+1)})$$

where, if $\bar{x}_i=0$, $\varphi''_{(p-1)\bar{x}_i+1}(c_{p,0}(\cdot)^{\bar{x}_i};\omega_{l(\bar{X}(i)+1)}\cdots\omega_{l(\bar{X}(i)+(p-1)\bar{x}_i+1)})=1$, otherwise,

$$\begin{aligned}&\varphi''_{(p-1)\bar{x}_i+1}(c_{p,0}(\cdot)^{\bar{x}_i};\omega_{l(\bar{X}(i)+1)}\cdots\omega_{l(\bar{X}(i)+(p-1)\bar{x}_i+1)})\\ &=\frac{(j\omega_{l(\bar{X}(i)+1)}+\cdots+j\omega_{l(\bar{X}(i)+(p-1)\bar{x}_i+1)})^k}{-L_{(p-1)\bar{x}_i+1}(j\omega_{l(\bar{X}(i)+1)}+\cdots+j\omega_{l(\bar{X}(i)+(p-1)\bar{x}_i+1)})}\cdot\\ &\sum_{\substack{\text{all the different combinations}\\ \text{of }\{x_1,x_2,\ldots,x_p\}\text{ satisfying}\\ x_1+\cdots+x_p=\bar{x}_i-1,\ 0\le x_j\le\bar{x}_i-1}} n_x^*(x_1,\cdots,x_p)\cdot\prod_{j=1}^{p}\varphi''_{(p-1)x_j+1}(c_{p,0}(\cdot)^{x_j};\omega_{l(\bar{X}'(j)+1)}\cdots\omega_{l(\bar{X}'(j)+(p-1)x_j+1)})\end{aligned}$$

The recursive terminal of $\varphi''_{(p-1)\bar{x}_i+1}(c_{p,0}(\cdot)^{\bar{x}_i};\omega_{l(\bar{X}(i)+1)}\cdots\omega_{l(\bar{X}(i)+(p-1)\bar{x}_i+1)})$ is $\bar{x}_i=1$.

Proof of Lemma 3.

$$\begin{aligned}&\varphi_{n(\bar{s})}(c_{p,0}(\cdot)^n;\omega_{l(1)}\cdots\omega_{l(n(\bar{s}))})=\varphi_{(p-1)n+1}(c_{p,0}(\cdot)c_{p,0}(\cdot)\cdots c_{p,0}(\cdot);\omega_{l(1)}\cdots\omega_{(l(p-1)n+1)})\\ &=\sum_{\substack{\text{all the 2-partitions}\\ \text{for }\bar{s}\text{ satisfying}\\ s_1(\bar{s})=c_{p,0}(\cdot)}}\Big\{f_1(c_{p,0}(\cdot),(p-1)n+1;\omega_{l(1)}\cdots\omega_{l((p-1)n+1)})\cdot\sum_{\substack{\text{all the p-partitions}\\ \text{for }\bar{s}/c_{3,0}(\cdot)}}\ \sum_{\substack{\text{all the different}\\ \text{permutations}\\ \text{of }\{s_{x_1},\cdots,s_{x_p}\}}}\\ &\Big[f_{2a}(s_{\bar{x}_1}\cdots s_{\bar{x}_p}(c_{p0}(\cdot)^{n-1});\omega_{l(1)}\cdots\omega_{l(n(\bar{s}))})\cdot\prod_{i=1}^{p}\varphi_{n(s_{\bar{x}_i}(c_{p,0}(\cdot)^{n-1}))}(s_{\bar{x}_i}(c_{p,0}(\cdot)^{n-1});\omega_{l(\bar{X}(i)+1)}\cdots\omega_{l(\bar{X}(i)+n(s_{\bar{x}_i}(c_{p,0}(\cdot)^{n-1})))})\Big]\Big\}\end{aligned}$$

$$\begin{aligned}&=\frac{1}{L_{(p-1)n+1}(j\omega_{l(1)}+\cdots+j\omega_{l((p-1)n+1)})}\cdot\sum_{\substack{\text{all the p-partitions}\\ \text{for }\bar{s}/c_{p,0}(\cdot)}}\ \sum_{\substack{\text{all the different}\\ \text{permutations}\\ \text{of }\{s_{x_1},\cdots,s_{x_p}\}}}\Big[\prod_{i=1}^{p}(j\omega_{l(\bar{X}(i)+1)}+\cdots+j\omega_{l(\bar{X}(i)+n(s_{\bar{x}_i}(c_{p,0}(\cdot)^{n-1})))})^{k_i}\\ &\cdot\prod_{i=1}^{p}\varphi_{n(s_{\bar{x}_i}(c_{p,0}(\cdot)^{n-1}))}(s_{\bar{x}_i}(c_{p,0}(\cdot)^{n-1});\omega_{l(\bar{X}(i)+1)}\cdots\omega_{l(\bar{X}(i)+n(s_{\bar{x}_i}(c_{p,0}(\cdot)^{n-1})))})\Big]\end{aligned}$$

$$\begin{aligned}&=\frac{1}{L_{(p-1)n+1}(j\omega_{l(1)}+\cdots+j\omega_{l((p-1)n+1)})}\cdot\sum_{\substack{\text{all the different combinations}\\ \text{of }\{\bar{x}_1,\bar{x}_2,\ldots,\bar{x}_p\}\text{ satisfying}\\ \bar{x}_1+\cdots+\bar{x}_p=n-1,\ 0\le\bar{x}_i\le n-1}}\ \sum_{\substack{\text{all the different}\\ \text{permutations of}\\ \text{each combination}}}\Big[\prod_{i=1}^{p}(j\omega_{l(\bar{X}(i)+1)}+\cdots+j\omega_{l(\bar{X}(i)+(p-1)\bar{x}_i+1)})^{k_i}\\ &\cdot\prod_{i=1}^{p}\varphi_{(p-1)\bar{x}_i+1}(c_{p,0}(\cdot)^{\bar{x}_i};\omega_{l(\bar{X}(i)+1)}\cdots\omega_{l(\bar{X}(i)+(p-1)\bar{x}_i+1)})\Big]\end{aligned}$$

Note that different permutations in each combination have no difference to $\prod_{i=1}^{p}\varphi_{(p-1)\bar{x}_i+1}(c_{p,0}(\cdot)^{\bar{x}_i};\omega_{l(\bar{X}(i)+1)}\cdots\omega_{l(\bar{X}(i)+(p-1)\bar{x}_i+1)})$, thus $\varphi_{(p-1)n+1}(c_{p,0}(\cdot)^n;\omega_1\cdots\omega_{(p-1)n+1})$ can be written as

$$\varphi_{(p-1)n+1}(c_{p,0}(\cdot)^n;\omega_1\cdots\omega_{(p-1)n+1})$$

$$=\frac{1}{L_{(p-1)n+1}(j\omega_{l(1)}+\cdots+j\omega_{l((p-1)n+1)})}\cdot\sum_{\substack{\text{all the different combinations}\\ \text{of }\{\bar{x}_1,\bar{x}_2,\ldots,\bar{x}_p\}\text{ satisfying}\\ \bar{x}_1+\cdots+\bar{x}_3=n-1,\ 0\le\bar{x}_i\le n-1}}\prod_{i=1}^{p}\varphi_{(p-1)\bar{x}_i+1}(c_{p,0}(\cdot)^{\bar{x}_i};\omega_{l(\bar{X}(i)+1)}\cdots\omega_{l(\bar{X}(i)+(p-1)\bar{x}_i+1)})$$

$$\cdot\sum_{\substack{\text{all the different}\\ \text{permutations of}\\ \text{each combination}}}\prod_{i=1}^{p}(j\omega_{l(\bar{X}(i)+1)}+\cdots+j\omega_{l(\bar{X}(i)+(p-1)\bar{x}_i+1)})^{k_i}$$

$$=\frac{1}{L_{(p-1)n+1}(j\omega_{l(1)}+\cdots+j\omega_{l((p-1)n+1)})}\cdot\sum_{\substack{\text{all the different combinations}\\ \text{of }\{\bar{x}_1,\bar{x}_2,\ldots,\bar{x}_p\}\text{ satisfying}\\ \bar{x}_1+\cdots+\bar{x}_p=n-1,\ 0\le\bar{x}_i\le n-1}}\prod_{i=1}^{p}\varphi_{(p-1)\bar{x}_i+1}(c_{p,0}(\cdot)^{\bar{x}_i};\omega_{l(\bar{X}(i)+1)}\cdots\omega_{l(\bar{X}(i)+(p-1)\bar{x}_i+1)})$$

$$\cdot\frac{n_x^*(\bar{x}_1,\cdots,\bar{x}_p)}{n_k^*(k_1,\cdots,k_p)}\cdot\sum_{\substack{\text{all the different}\\ \text{permutations of}\\ \{k_1,\cdots,k_p\}}}\prod_{i=1}^{p}(j\omega_{l(\bar{X}(i)+1)}+\cdots+j\omega_{l(\bar{X}(i)+(p-1)\bar{x}_i+1)})^{k_i}$$

$n_x^*(\bar{x}_1,\cdots,\bar{x}_p)$ and $n_k^*(k_1,\cdots,k_p)$ are the numbers of the corresponding combinations involved, which can be obtained from the combination theory and can also be referred to [14]. Inspection of the recursion in the equation above, it can be seen that there are $(p-1)n+1$ $H_1(j\omega_i)$ with different frequency variable at the end of the recursion. Thus they can be brought out as a common factor. This gives

$$\varphi_{(p-1)n+1}(c_{p,0}(\cdot)^n;\omega_{l(1)}\cdots\omega_{l((p-1)n+1)})=(-1)^n\prod_{i=1}^{(p-1)n+1}H_1(j\omega_{l(i)})\cdot\varphi'_{(p-1)n+1}(c_{p,0}(\cdot)^n;\omega_{l(1)}\cdots\omega_{l((p-1)n+1)})\quad\text{(A1)}$$

where,

$$\varphi'_{(p-1)n+1}(c_{p,0}(\cdot)^n;\omega_{l(1)}\cdots\omega_{l((p-1)n+1)})$$

$$=\frac{-1}{L_{(p-1)n+1}(j\omega_{l(1)}+\cdots+j\omega_{l((p-1)n+1)})}\cdot\sum_{\substack{\text{all the different combinations}\\ \text{of }\{\bar{x}_1,\bar{x}_2,\ldots,\bar{x}_p\}\text{ satisfying}\\ \bar{x}_1+\cdots+\bar{x}_p=n-1,\ 0\le\bar{x}_i\le n-1}}\prod_{i=1}^{p}\varphi'_{(p-1)\bar{x}_i+1}(c_{p,0}(\cdot)^{\bar{x}_i};\omega_{l(\bar{X}(i)+1)}\cdots\omega_{l(\bar{X}(i)+(p-1)\bar{x}_i+1)})\quad\text{(A2)}$$

$$\cdot\frac{n_x^*(\bar{x}_1,\cdots,\bar{x}_p)}{n_k^*(k_1,\cdots,k_p)}\cdot\sum_{\substack{\text{all the different}\\ \text{permutations of}\\ \{k_1,\cdots,k_p\}}}\prod_{i=1}^{p}(j\omega_{l(\bar{X}(i)+1)}+\cdots+j\omega_{l(\bar{X}(i)+(p-1)\bar{x}_i+1)})^{k_i}$$

the termination is $\varphi_1'(1;\omega_i)=1$. Note that when $\bar{x}_i=0$, there is a term $(j\omega_{l(\bar{X}(i)+1)})^{k_i}$ appearing from $\frac{n_x^*(\bar{x}_1,\cdots,\bar{x}_p)}{n_k^*(k_1,\cdots,k_p)}\cdot\sum_{\substack{\text{all the different}\\ \text{permutations of}\\ \{k_1,\cdots,k_p\}}}\prod_{i=1}^{p}(j\omega_{l(\bar{X}(i)+1)}+\cdots+j\omega_{l(\bar{X}(i)+(p-1)\bar{x}_i+1)})^{k_i}$. It can be verified that in each recursion of $\varphi'_{(p-1)n+1}(c_{p,0}(\cdot)^n;\omega_{l(1)}\cdots\omega_{l((p-1)n+1)})$, there may be some frequency variables appearing individually in the form of $(j\omega_{l(\bar{X}(i)+1)})^{k_i}$, and these variables will not appear individually in the same form in the subsequent recursion. At the end of the recursion, all the frequency variables should have appeared in this form. Thus these terms can also be brought out as common factors if $k_1=k_2=\ldots=k_p$. In the case of $k_1=k_2=\ldots=k_p=k$,

$$\frac{n_x^*(\overline{x}_1,\cdots,\overline{x}_p)}{n_k^*(k_1,\cdots,k_p)}\cdot\sum_{\substack{\text{all the different}\\ \text{permutations of}\\ \{k_1,\cdots,k_p\}}}\prod_{i=1}^{p}(j\omega_{l(\overline{X}(i)+1)}+\cdots+j\omega_{l(\overline{X}(i)+(p-1)\overline{x}_i+1)})^{k_i}$$
$$=n_x^*(\overline{x}_1,\cdots,\overline{x}_p)\cdot\prod_{i=1}^{p}(j\omega_{l(\overline{X}(i)+1)}+\cdots+j\omega_{l(\overline{X}(i)+(p-1)\overline{x}_i+1)})^{k_i}$$

Therefore (A1) and (A2) can be written, if $k_1=k_2=\ldots=k_p$, as

$$\begin{aligned}&\varphi_{(p-1)n+1}(c_{p,0}(\cdot)^n;\omega_{l(1)}\cdots\omega_{l((p-1)n+1)})\\&=(-1)^n\prod_{i=1}^{(p-1)n+1}[(j\omega_{l(i)})^kH_1(j\omega_{l(i)})]\cdot\varphi'_{(p-1)n+1}(c_{p,0}(\cdot)^n;\omega_{l(1)}\cdots\omega_{l((p-1)n+1)})\end{aligned}\tag{A3}$$

$$\begin{aligned}&\varphi'_{(p-1)n+1}(c_{p,0}(\cdot)^n;\omega_{l(1)}\cdots\omega_{l((p-1)n+1)})\\&=\frac{-1}{L_{(p-1)n+1}(j\omega_{l(1)}+\cdots+j\omega_{l((p-1)n+1)})}\cdot\sum_{\substack{\text{all the different combinations}\\ \text{of }\{\overline{x}_1,\overline{x}_2,\ldots,\overline{x}_p\}\text{ satisfying}\\ \overline{x}_1+\cdots+\overline{x}_p=n-1,\ 0\le\overline{x}_i\le n-1}}\prod_{i=1}^{p}\varphi'_{(p-1)\overline{x}_i+1}(c_{p,0}(\cdot)^{\overline{x}_i};\omega_{l(\overline{X}(i)+1)}\cdots\omega_{l(\overline{X}(i)+(p-1)\overline{x}_i+1)})\\&\quad\cdot n_x^*(\overline{x}_1,\cdots,\overline{x}_p)\cdot\prod_{i=1}^{p}(j\omega_{l(\overline{X}(i)+1)}+\cdots+j\omega_{l(\overline{X}(i)+(p-1)\overline{x}_i+1)})^{k_i(1-\delta(\overline{x}_i))}\end{aligned}\tag{A4}$$

(A4) can be further written as

$$\begin{aligned}&\varphi'_{(p-1)n+1}(c_{p,0}(\cdot)^n;\omega_{l(1)}\cdots\omega_{l((p-1)n+1)})\\&=\frac{-1}{L_{(p-1)n+1}(j\omega_{l(1)}+\cdots+j\omega_{l((p-1)n+1)})}\\&\quad\cdot\sum_{\substack{\text{all the different combinations}\\ \text{of }\{\overline{x}_1,\overline{x}_2,\ldots,\overline{x}_p\}\text{ satisfying}\\ \overline{x}_1+\cdots+\overline{x}_p=n-1,\ 0\le\overline{x}_i\le n-1}}n_x^*(\overline{x}_1,\cdots,\overline{x}_p)\cdot\prod_{i=1}^{p}\varphi''_{(p-1)\overline{x}_i+1}(c_{p,0}(\cdot)^{\overline{x}_i};\omega_{l(\overline{X}(i)+1)}\cdots\omega_{l(\overline{X}(i)+(p-1)\overline{x}_i+1)})\end{aligned}\tag{A5}$$

where, if $\overline{x}_i=0$, $\varphi''_{(p-1)\overline{x}_i+1}(c_{p,0}(\cdot)^{\overline{x}_i};\omega_{l(\overline{X}(i)+1)}\cdots\omega_{l(\overline{X}(i)+(p-1)\overline{x}_i+1)})=1$, otherwise,

$$\varphi''_{(p-1)\bar{x}_i+1}(c_{p,0}(\cdot)^{\bar{x}_i};\omega_{l(\bar{X}(i)+1)}\cdots\omega_{l(\bar{X}(i)+(p-1)\bar{x}_i+1)})$$
$$=(j\omega_{l(\bar{X}(i)+1)}+\cdots+j\omega_{l(\bar{X}(i)+(p-1)x_i+1)})^k\varphi'_{(p-1)\bar{x}_i+1}(c_{p,0}(\cdot)^{\bar{x}_i};\omega_{l(\bar{X}(i)+1)}\cdots\omega_{l(\bar{X}(i)+(p-1)\bar{x}_i+1)})$$
$$=\frac{(j\omega_{l(\bar{X}(i)+1)}+\cdots+j\omega_{l(\bar{X}(i)+(p-1)\bar{x}_i+1)})^k}{-L_{(p-1)\bar{x}_i+1}(j\omega_{l(\bar{X}(i)+1)}+\cdots+j\omega_{l(\bar{X}(i)+(p-1)\bar{x}_i+1)})}\cdot\sum_{\substack{\text{all the different combinations}\\ \text{of }\{x_1,x_2,\ldots,x_p\}\text{ satisfying}\\ x_1+\cdots+x_p=\bar{x}_i-1,\ 0\le x_i\le\bar{x}_i-1}} n_x^*(x_1,\cdots,x_p)$$
$$\cdot\prod_{i=1}^{p}(j\omega_{l(\bar{X}'(i)+1)}+\cdots+j\omega_{l(\bar{X}'(i)+(p-1)x_i+1)})^{k_i(1-\delta(x_i))}\varphi'_{(p-1)x_i+1}(c_{p,0}(\cdot)^{x_i};\omega_{l(\bar{X}'(i)+1)}\cdots\omega_{l(\bar{X}'(i)+(p-1)x_i+1)})$$
$$=\frac{(j\omega_{l(\bar{X}(i)+1)}+\cdots+j\omega_{l(\bar{X}(i)+(p-1)\bar{x}_i+1)})^k}{-L_{(p-1)\bar{x}_i+1}(j\omega_{l(\bar{X}(i)+1)}+\cdots+j\omega_{l(\bar{X}(i)+(p-1)\bar{x}_i+1)})}\cdot$$
$$\sum_{\substack{\text{all the different combinations}\\ \text{of }\{x_1,x_2,\ldots,x_p\}\text{ satisfying}\\ x_1+\cdots+x_p=\bar{x}_i-1,\ 0\le x_i\le\bar{x}_i-1}} n_x^*(x_1,\cdots,x_p)\cdot\prod_{i=1}^{p}\varphi''_{(p-1)x_i+1}(c_{p,0}(\cdot)^{x_i};\omega_{l(\bar{X}'(i)+1)}\cdots\omega_{l(\bar{X}'(i)+(p-1)x_i+1)})$$

The recursive terminal of (A6) is $\bar{x}_i=1$. Substituting (A2) into (A1) gives the first point of the lemma and substituting (A5) and (A6) into (A3) yields the first point of the lemma. This completes the proof. □

Now proceed with the proof of Theorem 2. For convenience, denote

$$\text{sgn}_c(\upsilon_1)*\text{sgn}_c(\upsilon_2)=\text{sgn}_c(\upsilon_1\upsilon_2)=\left[\text{sgn}_r(\text{Re}(\upsilon_1\upsilon_2))\quad \text{sgn}_r(\text{Im}(\upsilon_1\upsilon_2))\right]$$

for any $\upsilon_1,\upsilon_2\in\mathbb{C}$.

Proof of Theorem 2. (1). From Lemma 1, any asymmetric $\varphi_{(p-1)n+1}(c_{p,0}(\cdot)^n;\omega_{k_1},\cdots,\omega_{k_{(p-1)n+1}})$ is sufficient for the computation of $\tilde{F}_{(p-1)n+1}(j\Omega)$. It can be obtained that

$$\text{sgn}_c(\tilde{F}_{(p-1)n+1}(j\Omega))=\text{sgn}_c(-j(\frac{F_d}{2})^{(p-1)n+1})*\text{sgn}_c(\sum_{\omega_{k_1}+\cdots+\omega_{k_{(p-1)n+1}}=\Omega}\varphi_{(p-1)n+1}(c_{p,0}(\cdot)^n;\omega_{k_1},\cdots,\omega_{k_{(p-1)n+1}}))$$

From Lemma 2, $\text{sgn}_c(-j(\frac{F_d}{2})^{(p-1)n+1})$ has no effect on the alternating nature of the sequence $\tilde{F}_{(p-1)n+1}(j\Omega)$ for n=1,2,3,…. Hence, (21a-c) is an alternating series with respect to $c_{p,0}(.)$ if and only if the sequence $\sum_{\omega_{k_1}+\cdots+\omega_{k_{(p-1)n+1}}=\Omega}\varphi_{(p-1)n+1}(c_{p,0}(\cdot)^n;\omega_{k_1},\cdots,\omega_{k_{(p-1)n+1}})$ for n=1,2,3,… is alternating. This is equivalent to

$$\text{sgn}_c\left(\sum_{\omega_{k_1}+\cdots+\omega_{k_{(p-1)n+1}}=\Omega}(-1)^{n-1}\varphi_{(p-1)n+1}(c_{p,0}(\cdot)^n;\omega_{l(1)}\cdots\omega_{l((p-1)n+1)})\right)=const$$

In the equation above, replacing $\varphi_{(p-1)n+1}(c_{p,0}(\cdot)^n;\omega_{k_1},\cdots,\omega_{k_{(p-1)n+1}})$ by using the result in Lemma 3 and noting $(p-1)n+1$ is an odd integer, it can be obtained that

$$\operatorname{sgn}_c\left(\sum_{\omega_{k_1}+\cdots+\omega_{k_{(p-1)n+1}}=\Omega}\frac{\prod_{i=1}^{(p-1)n+1}H_1(j\omega_{l(i)})}{L_{(p-1)n+1}(j\omega_{l(1)}+\cdots+j\omega_{l((p-1)n+1)})}\cdot\sum_{\substack{\text{all the different combinations}\\ \text{of }\{\bar{x}_1,\bar{x}_2,\ldots,\bar{x}_p\}\text{ satisfying}\\ \bar{x}_1+\cdots+\bar{x}_p=n-1,\ 0\le\bar{x}_i\le n-1}}\left[\prod_{i=1}^{p}\varphi'_{(p-1)\bar{x}_i+1}(c_{p,0}(\cdot)^{\bar{x}_i};\omega_{l(\bar{X}(i)+1)}\cdots\omega_{l(\bar{X}(i)+(p-1)\bar{x}_i+1)})\cdot\frac{n_x^*(\bar{x}_1,\cdots,\bar{x}_p)}{n_k^*(k_1,\cdots,k_p)}\cdot\sum_{\substack{\text{all the different}\\ \text{permutations of}\\ \{k_1,\cdots,k_p\}}}\prod_{i=1}^{p}(j\omega_{l(\bar{X}(i)+1)}+\cdots+j\omega_{l(\bar{X}(i)+(p-1)\bar{x}_i+1)})^{k_i}\right]\right)$$

$$=\operatorname{sgn}_c\left(\frac{H_1(j\Omega)\prod_{i=1}^{(p-1)n/2}\left|H_1(j\Omega)\right|^2}{L_{(p-1)n+1}(j\Omega)}\cdot\sum_{\omega_{k_1}+\cdots+\omega_{k_{(p-1)n+1}}=\Omega}\ \sum_{\substack{\text{all the different combinations}\\ \text{of }\{\bar{x}_1,\bar{x}_2,\ldots,\bar{x}_p\}\text{ satisfying}\\ \bar{x}_1+\cdots+\bar{x}_p=n-1,\ 0\le\bar{x}_i\le n-1}}\left[\prod_{i=1}^{p}\varphi'_{(p-1)\bar{x}_i+1}(c_{p,0}(\cdot)^{\bar{x}_i};\omega_{l(\bar{X}(i)+1)}\cdots\omega_{l(\bar{X}(i)+(p-1)\bar{x}_i+1)})\cdot\frac{n_x^*(\bar{x}_1,\cdots,\bar{x}_p)}{n_k^*(k_1,\cdots,k_p)}\cdot\sum_{\substack{\text{all the different}\\ \text{permutations of}\\ \{k_1,\cdots,k_p\}}}\prod_{i=1}^{p}(j\omega_{l(\bar{X}(i)+1)}+\cdots+j\omega_{l(\bar{X}(i)+(p-1)\bar{x}_i+1)})^{k_i}\right]\right)=const$$

Note that $\prod_{i=1}^{(p-1)n/2}\left|H_1(j\Omega)\right|^2$ has no effect on the equality above according to Lemma 2, then the equation above is equivalent to (22).

(2). If additionally, $k_1=k_2=\ldots=k_p=k$ in $c_{p,0}(.)$, then using the result in Lemma 3, (22) can be written as

$$\operatorname{sgn}_c\left(\begin{array}{l}\dfrac{(j\Omega)^k H_1(j\Omega)}{L_{(p-1)n+1}(j\Omega)}\sum_{\omega_{k_1}+\cdots+\omega_{k_{(p-1)n+1}}=\Omega}\ \sum_{\substack{\text{all the different combinations}\\ \text{of } \{\bar{x}_1,\bar{x}_2,\ldots,\bar{x}_p\} \text{ satisfying}\\ \bar{x}_1+\cdots+\bar{x}_p=n-1,\ 0\le\bar{x}_i\le n-1}}\Big[n_x^*(\bar{x}_1,\cdots,\bar{x}_p)\\ \qquad\cdot\prod_{i=1}^{p}\varphi''_{(p-1)\bar{x}_i+1}(c_{p,0}(\cdot)^{\bar{x}_i};\omega_{l(\bar{X}(i)+1)}\cdots\omega_{l(\bar{X}(i)+(p-1)\bar{x}_i+1)})\Big]\end{array}\right)=const$$

From Lemma 2, $(j\Omega)^k$ has no effect on this equation. Then the equation above is equivalent to

$$\operatorname{sgn}_c\left(\begin{array}{l}\dfrac{H_1(j\Omega)}{L_{(p-1)n+1}(j\Omega)}\sum_{\omega_{k_1}+\cdots+\omega_{k_{(p-1)n+1}}=\Omega}\ \sum_{\substack{\text{all the different combinations}\\ \text{of } \{\bar{x}_1,\bar{x}_2,\ldots,\bar{x}_p\} \text{ satisfying}\\ \bar{x}_1+\cdots+\bar{x}_p=n-1,\ 0\le\bar{x}_i\le n-1}}\Big[n_x^*(\bar{x}_1,\cdots,\bar{x}_p)\\ \qquad\cdot\prod_{i=1}^{p}\varphi''_{(p-1)\bar{x}_i+1}(c_{p,0}(\cdot)^{\bar{x}_i};\omega_{l(\bar{X}(i)+1)}\cdots\omega_{l(\bar{X}(i)+(p-1)\bar{x}_i+1)})\Big]\end{array}\right)=const$$

If $\mathrm{Re}(\dfrac{H_1(j\Omega)}{L_{(p-1)n+1}(j\Omega)})\mathrm{Im}(\dfrac{H_1(j\Omega)}{L_{(p-1)n+1}(j\Omega)})=0$, then $\dfrac{H_1(j\Omega)}{L_{(p-1)n+1}(j\Omega)}$ has no effect, either. This gives Equation (23). The proof is completed.

7. References

[1] Billings S.A. and Peyton-Jones J.C., "Mapping nonlinear integro-differential equation into the frequency domain", International Journal of Control, Vol 54, 863-879, 1990

[2] Boyd, S. and Chua L., "Fading memory and the problem of approximating nonlinear operators with Volterra series". IEEE Trans. On Circuits and Systems, Vol. CAS-32, No 11, pp 1150-1160, 1985

[3] Bromwich T. J., An Introduction to the Theory of Infinite Series, American Mathematical Society, AMS Chelsea Publishing, 1991

[4] George D.A., "Continuous nonlinear systems", Technical Report 355, MIT Research Laboratory of Electronics, Cambridge, Mass. Jul. 24, 1959.

[5] Graham D. and McRuer D., Analysis of nonlinear control systems. New York; London : Wiley, 1961.

[6] Jing X.J., Lang Z.Q., Billings S. A. and Tomlinson G. R., "The Parametric Characteristics of Frequency Response Functions for Nonlinear Systems", International Journal of Control, Vol. 79, No. 12, December, pp 1552 - 1564, 2006

[7] Jing X.J., Lang Z.Q., and Billings S. A., "Mapping from parametric characteristics to generalised frequency response functions of nonlinear systems", International Journal of Control, 81(7), 1071-1088, 2008a

[8] Jing X.J., Lang Z.Q., Billings S. A. and Tomlinson G. R., Frequency domain analysis for suppression of output vibration from periodic disturbance using nonlinearities. Journal of Sound and Vibration, 314, 536-557, 2008b

[9] Lang Z.Q., and Billings S. A. "Output frequency characteristics of nonlinear systems". International Journal of Control, Vol. 64, 1049-1067, 1996

[10] Leonov G.A., Ponomarenko D.V. and Smirnova V.B. Frequency-domain methods for nonlinear analysis, theory and applications. World Scientific Publishing Co Pte Ltd, Singapore, 1996

[11] Ljung, L. System Identification: Theory for the User (second edition). Prentice Hall, Upper Saddle River. 1987 and 1999

[12] Pavlov A., van de Wouw N., and Nijmeijer H., Frequency Response Functions for Nonlinear Convergent Systems, IEEE Trans. Automatic Control, Vol 52, No 6, 1159-1165, 2007

[13] Nuij P.W.J.M., Bosgra O.H., Steinbuch M. "Higher-order sinusoidal input describing functions for the analysis of non-linear systems with harmonic responses". Mechanical Systems and Signal Processing Vol 20, pp1883 - 1904, 2006

[14] Peyton-Jones J.C. "Simplified computation of the Volterra frequency response functions of nonlinear systems". Mechanical systems and signal processing, Vol 21, Issue 3, pp 1452-1468, April 2007

[15] Pintelon R. and Schoukens J., System Identification: A Frequency Domain Approach, IEEE Press, Piscataway, NJ, 2001

[16] Rugh W.J., Nonlinear System Theory: the Volterra/Wiener Approach, Baltimore, Maryland, U.S.A.: Johns Hopkins University Press, 1981

[17] Sandberg I. W., The mathematical foundations of associated expansions for mildly nonlinear systems, IEEE Trans. Circuits Syst., CAS-30, pp441-455, 1983

[18] Solomou, M. Evans, C. Rees, D. Chiras, N. "Frequency domain analysis of nonlinear systems driven by multiharmonic signals", Proceedings of the 19th IEEE conference on Instrumentation and Measurement Technology, Vol 1, pp: 799- 804, 2002

[19] Schetzen M., The Volterra and Wiener Theory of Nonlinear Systems, J. Wiley and Sons, 1980

[20] Volterra V., Theory of Functionals and of Integral and Integrodifferential Equations, Dover, New York, 1959

[21] Orlowski P., Frequency domain analysis of uncertain time-varying discrete-time systems, Circuits Systems Signal Processing, Vol. 26, No. 3, 2007, PP. 293–310, 2007

[22] Jing X.J., Lang Z.Q., and Billings S. A., Parametric Characteristic Analysis for Generalized Frequency Response Functions of Nonlinear Systems. Circuits Syst Signal Process, DOI: 10.1007/s00034-009-9106-7, 2009

[23] Jing X.J., Lang Z.Q., and Billings S. A., Determination of the analytical parametric relationship for output spectrum of Volterra systems based on its parametric characteristics, Journal of Mathematical Analysis and Application, 351, 694 - 706, 2009

Chapter 3

LPV Gain-Scheduled Observer-Based State Feedback for Active Control of Harmonic Disturbances with Time-Varying Frequencies

Wiebke Heins, Pablo Ballesteros, Xinyu Shu and Christian Bohn

Additional information is available at the end of the chapter

1. Introduction

The design of controllers for the rejection of multisine disturbances with time-varying frequencies is considered. The frequencies are assumed to be known. Such a control problem frequently arises in active noise and vibration control (ANC/AVC) applications where the disturbances are caused by imbalances due to rotating or oscillating masses or periodically fluctuating excitations, for example the torque of a combustion engine, and the rotational speed is measured. Application examples are automobiles and aircrafts.

For the rejection of disturbances with time-varying frequencies, time-varying controllers that are automatically adjusted to the disturbance frequencies are usually used. Although time-invariant controllers might be sufficient in some applications [22], time-varying controllers usually result in a much better performance, particularly if the disturbance frequencies vary over fairly wide ranges. Such a controller can be constructed in several ways (see Sec. 2.1). Two observer-based state-feedback controllers are presented in this chapter (see Sec. 3). General output-feedback controllers are treated in the next chapter. The approaches presented in this chapter use state augmentation in order to achieve disturbance rejection. One consists of a time-invariant plant observer and a time-varying state-feedback gain for the state-augmented plant, where the state augmenation is based on a time-varying error filter, as proposed by Kinney & de Callafon [19]. The other controller approach is based on the disturbance observer of Bohn et al. [7], where the plant is augmented with a time-varying disturbance model. A time-varying observer for the overall system and a time-invariant state-feedback gain are used to track and reject the disturbance.

The remainder of this chapter is organized as follows. In Sec. 2, existing approaches to the problem are classified and some general control design considerations are discussed. The state-augmented observer-based state-feedback approaches are described in Sec. 3. In Sec. 4, the calculation of stabilizing state-feedback gains for time-varying systems in polytopic linear parameter-varying (pLPV) form is discussed. The application of this method for the rejection of harmonic disturbances is treated in Sec. 5. Real-time results are presented in Sec. 6. A discussion and conclusions are given in Sec. 7.

2. Controller design: Overview, stability and implementation aspects

In this section, an overview of control approaches for the rejection of harmonic disturbances with time-varying frequencies is given (Sec. 2.1) and stability and implementation aspects are discussed (Secs. 2.2 and 2.3, respectively).

2.1. Overview and classification of control approaches

A common approach in ANC/AVC is adaptive filtering, where the filter weights are usually updated with the FxLMS algorithm [24]. In most cases, disturbance feedforward is used, although it is possible to use an adaptive filter in feedback control with a technique called "secondary path neutralization" [24], which is equivalent to internal model control [25].

Another approach is to use gain scheduling, where the scheduling parameters are calculated from the disturbance frequencies. This can be further subdivided into indirect and direct scheduling methods. In indirect scheduling, the controller, or part of it, for example a state-feedback or observer gain, is determined from a set of pre-computed quantities through interpolation or switching. For example, for linear parameter-varying (LPV) systems, where the uncertain parameters are contained in a polytope, one controller is calculated for each vertex and the resulting controller is obtained from interpolation [2]. For the rejection of harmonic disturbances, continuous-time LPV approaches have been suggested by Darengosse & Chevrel [10], Du & Shi [11], Du et al. [12], Witte et al. [28], Balini et al. [6] and tested for a single sinusoidal disturbance by Darengosse & Chevrel [10], Du et al. [12], Witte et al. [28] and Balini et al. [6]. Methods based on observer-based state-feedback controllers are presented by Bohn et al. [7, 8], Kinney & de Callafon [19, 20, 21] and Heins et al. [16, 17]. In the approach of Bohn et al. [7, 8], the observer gain is selected from a set of pre-computed gains by switching. In the other approaches of Kinney & de Callafon [21], Heins et al. [17] and in this chapter, the observer gain is calculated by interpolation. In the other approach presented here, which is also used by Kinney & de Callafon [19, 20] and Heins et al. [16], the state-feedback gain is scheduled using interpolation.

In direct scheduling, the dependence of the controller on the scheduling parameter does not correspond to a simple interpolation or switching law [2, 4, 5, 21, 23, 26]. For example, for LPV systems where the parameter dependence is expressed as a linear fractional transformation (LFT), the uncertain parameters also enter the controller through an LFT [2]. For harmonic disturbances, an LPV-LFT approach has been suggested and applied in real time by Ballesteros & Bohn [4, 5] and Shu et al. [26]. Another example for direct scheduling is a controller based on a time-varying state estimator, for example a Kalman filter, where the scheduling parameters enter the controller through the recursive equations for the state estimate and the error covariance matrix. Such a controller is presented and compared to an indirect (interpolation) approach by Kinney & de Callafon [21].

2.2. Stability considerations

The existing design approaches can be classified as approaches that take stability into consideration and such that do not. In indirect scheduling, for example, the controllers or gains are sometimes pre-computed for fixed operating points and then interpolated in an ad-hoc fashion [7, 8, 20]. Stability is then not guaranteed, although it might be expected that the system is stable for slow variations of the scheduling parameter. For the adaptive filtering approaches, only approximate stability results seem available to date [13, 24]. To take stability into consideration, it is attractive to model the control problem as an LPV system and then use suitable gain-scheduling techniques [4–6, 10, 16, 17, 19, 23, 26, 28]. If parameter-independent Lyapunov functions are used, this guarantees closed-loop stability even for arbitrarily fast changes of the scheduling parameters at the expense of conservatism. The methods presented in this chapter guarantee closed-loop stability. In designs based on parameter-dependent Lyapunov functions, limits on the rate of change of the parameters can be incorporated.

2.3. Implementation aspects

For a practical application, the resulting controller has to be implemented in discrete time. In applications of ANC/AVC, the plant model is often obtained through system identification. This usually gives a discrete-time plant model. It is therefore most natural to carry out the whole design in discrete time. If a continuous-time controller is computed, the controller has to be discretized. Since the controllers considered here are time-varying, the discretization would have to be carried out at each sampling instant. An exact discretization involves the calculation of a matrix exponential, which is computationally too expensive. Particularly in LPV gain-scheduling control, an approximate discretization is proposed by Apkarian [3]. However, this leads to a distortion of the frequency scale. Usually, this can be tolerated, but not for the suppression of harmonic disturbances. It is therefore not surprising that the continuous-time controllers of Darengosse & Chevrel [10], Du et al. [12], Kinney & de Callafon [19] and Köroğlu & Scherer [23] are only tested in simulation studies with a very simple system as a plant and a single frequency in the disturbance signal. The design methods that are tested in real time are usually formulated in discrete time [4, 5, 7, 8, 16, 17, 20, 21, 26]. Exceptions are Witte et al. [28] and Balini et al. [6], who designed continuous-time controllers which then are approximately discretized. However, Witte et al. [28] use a very high sampling frequency of 40 kHz to reject a harmonic disturbance with a frequency up to 48 Hz (in fact, the authors state that they chose "the smallest [sampling time] available by the hardware") and Balini et al. [6] use a maximal sampling frequency of 50 kHz. It seems more natural to directly carry out the design in discrete time to avoid discretization issues.

3. State-augmented observer-based state-feedback control

Usually, disturbances act somewhere on the plant under consideration. The control objective is then to reject this disturbance. For the design methods considered in this paper, it is assumed that the disturbance acts on the plant input. For linear systems, a disturbance acting somewhere in the plant can be represented as a disturbance acting on the plant input under very mild assumptions. Therefore, assuming an input disturbance does not mean that the true disturbance has to act on the plant input (in fact, it usually does not). The design method used in this chapter is based on the internal model principle [14]. Applying this principle results in controllers with high gain in the frequency ranges where disturbances are assumed. These disturbances are then rejected no matter where they act upon the plant.

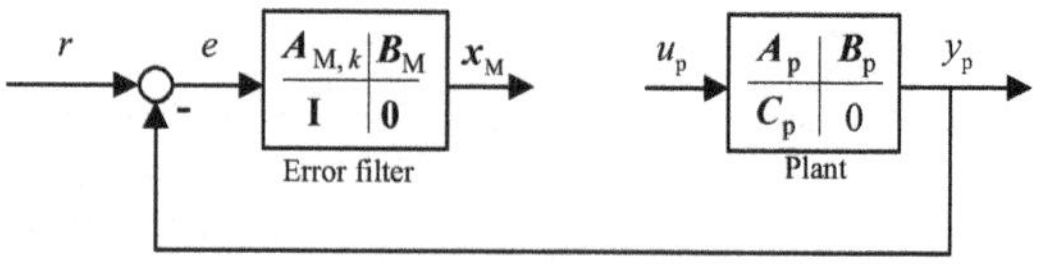

Figure 1. Interconnection of plant model and error filter

A discrete-time model for the plant with a disturbance $y_{d,k}$ acting on the input is given by

$$\begin{aligned} x_{p,k+1} &= A_p x_{p,k} + B_p(u_{p,k} + y_{d,k}), \\ y_{p,k} &= C_p x_{p,k}. \end{aligned} \tag{1}$$

In the approaches presented in the following, the disturbance rejection will be achieved with observer-based state-feedback controllers. The design methods presented here use state augmentation to add certain desired dynamics to the controller. In the time-invariant case, such a state augmentation can be used to prescribe controller poles. It can easily be verified that controller poles show up as zeros in the closed-loop disturbance transfer functions. Therefore, controller poles can be chosen to correspond to disturbances that are to be suppressed.

This argumentation, of course, only holds in the time-invariant case for which transfer functions and poles are defined (and only for controller poles that are not cancelled by plant zeros). If a time-varying internal model is used, it is not easy to interpret what happens when the internal model changes. Conceptually, the argument that the controller has high (infinite) gain at the disturbance frequencies that are to be rejected should still hold. It is confirmed by experiments (see Sec. 6) that even for fairly fast changes of the disturbance frequencies, excellent disturbance rejection is achieved.

3.1. State augmentation through an error filter

In this approach, disturbance rejection is achieved by including additional dynamics in a state-feedback controller through error filtering. A general error filter in state-space representation is given by

$$x_{M,k+1} = A_M x_{M,k} + B_M \left(r_k - y_{p,k} \right) \tag{2}$$

and interconnected with the plant model (without disturbance input)

$$\begin{aligned} x_{p,k+1} &= A_p x_{p,k} + B_p u_{p,k}, \\ y_{p,k} &= C_p x_{p,k}. \end{aligned} \tag{3}$$

as shown in Fig. 1. Usually, the controller error $e = r - y_p$ is used as input for the error filter. In the application considered here, however, no set-point r will be given and the error filter basically acts as an output filter with input $-y_p$. Nonetheless it will be referred to as error filter in the following.

The dynamics of the error filter can be used to include additional desired dynamics in a state-feedback controller for the overall system given by the plant and the filter. The additional dynamics can be chosen such that they describe the disturbances that the controller should reject, which corresponds to the internal model principle [14]. The additional dynamics can therefore be referred to as the "internal (disturbance) model." If, as in the control problem considered in this chapter, the disturbance characteristics change over time, a time-varying internal model

$$x_{\mathrm{M},k+1} = A_{\mathrm{M},k} x_{\mathrm{M},k} - B_{\mathrm{M}} y_{\mathrm{p},k} \tag{4}$$

can be used.

In order to design a state-feedback controller for the overall system, first the overall system model is formed by combining the plant model (3) and the error filter dynamics (4) through introduction of an augmented state vector. This yields

$$\begin{aligned} \begin{bmatrix} x_{\mathrm{M},k+1} \\ x_{\mathrm{p},k+1} \end{bmatrix} &= \begin{bmatrix} A_{\mathrm{M},k} & -B_{\mathrm{M}} C_{\mathrm{p}} \\ 0 & A_{\mathrm{p}} \end{bmatrix} \begin{bmatrix} x_{\mathrm{M},k} \\ x_{\mathrm{p},k} \end{bmatrix} + \begin{bmatrix} 0 \\ B_{\mathrm{p}} \end{bmatrix} u_{\mathrm{p},k}, \\ y_{\mathrm{p},k} &= \begin{bmatrix} 0 & C_{\mathrm{p}} \end{bmatrix} \begin{bmatrix} x_{\mathrm{M},k} \\ x_{\mathrm{p},k} \end{bmatrix}. \end{aligned} \tag{5}$$

Then, a state-feedback gain K_k for the overall system (5) can be designed. Due to the time-varying dynamics of the error filter, the state-feedback gain may be time-varying.

Usually the plant states are not available for feedback. Therefore, an identity observer

$$\hat{x}_{\mathrm{p},k+1} = (A_{\mathrm{p}} - L_{\mathrm{p},k} C_{\mathrm{p}}) \hat{x}_{\mathrm{p},k} + B_{\mathrm{p}} u_{\mathrm{p},k} + L_{\mathrm{p},k} y_{\mathrm{p},k} \tag{6}$$

is used to obtain an estimate $\hat{x}_{\mathrm{p}}$ of the plant states x_{p}. The estimate is then used for feedback instead of the unknown plant states.

For an overall state-feedback gain

$$K_k = \begin{bmatrix} K_{\mathrm{M},k} & K_{\mathrm{p},k} \end{bmatrix}, \tag{7}$$

where $K_{\mathrm{M},k}$ and $K_{\mathrm{p},k}$ denote the parts which feed back the error filter states x_{M} and the plant states x_{p} (or in this case, the estimated plant states $\hat{x}_{\mathrm{p}}$), respectively, the state-feedback law is given by

$$u_{\mathrm{p},k} = -K_{\mathrm{M},k} x_{\mathrm{M},k} - K_{\mathrm{p},k} \hat{x}_{\mathrm{p},k}. \tag{8}$$

The structure of the overall closed-loop system is shown in Fig. 2. The plant is affected by a disturbance y_{d} at the plant input. The plant output y_{p} is used as input for the error filter (with a sign reversal), which outputs are its states x_{M}, and for the identity observer for the plant. The latter gives an estimate $\hat{x}_{\mathrm{p}}$ of the plant states computed from y_{p} and u_{p}. The states are then combined and fed back with the overall state-feedback gain consisting of $K_{\mathrm{M},k}$ and $K_{\mathrm{p},k}$.

For stability analysis of the closed-loop system it is convenient to separate the state-space representation of the overall closed-loop dynamics including plant, observer, error filter, state-feedback and disturbance input in several parts as derived in the following. First, the controlled plant under disturbance at the input can be derived by combining (1), (6) and (8) to

$$x_{\mathrm{p},k+1} = (A_{\mathrm{p}} - B_{\mathrm{p}} K_{\mathrm{p},k}) x_{\mathrm{p},k} - B_{\mathrm{p}} K_{\mathrm{M},k} x_{\mathrm{M},k} - B_{\mathrm{p}} K_{\mathrm{p},k} (\hat{x}_{\mathrm{p},k} - x_{\mathrm{p},k}) + B_{\mathrm{p}} y_{\mathrm{d},k}. \tag{9}$$

It is driven by the error filter states x_{M}, the observer error $\hat{x}_{\mathrm{p}} - x_{\mathrm{p}}$ and the real unknown disturbance y_{d}.

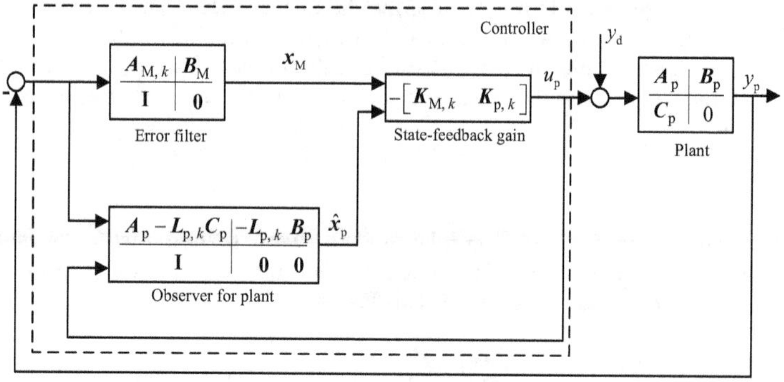

Figure 2. Structure of the resulting controller based on error filtering

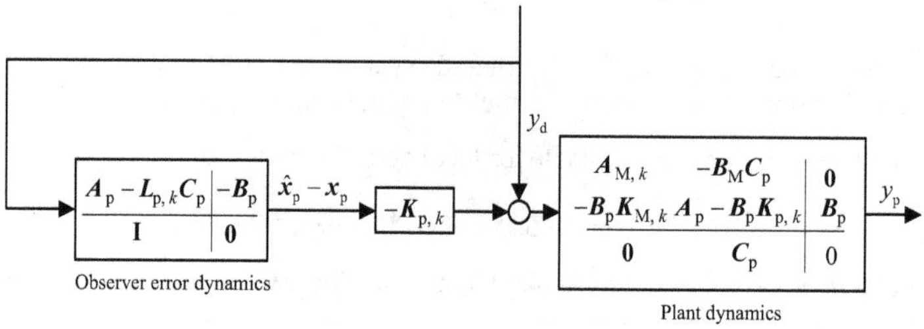

Figure 3. Dynamics of the overall closed-loop system for stability analysis

The dynamics of the identity observer for the plant states are described by (6). Since the observer does not include a model of the disturbance, the dynamics of the observer error, from now on denoted by $\tilde{x}_p$, are given as

$$\tilde{x}_{p,k+1} = (A_p - L_{p,k}C_p)\tilde{x}_{p,k} - B_p y_{d,k}, \tag{10}$$

which means that the observer is driven by the unknown disturbance y_d. With the error filter dynamics given by (4), the overall system can be described as

$$\begin{bmatrix} \tilde{x}_{p,k+1} \\ x_{M,k+1} \\ x_{p,k+1} \end{bmatrix} = \begin{bmatrix} A_p - L_{p,k}C_p & 0 & 0 \\ 0 & A_{M,k} & -B_MC_p \\ -B_pK_{p,k} & -B_pK_{M,k} & A_p - B_pK_{p,k} \end{bmatrix} \begin{bmatrix} \tilde{x}_{p,k} \\ x_{M,k} \\ x_{p,k} \end{bmatrix} + \begin{bmatrix} -B_p \\ 0 \\ B_p \end{bmatrix} y_{d,k}. \tag{11}$$

The overall system behavior is shown in Fig. 3. From this representation it can be seen that the time-varying dynamics of the augmented plant under state feedback are driven by the time-varying dynamics of the observer error for the plant states. The augmented plant under state feedback and the dynamics of the observer error for the plant states can therefore be interpreted as two distinct systems connected in series. From this it follows that choosing an observer gain $L_{p,k}$ such that $A_p - L_{p,k}C_p$ is stable and a state-feedback gain such that the time-varying dynamics of the augmented plant under state feedback are stable guarantees overall stability of the closed-loop system.

3.2. State augmentation through a disturbance model

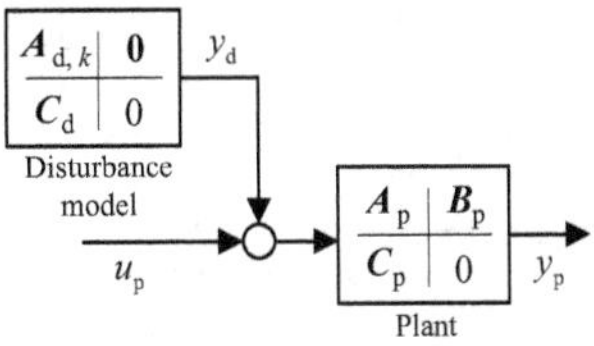

Figure 4. System model for the disturbance observer approach

In this approach, the disturbance y_d acting on the plant input u_p is explicitly modeled as the output of an unforced linear time-varying system

$$\begin{aligned} x_{d,k+1} &= A_{d,k}x_{d,k}, \\ y_{d,k} &= C_d x_{d,k} \end{aligned} \tag{12}$$

as shown in Fig. 4. An overall linear time-varying system model is then given by the combination of the plant model (1), which is affected by the disturbance $y_{d,k}$ at the plant input, and disturbance through

$$\begin{aligned} \begin{bmatrix} x_{d,k+1} \\ x_{p,k+1} \end{bmatrix} &= \begin{bmatrix} A_{d,k} & 0 \\ B_p C_d & A_p \end{bmatrix} \begin{bmatrix} x_{d,k} \\ x_{p,k} \end{bmatrix} + \begin{bmatrix} 0 \\ B_p \end{bmatrix} u_{p,k}, \\ y_{p,k} &= \begin{bmatrix} 0 & C_p \end{bmatrix} \begin{bmatrix} x_{d,k} \\ x_{p,k} \end{bmatrix}. \end{aligned} \tag{13}$$

With

$$x_k = \begin{bmatrix} x_{d,k} \\ x_{p,k} \end{bmatrix}, A_k = \begin{bmatrix} A_{d,k} & 0 \\ B_p C_d & A_p \end{bmatrix}, B = \begin{bmatrix} 0 \\ B_p \end{bmatrix}, C = \begin{bmatrix} 0 & C_p \end{bmatrix}, \tag{14}$$

the overall system can be written as

$$\begin{aligned} x_{k+1} &= A_k x_k + B u_{p,k}, \\ y_{p,k} &= C x_k. \end{aligned} \tag{15}$$

Estimates for the states of the disturbance and the plant model can then be obtained by application of a linear time-varying identity observer with time-varying observer gain L_k

$$\hat{x}_{k+1} = (A_k - L_k C)\hat{x}_k + B u_{p,k} + L_k y_{p,k}. \tag{16}$$

A state-feedback control for the overall system is given by

$$u_{p,k} = -K_k \hat{x}_k. \tag{17}$$

This leads to the typical structure of an observer-based state-feedback controller, but due to the time-varying disturbance model, the observer gain as well as state-feedback gain might be time-varying. Fig. 5 shows the general structure. The idea behind the disturbance observer is to use the estimate of the disturbance states (with a sign reversal) as control input. This should

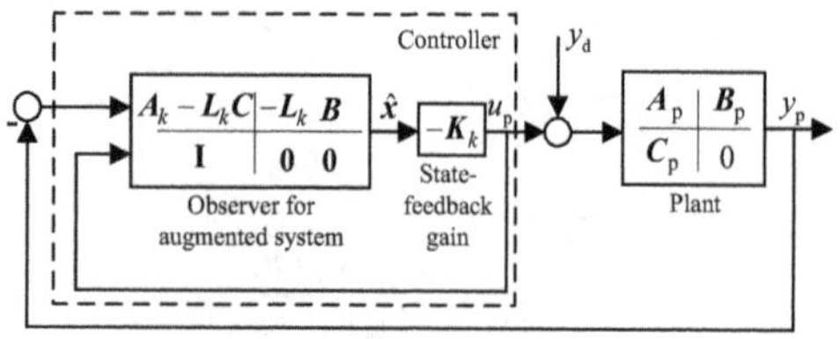

Figure 5. Observer-based state-feedback control in the disturbance observer approach

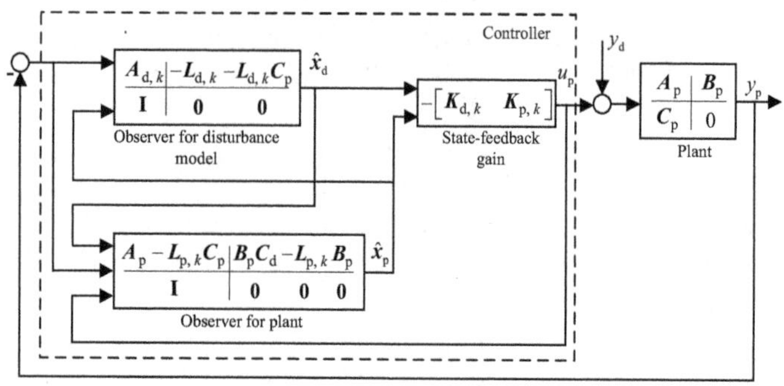

Figure 6. Detailed illustration of observer structure

lead to perfect disturbance cancellation if the disturbance model is chosen appropriately. For a more detailed analysis of this aspect, in the following observer gains and state-feedback gains are seperated into a part that feeds back the disturbance states via $L_{d,k}$ and $K_{d,k}$ and a part that feeds back the plant states or their estimates via $L_{p,k}$ and $K_{p,k}$. The estimated disturbance states can be described by

$$\hat{x}_{d,k+1} = A_{d,k}\hat{x}_{d,k} - L_{d,k}C_p\hat{x}_{p,k} + L_{d,k}y_{p,k}, \tag{18}$$

and the estimated plant states are given through

$$\hat{x}_{p,k+1} = (A_p - L_{p,k}C_p)\hat{x}_{p,k} + B_pC_d\hat{x}_{d,k} + L_{p,k}y_{p,k} + B_pu_{p,k}, \tag{19}$$

where

$$L_k = \begin{bmatrix} L_{d,k} & L_{p,k} \end{bmatrix}. \tag{20}$$

In Fig. 6, this more detailed representation is illustrated.

For stability analysis of the overall system including overall observer dynamics and state feedback, the dynamics of the observer error for the plant states

$$\hat{x}_{p,k+1} - x_{p,k+1} = (A_p - L_{p,k}C_p)(\hat{x}_{p,k} - x_{p,k}) + B_pC_d\hat{x}_{d,k} - B_py_{d,k} \tag{21}$$

are considered.

The dynamics of the states of the plant under state feedback are given by

$$x_{p,k+1} = (A_p - B_pK_{p,k})x_{p,k} - B_pK_{d,k}\hat{x}_{d,k} + B_pK_{p,k}(\hat{x}_{p,k} - x_{p,k}) + B_py_{d,k}. \tag{22}$$

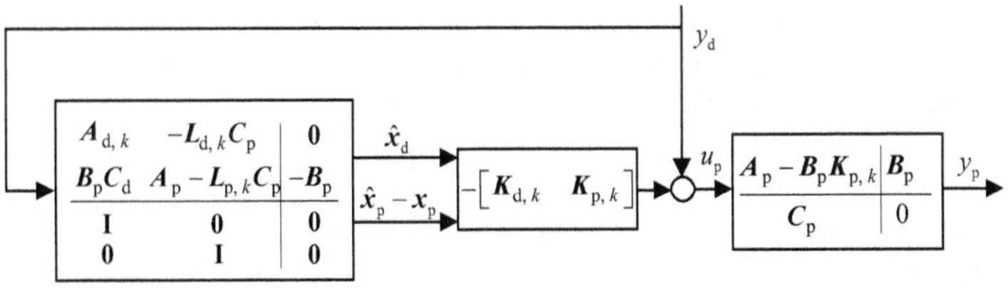

Figure 7. Dynamics of the overall closed-loop system for stability analysis

This yields the overall closed-loop system dynamics given by

$$\begin{bmatrix} \hat{x}_{d,k+1} \\ \tilde{x}_{p,k+1} \\ x_{p,k+1} \end{bmatrix} = \begin{bmatrix} A_{d,k} & -L_{d,k}C_p & 0 \\ B_pC_d & A_p - L_{p,k}C_p & 0 \\ -B_pK_{d,k} & B_pK_{p,k} & A_p - B_pK_{p,k} \end{bmatrix} \begin{bmatrix} \hat{x}_{d,k} \\ \tilde{x}_{p,k} \\ x_{p,k} \end{bmatrix} + \begin{bmatrix} 0 \\ -B_p \\ B_p \end{bmatrix} y_{d,k}. \tag{23}$$

Fig. 7 illustrates the structure of the overall system dynamics. This representation shows, similiar to the error filtering approach, that the dynamics of the plant under state feedback are driven by the time-varying dynamics of the observer for the augmented system, where in this case the outputs are given by the estimate of the disturbance states and the observer error for the plant states. The plant under state feedback and the observer of the augmented system can therefore be considered as two distinct systems in series connection. From this it follows that as long as a stabilizing state-feedback gain $K_{p,k}$ for the linear time-invariant plant and a stabilizing observer gain for the augmented system model is chosen, the overall closed-loop system is stable. It is also evident that the choice of the disturbance model does not influence the overall system stability, as long as the disturbance model itself is stable. Also, the choice of $K_{d,k}$ has no effect on the overall system stability (as long as $K_{d,k}\hat{x}_{d,k}$ remains bounded). Therefore, the overall state-feedback gain K does not have to be time-varying since only the linear time-invariant plant must be stabilized by K_p and the choice of K_d is not important for system stability. Furthermore, it can be seen that, assuming a perfect disturbance model, complete disturbance cancellation will be achieved after some transient if $K_d = C_d$.

4. Gain-scheduled state-feedback and observer design for pLPV systems

In this section, design methods for time-varying state-feedback and observer gains for pLPV systems based on quadratic stability theory are reviewed.

An LPV system is a linear system described by

$$\begin{aligned} x_{k+1} &= A(\theta)x_k + B(\theta)u_k \\ y_k &= C(\theta)x_k + D(\theta)u_k \end{aligned} \tag{24}$$

with $x_k \in \mathbb{R}^n$, where the state-space matrices depend on a possibly time-varying parameter vector

$$\theta = \begin{bmatrix} \theta_1 \ \theta_2 \ \cdots \ \theta_N \end{bmatrix}^T \in \Theta \subset \mathbb{R}^N. \tag{25}$$

In the following, only unforced systems of the form

$$x_{k+1} = A(\theta)x_k \tag{26}$$

are considered. In general, an LPV system is called a pLPV system, if the matrices depend affinely on the varying parameter. For the unforced system given in (26), this means that it is a pLPV system, if the system matrix $A(\cdot)$ depends affinely on $\boldsymbol{\theta}$, that is

$$A(\boldsymbol{\theta}) = \mathcal{A}_0 + \theta_1 \mathcal{A}_1 + \theta_2 \mathcal{A}_2 + \cdots + \theta_N \mathcal{A}_N, \tag{27}$$

where $\mathcal{A}_i \in \mathbb{R}^{n \times n}$ for all $i = 0, 1, ..., N$ are constant matrices and $\boldsymbol{\Theta}$ is a convex polytope in $\mathbb{R}^N$ with a finite set of vertices $V = \{v_1, v_2, ..., v_M\} \subset \mathbb{R}^N$. A point $\boldsymbol{\theta} \in \boldsymbol{\Theta}$ can be written as a convex combination of vertices, which means there exists a coordinate vector $\boldsymbol{\lambda} = [\lambda_1 \, \lambda_2 \, \ldots \, \lambda_M]^{\mathrm{T}} \in \mathbb{R}^M$ such that $\boldsymbol{\theta}$ can be written as

$$\boldsymbol{\theta} = \sum_{j=1}^{M} \lambda_j v_j, \quad \lambda_j \geq 0 \quad \forall j = 1, ..., M, \quad \sum_{j=1}^{M} \lambda_j = 1. \tag{28}$$

Defining $A_{\mathrm{v},j} = A(v_j)$ for $j = 1, \ldots, M$, $A(\boldsymbol{\theta})$ can be represented as

$$A(\boldsymbol{\theta}) = A(\boldsymbol{\lambda}) = \lambda_1 A_{\mathrm{v},1} + \lambda_2 A_{\mathrm{v},2} + \cdots + \lambda_M A_{\mathrm{v},M}. \tag{29}$$

The matrices $A_{\mathrm{v},j}$ can be considered as system matrices of linear time-invariant vertex systems

$$x_{k+1} = A_{\mathrm{v},j} x_k, \quad j = 1, \ldots, M, \tag{30}$$

and the current system (26) is a convex combination of the vertex systems in (30).

The pLPV system (26) is called *quadratically stable* in $\boldsymbol{\Theta}$ if and only if there exists a positive definite matrix $\boldsymbol{P}$ such that for all $\boldsymbol{\theta} \in \boldsymbol{\Theta}$

$$A^{\mathrm{T}}(\boldsymbol{\theta}) \boldsymbol{P} A(\boldsymbol{\theta}) - \boldsymbol{P} < 0, \tag{31}$$

or equivalently via the Schur complement [18]

$$\begin{bmatrix} \boldsymbol{P} & \boldsymbol{P} A(\boldsymbol{\theta}) \\ A^{\mathrm{T}}(\boldsymbol{\theta}) \boldsymbol{P} & \boldsymbol{P} \end{bmatrix} > 0, \tag{32}$$

where "> 0" and "< 0" indicate positive and negative definiteness, respectively.

Since (32) has to hold for every $\boldsymbol{\theta}$, this constitutes an infinite set of linear matrix inequalities (LMIs), which is computationally intractable. By the following result, it is possible to use a finite set of inequalities. It holds that a pLPV system in the form of (26) is quadratically stable for every $\boldsymbol{\theta} \in \boldsymbol{\Theta}$ if and only if there exists a symmetric positive definite matrix $\boldsymbol{P}$ such that

$$A_{\mathrm{v},j}^{\mathrm{T}} \boldsymbol{P} A_{\mathrm{v},j} - \boldsymbol{P} < 0, \tag{33}$$

or equivalently

$$\begin{bmatrix} \boldsymbol{P} & \boldsymbol{P} A_{\mathrm{v},j} \\ A_{\mathrm{v},j}^{\mathrm{T}} \boldsymbol{P} & \boldsymbol{P} \end{bmatrix} > 0, \tag{34}$$

for all $j = 1, \ldots, M$. This result is based on one single quadratic Lyapunov function that assures stability for the whole parameter space $\boldsymbol{\Theta}$. Therefore, $\boldsymbol{\theta}$ is allowed to change arbitrarily fast with time and $\boldsymbol{\theta}$ and $\boldsymbol{\lambda}$ can explicitly be assumed time-varying and denoted as $\boldsymbol{\theta}_k$ and $\boldsymbol{\lambda}_k$, respectively. A proof is given in Amato [1].

These results can be applied to the design of time-varying state-feedback and observer gains, as reviewed in the following sections.

4.1. State-feedback design for pLPV systems based on quadratic stability

A system of the form

$$x_{k+1} = A(\theta_k)x_k + Bu_k \tag{35}$$

is considered and the objective is to find a state-feedback gain $K(\theta_k)$ that quadratically stabilizes the closed-loop system

$$x_{k+1} = (A(\theta_k) - BK(\theta_k))x_k. \tag{36}$$

From the above result it follows that it suffices to find a single symmetric positive definite matrix P and a finite set of matrices $K_{\mathrm{v},j}$ such that

$$\begin{bmatrix} P & P(A_{\mathrm{v},j} - BK_{\mathrm{v},j}) \\ (A_{\mathrm{v},j} - BK_{\mathrm{v},j})^{\mathrm{T}}P & P \end{bmatrix} > 0, \quad j = 1, \ldots, M. \tag{37}$$

Then, $K(\theta_k)$ has to be chosen as

$$K(\theta_k) = \sum_{j=1}^{M} \lambda_{j,k} K_{\mathrm{v},j} = K(\lambda_k), \tag{38}$$

because then it holds that

$$A(\theta_k) - BK(\theta_k) = \sum_{j=1}^{M} \lambda_{j,k} A_{\mathrm{v},j} - B \sum_{j=1}^{M} \lambda_{j,k} K_{\mathrm{v},j} = \sum_{j=1}^{M} \lambda_{j,k}(A_{\mathrm{v},j} - BK_{\mathrm{v},j}), \tag{39}$$

and therefore quadratic stability of (36) is implied due to the results presented above.

4.2. Computation of state-feedback gains for the vertex systems

In this section, sufficient conditions that guarantee closed-loop stability and a certain H_2 performance level for the controlled vertex systems that can be used to calculate state-feedback gains for the vertex systems are derived. To use an H_2 performance level, an additional performance input is introduced that enters the state update equation for each vertex system of the pLPV system in (35). This input can be interpreted as process noise. For every $j = 1, \ldots, M$, this yields

$$x_{k+1} = A_{\mathrm{v},j}x_k + Bu_k + w_k. \tag{40}$$

As a performance output, the artificial signal

$$z_k = \begin{bmatrix} Q^{\frac{1}{2}} x_k \\ R^{\frac{1}{2}} u_k \end{bmatrix} \tag{41}$$

is defined that weighs the states and the control signal with matrices $Q^{\frac{1}{2}}$ and $R^{\frac{1}{2}}$, respectively. The objective then is to find state-feedback gains $K_{\mathrm{v},j}$ that stabilize the system and minimize the H_2 norm of the transfer path from w_k to z_k, when the control signal u_k in (40) and (41) is chosen as

$$u_k = -K_{v,j} x_k. \tag{42}$$

This objective can be stated as the minimization of the H_2 norm of the system

$$z = Gw \tag{43}$$

which has the state-space representation

$$\begin{aligned} x_{k+1} &= \widetilde{A} x_k + \widetilde{B} w_k, \\ z_k &= \widetilde{C} x_k, \end{aligned} \tag{44}$$

with

$$\widetilde{A} = A_{v,j} - BK_{v,j}, \quad \widetilde{B} = \mathbf{I}, \quad \widetilde{C} = \begin{bmatrix} Q^{\frac{1}{2}} \\ -R^{\frac{1}{2}} K_{v,j} \end{bmatrix}. \tag{45}$$

If the system is stable, the H_2 norm of this discrete-time linear time-invariant system is given by

$$\|G\|_2^2 = \text{trace}\left(\widetilde{C} W_c \widetilde{C}^{\mathrm{T}}\right), \tag{46}$$

where the controllability gramian W_c satisfies the discrete-time Lyapunov equation

$$\widetilde{A} W_c \widetilde{A}^{\mathrm{T}} - W_c + \widetilde{B}\widetilde{B}^{\mathrm{T}} = 0. \tag{47}$$

Therefore, if there exist symmetric positive definite matrices P and W such that

$$\widetilde{A} P \widetilde{A}^{\mathrm{T}} - P + \widetilde{B}\widetilde{B}^{\mathrm{T}} < 0, \tag{48}$$

$$W - \widetilde{C} P \widetilde{C}^{\mathrm{T}} > 0, \tag{49}$$

$$\text{trace}(W) < \gamma^2, \tag{50}$$

then it follows that $\|G\|_2 < \gamma$ for a $\gamma > 0$. Through the Schur complement, (48) and (49) can be transformed to

$$\begin{bmatrix} P & P\widetilde{A}^{\mathrm{T}} \\ \widetilde{A}P & P - \widetilde{B}\widetilde{B}^{\mathrm{T}} \end{bmatrix} > 0, \tag{51}$$

$$\begin{bmatrix} W & \widetilde{C}P \\ P\widetilde{C}^{\mathrm{T}} & P \end{bmatrix} > 0. \tag{52}$$

Introducing

$$\widetilde{Q} = \begin{bmatrix} Q^{\frac{1}{2}} \\ 0 \end{bmatrix}, \quad \widetilde{R} = \begin{bmatrix} 0 \\ R^{\frac{1}{2}} \end{bmatrix}, \quad Y_{v,j} = K_{v,j} P, \tag{53}$$

it follows that if solutions for the matrix variables P, W and $Y_{v,j}$, $j = 1, \dots, M$, can be found that satisfy the $2M + 1$ LMIs

$$\begin{bmatrix} P & \left(A_{v,j}P - BY_{v,j}\right)^{\mathrm{T}} \\ A_{v,j}P - BY_{v,j} & P - \mathbf{I} \end{bmatrix} > 0, \ j = 1, \dots, M, \tag{54}$$

$$\begin{bmatrix} W & \widetilde{Q}P - \widetilde{R}Y_{\mathrm{v},j} \\ \left(\widetilde{Q}P - \widetilde{R}Y_{\mathrm{v},j}\right)^{\mathrm{T}} & P \end{bmatrix} > 0, \; j = 1, \ldots, M, \tag{55}$$

$$\operatorname{trace}(W) < \gamma^2, \tag{56}$$

then the system G has an H_2 norm bounded by γ. From the solutions P and $Y_{\mathrm{v},j}$, the state-feedback gain for each vertex system can be calculated as

$$K_{\mathrm{v},j} = Y_{\mathrm{v},j}P^{-1}. \tag{57}$$

Quadratic stability is then implied for each closed-loop vertex system because of (48). In order to guarantee quadratic stability for the whole parameter space, solutions for the matrix variables P and W in (54)-(56) have to be the same for all vertex systems. Therefore, if solutions are found, also the performance bound γ is guaranteed for every fixed θ in the parameter space Θ, since it depends only on P and W.

Once state-feedback gains are found for the vertex systems, in any instant of time of a realization of the pLPV system (35) with $\theta_k \in \Theta$, a state-feedback gain can be found via interpolation according to (38). Quadratic stability of the closed-loop system (36) is guaranteed. The question of how to compute the coordinates λ_k with the properties described in (28) depends on the specific polytope Θ. The case of an N-dimensional hyper box will be considered in detail in Sec. 4.4.

4.3. Observer design for pLPV systems based on quadratic stability

Although the observer design is dual to the controller design, the LMIs for observer design are shortly presented here for completeness.

For observer design, a system of the form

$$\begin{aligned} x_{k+1} &= A(\theta_k)x_k + Bu_k, \\ y_k &= Cx_k \end{aligned} \tag{58}$$

is considered and the objective is to find an observer gain $L(\theta_k)$ that quadratically (and therefore, asymptotically) stabilizes the dynamics of the observer error

$$\widetilde{x}_{k+1} = (A(\theta_k) - L(\theta_k)C)\widetilde{x}_k. \tag{59}$$

From the above result it follows that it suffices to find a single symmetric positive definite matrix P and a finite set of matrices $L_{\mathrm{v},j}$ such that

$$\begin{bmatrix} P & P(A_{\mathrm{v},j} - L_{\mathrm{v},j}C) \\ (A_{\mathrm{v},j} - L_{\mathrm{v},j}C)^{\mathrm{T}}P & P \end{bmatrix} > 0, j = 1, \ldots, M, \tag{60}$$

and $L(\theta_k)$ has to be chosen as

$$L(\theta_k) = \sum_{j=1}^{M} \lambda_{j,k}L_{\mathrm{v},j} = L(\lambda_k). \tag{61}$$

As in the design of the state-feedback gain in the previous section, sufficient conditions for quadratic stability of (59) and a certain H_2 performance level can be derived. It is assumed that white noise of unit covariance, given by w_1 and w_2 and weighted by matrices $Q^{\frac{1}{2}}$ and $R^{\frac{1}{2}}$ respectively, affects the states and the outputs of the vertex systems of (58). This gives

$$\begin{aligned} x_{k+1} &= A(\theta_k)x_k + Cu_k + Q^{\frac{1}{2}}w_{1,k} \\ y_k &= Cx_k + R^{\frac{1}{2}}w_{2,k}. \end{aligned} \tag{62}$$

The H_2-norm of the transfer path from this noise input to the observer error is chosen as the performance measure. Defining

$$w_k = \begin{bmatrix} w_{1,k} \\ w_{2,k} \end{bmatrix}, \tag{63}$$

this transfer path is described by

$$\tilde{x}_{k+1} = \left(A_{\mathrm{v},j} - L_{\mathrm{v},k}C\right)\tilde{x}_k + \left[\, Q^{\frac{1}{2}} \;\; -L_{\mathrm{v},k}R^{\frac{1}{2}} \,\right] w_k. \tag{64}$$

Therefore the objective is to minimize the H_2-norm of the system $z = Gw$ with state-space representation

$$\begin{aligned} x_{k+1} &= \tilde{A}x_k + \tilde{B}w_k, \\ z_k &= \tilde{C}x_k, \end{aligned} \tag{65}$$

with

$$\tilde{A} = A_{\mathrm{v},j} - L_{\mathrm{v},j}C, \quad \tilde{B} = \left[\, Q^{\frac{1}{2}} \;\; -L_{\mathrm{v},k}R^{\frac{1}{2}} \,\right], \quad \tilde{C} = \mathbf{I}. \tag{66}$$

If the system is stable, the H_2 norm of this discrete-time linear time-invariant system is given by

$$\|G\|_2^2 = \mathrm{trace}\left(\tilde{B}^{\mathrm{T}} W_{\mathrm{o}} \tilde{B}\right), \tag{67}$$

where the observability gramian W_{o} satisfies the discrete-time Lyapunov equation

$$\tilde{A} W_{\mathrm{o}} \tilde{A}^{\mathrm{T}} - W_{\mathrm{o}} + \tilde{C}^{\mathrm{T}}\tilde{C} = 0. \tag{68}$$

For the same arguments as in Sec. 4.1 and with the introduction of

$$\tilde{Q} = \begin{bmatrix} Q^{\frac{T}{2}} \\ 0 \end{bmatrix}, \quad \tilde{R} = \begin{bmatrix} 0 \\ R^{\frac{T}{2}} \end{bmatrix}, \quad Y_{\mathrm{v},j} = L_{\mathrm{v},j}^{\mathrm{T}}P, \;\; j = 1, ..., M, \tag{69}$$

it follows that if solutions for the matrix variables P, W and $Y_{\mathrm{v},j}$, $j = 1, \ldots, M$, can be found that satisfy the $2M + 1$ LMIs

$$\begin{bmatrix} P & PA_{\mathrm{v},j} - Y_{\mathrm{v},j}^{\mathrm{T}}C \\ (PA_{\mathrm{v},j} - Y_{\mathrm{v},j}^{\mathrm{T}}C)^{\mathrm{T}} & P - \mathbf{I} \end{bmatrix} > 0, \;\; j = 1, ..., M, \tag{70}$$

$$\begin{bmatrix} W & \tilde{Q}P - \tilde{R}Y_{\mathrm{v},j} \\ (\tilde{Q}P - \tilde{R}Y_{\mathrm{v},j})^{\mathrm{T}} & P \end{bmatrix} > 0, \;\; j = 1, ..., M, \tag{71}$$

$$\text{trace}\left(\boldsymbol{W}\right) < \gamma^2, \tag{72}$$

then the system $\boldsymbol{G}$ has an H_2 norm bounded by γ. From the solutions $\boldsymbol{P}$ and $\boldsymbol{Y}_{\mathrm{v},j}$, the observer gain for each vertex system can be calculated as

$$\boldsymbol{L}_{\mathrm{v},j} = \boldsymbol{P}^{-\mathrm{T}}\boldsymbol{Y}_{\mathrm{v},j}^{\mathrm{T}}. \tag{73}$$

In order to guarantee quadratic stability for the whole parameter space, solutions for the matrix variables $\boldsymbol{P}$ and $\boldsymbol{W}$ in (70)-(72) have to be the same for all vertex systems. Therefore, if solutions are found, also the performance bound γ is guaranteed for every fixed $\boldsymbol{\theta}$ in the parameter space $\boldsymbol{\Theta}$.

Once observer gains are found for the vertex systems, in any instant of time an observer gain for the realization of the pLPV system (58) with $\boldsymbol{\theta}_k \in \boldsymbol{\Theta}$ can be found via interpolation according to (61).

4.4. Computation of the coordinates for interpolation

The computation of the coordinates $\boldsymbol{\lambda}_k \in \mathbb{R}^M$ required for the interpolation depends on the specific polytope $\boldsymbol{\Theta} \subset \mathbb{R}^N$ that is used to cover the parameter space. In general, for a parameter vector $\boldsymbol{\theta}_k \in \boldsymbol{\Theta}$, where $\boldsymbol{\Theta}$ denotes a polytope with M vertices $v_1, v_2, \ldots, v_M \in \mathbb{R}^N$, the coordinate vector $\boldsymbol{\lambda}_k$ as required for the interpolation is given by the solution to the constrained system of linear equations

$$\begin{bmatrix} v_1 \; v_2 \; \ldots \; v_M \\ 1 \;\; 1 \; \ldots \;\; 1 \end{bmatrix} \boldsymbol{\lambda}_k = \begin{bmatrix} \boldsymbol{\theta}_k \\ 1 \end{bmatrix}, \; 0 \leq \lambda_{j,k} \leq 1, \; j = 1, \ldots, M. \tag{74}$$

Existence of the solution is guaranteed as long as $\boldsymbol{\theta}_k \in \boldsymbol{\Theta}$. If the polytope considered has less than or exactly $N+1$ vertices, the constrained linear equation system admits exactly one solution. If the polytope is given by more than $N+1$ vertices, the system is underdetermined and more than one solution is possible. Since the controller design is not a linear operation, different choices of coordinates might lead to different properties of the resulting controller.

In most cases, ranges of components of the parameter vector are given by closed intervals in $\mathbb{R}$ determined by upper and lower bounds, such that

$$\theta_{i,k} \in \left[\theta_{i,\min}, \theta_{i,\max}\right], i = 1, \ldots, N. \tag{75}$$

If no further information on the relations between parameters are known, the polytope $\boldsymbol{\Theta}$ has to be chosen as the N-dimensional hyper box, also referred to as the parameter box, given by

$$\boldsymbol{\Theta} = \left[\theta_{1,\min}, \theta_{1,\max}\right] \times \left[\theta_{2,\min}, \theta_{2,\max}\right] \times \ldots \times \left[\theta_{N,\min}, \theta_{N,\max}\right]. \tag{76}$$

One way to compute a set of coordinates $\boldsymbol{\lambda}_k$ for this case is introduced by Apkarian et al. [2] and generalized and implemented for an arbitrary number of vertices in the LMI Control Toolbox for Matlab [15]. Another implementation of this approach is proposed here that is suitable for real-time implementation purposes where variables have to be pre-initialized with fixed dimensions. Daafouz et al. [9] presented a compact way of writing the calculation scheme, on which the scheme proposed here and in [17] is based. If the order of vertices is not changed, any of the mentioned approaches leads to the same coordinates.

The i-th entry $v_{j,i}$ of a vertex j of the parameter box $\boldsymbol{\Theta}$ is either the lower bound $\theta_{i,\min}$ or the upper bound $\theta_{i,\max}$ of $\theta_{i,k}$. Now, $2N$ vectors

$$\boldsymbol{b}_{i_{\min}} = \begin{bmatrix} b_{i_{\min},1} \\ \vdots \\ b_{i_{\min},M} \end{bmatrix}, \boldsymbol{b}_{i_{\max}} = \begin{bmatrix} b_{i_{\max},1} \\ \vdots \\ b_{i_{\max},M} \end{bmatrix} \tag{77}$$

are pre-computed such that

$$b_{i_{\max},j} = \begin{cases} \frac{1}{\theta_{i,\max}-\theta_{i,\min}}, & \text{if } v_{j,i} = \theta_{i,\max}, \\ 0, & \text{if } v_{j,i} = \theta_{i,\min}, \end{cases} \tag{78}$$

$$b_{i_{\min},j} = \begin{cases} \frac{1}{\theta_{i,\max}-\theta_{i,\min}}, & \text{if } v_{j,i} = \theta_{i,\min}, \\ 0, & \text{if } v_{j,i} = \theta_{i,\max}. \end{cases} \tag{79}$$

The following steps are then carried out in every sampling instant:

1. $\theta_{i,k} = \cos(2\pi f_{i,k} T),\ i = 1, ..., N,$ (80)

2. $c_{i_{\max},k} = \theta_{i,k} - \theta_{i,\min},\ \ c_{i_{\min},k} = \theta_{i,\max} - \theta_{i,k},\ i = 1, ..., N,$ (81)

3. $\lambda_{j,k} = \prod_{i=1}^{N} (b_{i_{\max},j} c_{i_{\max},k} + b_{i_{\min},j} c_{i_{\min},k}), j = 1, ..., M.$ (82)

5. Application for the rejection of harmonic disturbances

In this section, the methods described in Sec. 4 are applied to the case of a harmonic multisine disturbance. Specific system models and the transformation that leads to a pLPV system are presented.

5.1. Internal model for harmonic disturbances

The disturbance is assumed to be a multisine. Let N be the number of components of the disturbance and $f_{i,k}$ the frequency of the i-th component at sampling instant k. The frequencies are assumed to vary in intervals $[f_{i,\min}, f_{i,\max}] \subset [0, 0.5 f_{\mathrm{s}}]$, where f_{s} denotes the sampling frequency.

As discussed in Sec. 3, an error filter or a disturbance model can be used to include desired dynamics in the controller according to the internal model principle [14]. Since the following relations are used for either $\boldsymbol{A}_{\mathrm{M},k}$ or $\boldsymbol{A}_{\mathrm{d},k}$ depending on the control approach, the notation $\boldsymbol{A}_{\mathrm{M/d},k}$ is employed. The dynamics for a multisine disturbance with time-varying frequency are given by a discrete-time state-space model with system matrix

$$\boldsymbol{A}_{\mathrm{M/d},k} = \begin{bmatrix} \boldsymbol{A}_{\mathrm{M}_1/\mathrm{d}_1,k} & \cdots & \mathbf{0} \\ \vdots & \ddots & \vdots \\ \mathbf{0} & \cdots & \boldsymbol{A}_{\mathrm{M}_N/\mathrm{d}_N,k} \end{bmatrix}, \tag{83}$$

where

$$A_{\mathrm{M}_i/\mathrm{d}_i,k} = \begin{bmatrix} \cos(2\pi f_{i,k}T) & \sin(2\pi f_{i,k}T) \\ -\sin(2\pi f_{i,k}T) & \cos(2\pi f_{i,k}T) \end{bmatrix} \tag{84}$$

with sampling time $T = 1/f_{\mathrm{s}}$. This time-varying matrix might therefore be used as system matrix $A_{\mathrm{M},k}$ of the error filter described in Sec. 3.1 or $A_{\mathrm{d},k}$ of the disturbance model described in Sec. 3.2. In order to find a representation of $A_{\mathrm{M/d},k}$ that depends affinely on a parameter vector $\boldsymbol{\theta}$, one approach could be to choose

$$\theta_{i,k} = \cos(2\pi f_{i,k}T),\ i = 1,...,N. \tag{85}$$

Unfortunately, through

$$\sin(2\pi f_{i,k}T) = \sqrt{1-\theta_{i,k}^2},\ i = 1,...,N, \tag{86}$$

a non-affine dependence is introduced into the model. To circumvent this, additional parameters

$$\theta_{i,k} = \sin(2\pi f_{i,k}T),\ i = N+1,...,2N \tag{87}$$

could be introduced. The use of this internal model leads to a large dimension of the parameter space and a polytope with many vertices. It can be expected that controllers based on this model are quite conservative, if there exists a solution to the LMIs at all. Therefore, a simplified disturbance model that leads to less conservative conditions is used for the controller design.

In the case of constant frequencies, instead of (84), the matrices

$$A_{\mathrm{M}_i/\mathrm{d}_i} = \begin{bmatrix} 0 & 1 \\ -1 & 2\cos(2\pi f_i T) \end{bmatrix},\ i = 1,...,N \tag{88}$$

can be used in the system matrix of the internal model (83). This leads to a time-invariant overall system model. As discussed in Sec. 3, the internal model then is a way to determine controller poles (that show up as zeros in the disturbance transfer functions). With the above approach, for a frequency f, a complex conjugate controller pole pair is placed at $\exp(j2\pi fT)$ and $\exp(-j2\pi fT)$, which causes complete asymptotic rejection of harmonic disturbances with this frequency.

Although this model does not take into account the rate of change of the disturbance frequencies and the argumentation based on controller poles is valid only in the time-invariant case, this simpler model is used here even for the case of time-varying frequencies, since it reduces the dimension of the parameter space. Thus, the matrices

$$A_{\mathrm{M}_i/\mathrm{d}_i,k} = \begin{bmatrix} 0 & 1 \\ -1 & 2\cos(2\pi f_{i,k}T) \end{bmatrix},\ i = 1,...,N \tag{89}$$

are used for the time-varying disturbance model. This might have an effect on the disturbance attenuation for fast changes of the frequencies, but it is expected that other effects are more dominant than the "incorrect" disturbance model. There will always be delay between the measured frequency used in the controller and the true frequency. Also, if the disturbance and the control signal do not enter at exactly the same point and the disturbance frequency varies, it is unclear which frequency "is present" at the point where the control signal enters the plant at a certain time. As shown in Sec. 3, closed-loop stability is not affected by the choice of the internal model.

5.2. Transformation to pLPV form

If the internal model can be written in the form of a pLPV system, it is possible to find a representation of the system matrix A_k of the overall system of the error filter approach (5) and the disturbance observer approach (15) in the form of a pLPV system as well. Then, the controller design and the gain-scheduling procedure can be carried out based on quadratic stability theory of pLPV systems, as has been presented in Sec. 4.

If the sampling theorem applies (which should be the case for any practical application), it holds that $0 \leq 2\pi f_{i,k}T \leq \pi$ for all $i = 1, ..., N$. Since the function $f \mapsto \cos(2\pi f T)$ is monotonically decreasing on $[0, \pi]$, the frequencies under consideration have to fulfill $0 \leq 2\pi f_{i,k}T \leq \pi$ for all $i = 1, ..., N$ in order to guarantee a unique mapping between frequencies and parameters. This can usually be achieved by choosing a sufficiently small value for T. Then, parameters $\boldsymbol{\theta}_i$ as defined in (85) are bounded by

$$\theta_{i,\min} = \cos(2\pi f_{i,\max}T),\ \theta_{i,\max} = \cos(2\pi f_{i,\min}T),\ i = 1, ..., N, \tag{90}$$

respectively. Defining

$$\boldsymbol{\Theta} = [\theta_{1,\min}, \theta_{1,\max}] \times [\theta_{2,\min}, \theta_{2,\max}] \times ... \times [\theta_{N,\min}, \theta_{N,\max}] \subset \mathbb{R}^N \tag{91}$$

with vertices $v_1, v_2, ..., v_M \in \mathbb{R}^N$ it follows that the parameter vector

$$\boldsymbol{\theta}_k = \begin{bmatrix} \theta_{1,k} \\ \theta_{2,k} \\ \vdots \\ \theta_{N,k} \end{bmatrix} \in \mathbb{R}^N \tag{92}$$

is always inside the hyper box $\boldsymbol{\Theta}$. Then, the system matrix $A_{\mathrm{M/d},k}$ of the internal model can be written as

$$A_{\mathrm{M/d},k} = A_{\mathrm{M/d}}(\boldsymbol{\theta}_k) = \mathcal{A}_{\mathrm{M_0/d_0}} + \theta_{1,k}\mathcal{A}_{\mathrm{M_1/d_1}} + \theta_{2,k}\mathcal{A}_{\mathrm{M_2/d_2}} + ... + \theta_{N,k}\mathcal{A}_{\mathrm{M_N/d_N}} \tag{93}$$

with

$$\mathcal{A}_{\mathrm{M_0/d_0}} = \begin{bmatrix} \mathcal{A}_{\mathrm{M_0/d_0},1} & \cdots & \mathbf{0} \\ \vdots & \ddots & \vdots \\ \mathbf{0} & \cdots & \mathcal{A}_{\mathrm{M_0/d_0},N} \end{bmatrix}, \tag{94}$$

where

$$\mathcal{A}_{\mathrm{M_0/d_0},i} = \begin{bmatrix} 0 & 1 \\ -1 & 0 \end{bmatrix},\ i = 1, ..., N, \tag{95}$$

and $\mathcal{A}_{\mathrm{M_i/d_i}}$ matrices with zero entries only except for

$$\mathcal{A}_{\mathrm{M_i/d_i}}(2i, 2i) = 2,\ i = 1, ..., N. \tag{96}$$

Therefore, the system matrix $A_{\mathrm{M/d},k}$ of the internal model used for the error filter from Sec. 3.1 and the disturbance model in Sec. 3.2 can be written in pLPV form based on the

N-dimensional hyper box $\boldsymbol{\Theta}$ with vertices $v_1, v_2, ..., v_M \in \mathbb{R}^N$. The system matrices A_k of (5) in the error filter approach can then be written as

$$A_k = A(\theta_k) = \mathcal{A}_0 + \theta_{1,k}\mathcal{A}_1 + ... + \theta_{N,k}\mathcal{A}_N \tag{97}$$

with

$$\mathcal{A}_0 = \begin{bmatrix} \mathcal{A}_{\mathrm{M}_0} & B_{\mathrm{M}}C_{\mathrm{p}} \\ 0 & A_{\mathrm{p}} \end{bmatrix}, \tag{98}$$

$$\mathcal{A}_i = \begin{bmatrix} \mathcal{A}_{\mathrm{M}_i} & 0 \\ 0 & 0 \end{bmatrix}, i = 1, ..., N, \tag{99}$$

and analogously in (15) for the disturbance observer approach

$$A_k = A(\boldsymbol{\theta}_k) = \mathcal{A}_0 + \theta_{1,k}\mathcal{A}_1 + ... + \theta_{N,k}\mathcal{A}_N \tag{100}$$

with

$$\mathcal{A}_0 = \begin{bmatrix} A_{\mathrm{p}} & B_{\mathrm{p}}C_{\mathrm{d}} \\ 0 & \mathcal{A}_{\mathrm{d}_0} \end{bmatrix}, \tag{101}$$

$$\mathcal{A}_i = \begin{bmatrix} 0 & 0 \\ 0 & \mathcal{A}_{\mathrm{d}_i} \end{bmatrix}, i = 1, ..., N. \tag{102}$$

Since $\boldsymbol{\Theta}$ is a convex polytope, in every instant there exist coordinates $\lambda_{j,k}$, $j = 1, ..., M$, such that

$$\boldsymbol{\theta}_k = \sum_{j=1}^{M} \lambda_{j,k} \boldsymbol{v}_j. \tag{103}$$

Since A_k depends affinely on $\boldsymbol{\theta}_k$ in both approaches, it follows that the system matrix can be expressed as

$$A(\boldsymbol{\theta}_k) = \sum_{j=1}^{M} \lambda_{j,k} A_{\mathrm{v},j} = A(\boldsymbol{\lambda}_k), \tag{104}$$

where

$$A_{\mathrm{v},j} = A(\boldsymbol{v}_j), j = 1, \ldots, M. \tag{105}$$

The required time-varying state-feedback and observer gains for the overall systems (5) and (15), respectively, can therefore be obtained by application of the design methods based on quadratic stability theory as reviewed in Sec. 4.

5.3. Controller implementation

Once the pLPV representation of the considered system is found, the required time-varying state-feedback and observer gains for the overall systems (5) and (15), respectively, can be obtained by application of the design methods based on quadratic stability theory as reviewed in Sec. 4.

The time-invariant vertex state-feedback or vertex observer gains are calculated in advance. Since the plant itself is linear time-invariant, also the time-invariant observer required for the estimation of the plant states in the error filter approach as well as the time-invariant state-feedback gain of the disturbance observer approach can be calculated off-line with

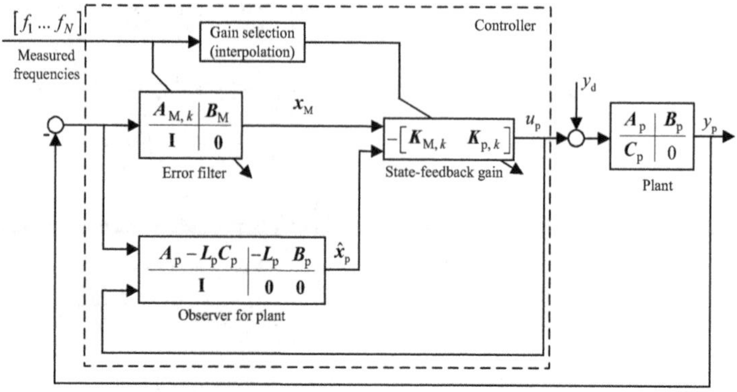

Figure 8. Control structure and interpolation for the error filter approach

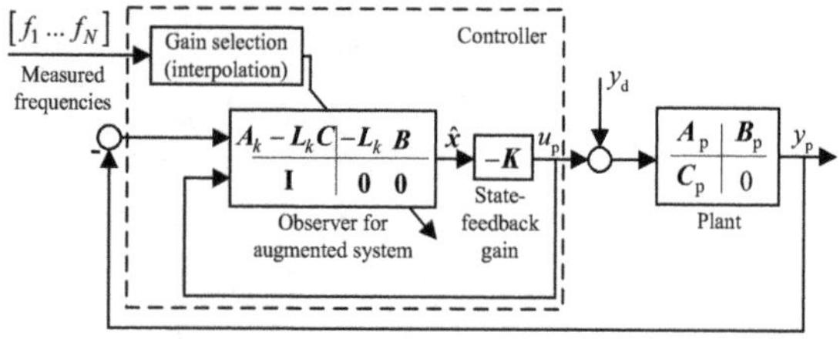

Figure 9. Control structure and interpolation the disturbance model approach

a standard method (e.g. LQR or pole placement). The controllers are then implemented according to the structures given in Sec. 3 and shown in Figs. 8 and 9 with the interpolation. Wherever possible, system matrices are updated in every sampling instant directly with the measured frequencies, while the time-varying state-feedback and observer gains are obtained by interpolation with the coordinates obtained by the scheme given in Sec. 4.4.

6. Real-time results

The gain-scheduled observer-based state-feedback controllers obtained through the design procedures presented in this chapter are validated with experimental results. Both controllers have been tested on an AVC and an ANC system and were found to work well. Results are presented for the test of the error filter with time-varying state-feedback gain on an ANC headset and the controller based on a disturbance observer with time-varying observer gain on an AVC test bed.

6.1. Real-time results for state-feedback gain-scheduling (error filtering) approach

The controller based on a disturbance observer with time-varying observer gain is tested experimentally on a Sennheiser PXC 300 headset. The experimental setup is shown schematically in Fig. 10. An external loudspeaker is used to generate a harmonic disturbance. The headset has one microphone on each loudspeaker. The objective is to cancel the

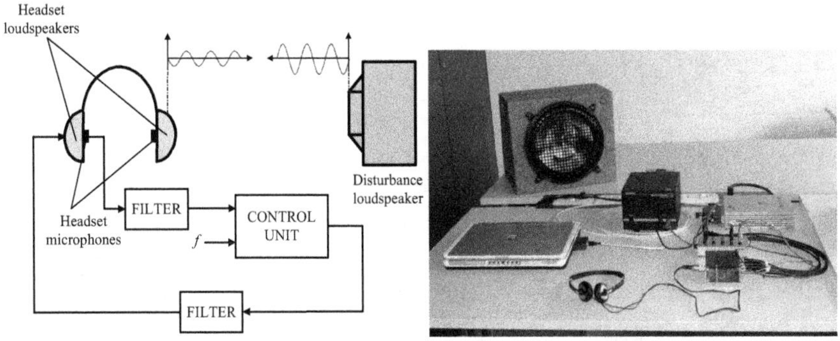

Figure 10. Block diagramm (left) and photograph (right) of the ANC system

disturbance with the loudspeakers of the headset. An anti-aliasing filter is applied to the output signal and a reconstruction filter to the control input. The control algorithm is implemented on a rapid control prototyping unit (dSpace MicroAutoBox). A sampling frequency of 1 kHz was used.

Standard black-box system identification techniques were used to obtain the transfer function between output and input of the control unit. The identified transfer function is of 12th order and the resulting controller of 20th order. As a disturbance signal, a sum of four harmonically related sine signals with fundamental frequency between 90 Hz and 100 Hz is used. The control algorithm has been implemented on both sides and results for the right side are shown.

Fig. 11 shows the amplitude frequency responses for constant fundamental frequencies of 90 Hz and 100 Hz. The amplitude frequency response plots show that amplification takes place in some frequency ranges, which is the known "waterbed" effect. It might cause problems in practical applications where significant background noise, e.g. broadband stochastic disturbances, is present. In Fig. 12, results are shown for a case where the fundamental frequency suddenly jumps from 90 Hz to 100 Hz. The controller remains stable even for such drastic variations of the fundamental frequency. The transient spike might be undesirable, particularly in ANC applications where it could be audible. However, a step change in the frequency does not commonly occur in real applications.

In Fig. 13, results for a gradually changing fundamental frequency are shown. The measurements show that excellent disturbance attenuation is achieved. The performance of the controller is tested with fast variations of the disturbance frequencies. Results for this experiment are shown in Fig. 14. It can be observed that the attenuation performance decreases, but the system remains stable.

6.2. Real-time results for observer gain-scheduling (disturbance model) approach

The scheme and a photograph of the AVC test bed are shown in Fig. 15. Two shakers (inertia mass actuators) are attached to a steel cantilever beam. One shaker acts as the disturbance source and the other shaker is driven by the control signal to counteract this disturbance. An accelerometer is used to measure the output signal. An anti-aliasing filter is applied to the output signal and a reconstruction filter to the control input.

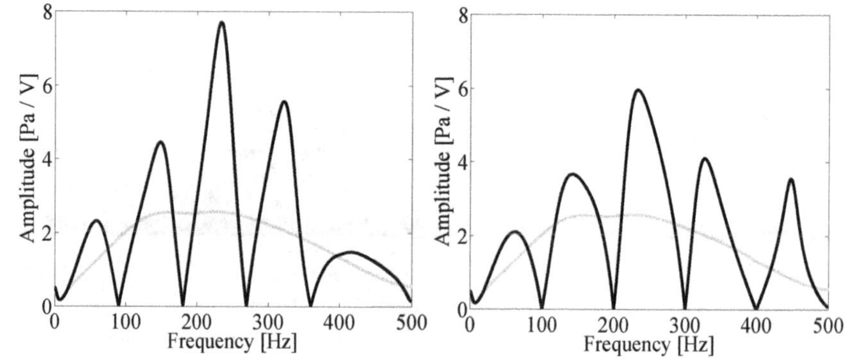

Figure 11. Open-loop (gray) and closed-loop (black) amplitude frequency responses for fixed disturbance frequencies of 90, 180, 270 and 360 Hz (left) and of 100, 200, 300 and 400 Hz (right)

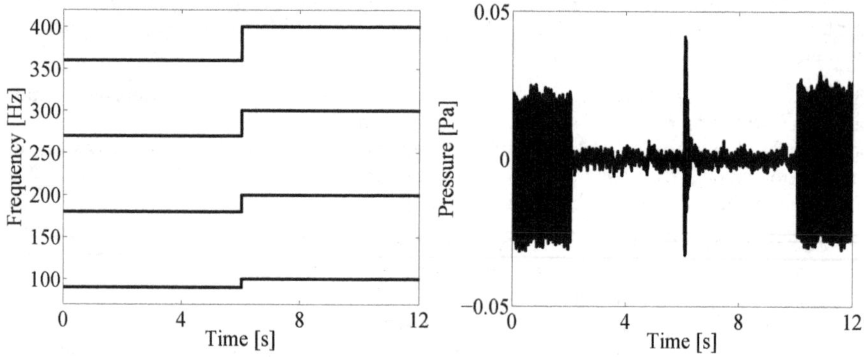

Figure 12. Results for a disturbance with time-varying frequencies. Variation of the frequencies (left) and measured sound pressure (right). The control sequence is off/on/off

A multisine test signal is used to identify the transfer function between output and input of the control unit using standard black-box system identification techniques. The controller is implemented on a rapid control prototyping unit (dSpace MicroAutoBox). A sampling frequency of 1 kHz was chosen. The identified system is of 10th order and the controller of 16th order. As a disturbance signal, a sum of three harmonically related sine signals with fundamental frequency between 110 and 120 Hz is used. The amplitude frequency responses of the open-loop and closed-loop systems for fixed disturbance frequencies of 110 Hz, 220 Hz and 330 Hz and 120 Hz, 240 Hz and 360 Hz are shown in Fig. 16. The resonance frequencies are damped by a factor 0.9 using pole placement with the state-feedback gain K_p. In Fig. 17, results are shown for a (rather unrealistic) case where the fundamental frequency suddenly jumps from 110 Hz to 120 Hz. After a short transient spike the disturbance rejection resumes quickly. The transient spike might be undesirable but a step change in the frequency does not commonly occur in real applications. The measurements confirm the excellent disturbance rejection that can be expected from the amplitude frequency responses. The amplitude response plots show that amplification takes place in frequency ranges between the rejected

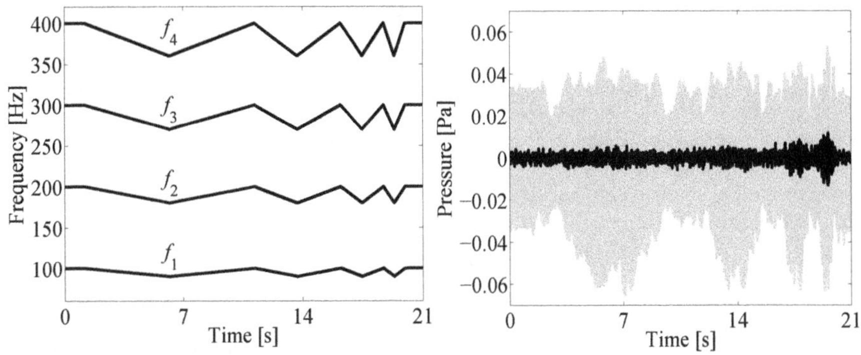

Figure 13. Results for a disturbance with time-varying frequencies. Variation of the frequencies (left) and measured sound pressure (right) in open loop (gray) and closed loop (black)

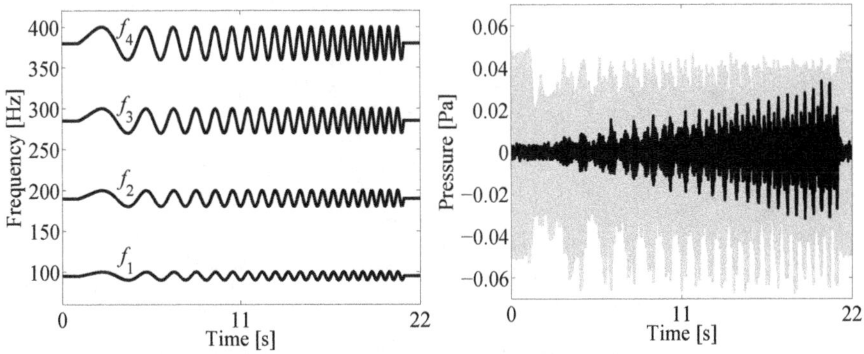

Figure 14. Results for a disturbance with time-varying frequencies. Variation of the frequencies (left) and measured sound pressure (right) in open loop (gray) and closed loop (black)

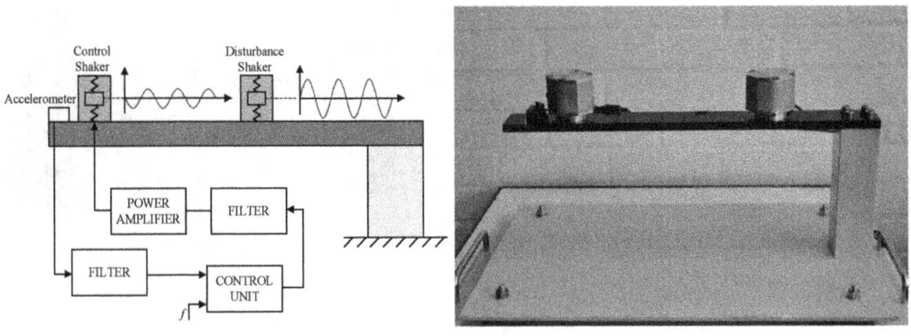

Figure 15. Schematic representation (left) and photograph (right) of the AVC system

frequencies. This is due to Bode's sensitivity integral ("waterbed" effect). Whether this is tolerable or not depends on the application. It is possible to limit the maximum disturbance amplification by a suitable overall controller design (which need not be observer based), but at the expense of either not fully suppressing the harmonic disturbances (that is, the notches in the frequency response would not get to zero) or requiring the frequency measurements to be very exact (that is, making the notches narrower) or worsening the transient behavior (that is, it would take longer before a harmonic disturbance is suppressed to a certain level). However, in this chapter only harmonic disturbances are considered, so these issues are not addressed in the design.

In Fig. 18, results for a gradually changing fundamental frequency are shown. It decreases linearly from 120 Hz to 110 Hz, remains constant for a while and then rises back to 120 Hz. This is repeated four times, every time in a shorter time interval. At the end the fourth harmonic rises from 330 Hz to 360 Hz and decreases back to 330 Hz in less than two seconds. The effect of fast variations of the disturbance frequencies has been further investigated in another experiment. The results are shown in Fig. 19. The disturbance frequency varies

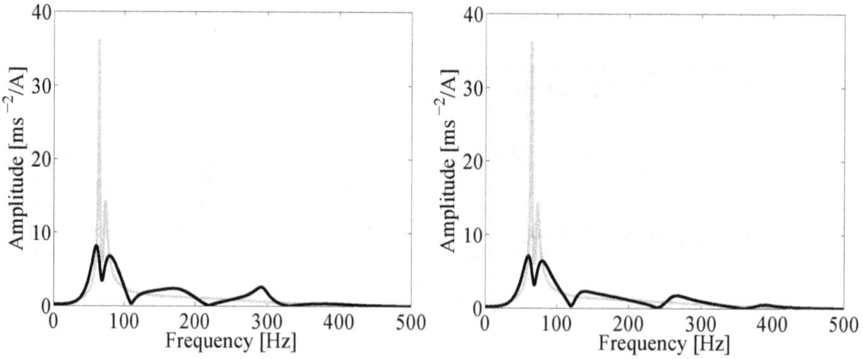

Figure 16. Open-loop (gray) and closed-loop (black) amplitude frequency responses for fixed disturbance frequencies of 110, 220 and 330 Hz (left) and of 120, 240 and 360 Hz (right)

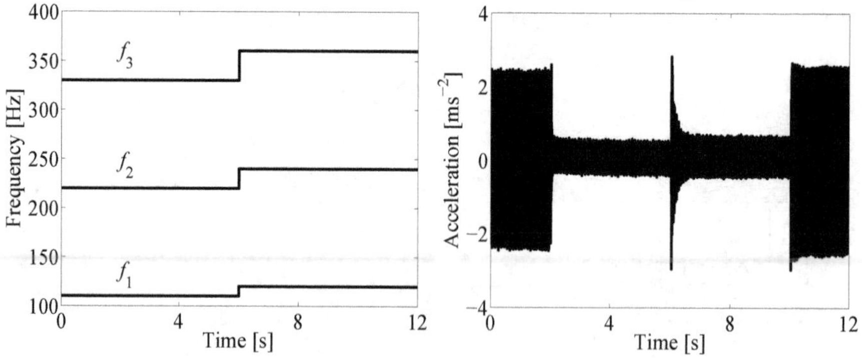

Figure 17. Results for a disturbance with time-varying frequencies. Variation of the frequencies (left) and measured acceleration (right). The control sequence is off/on/off

sinusoidally between the minimum and the maximum value with a period that decays from 10 to 0.5 seconds. It is seen that for very fast frequency variations, the attenuation performance decreases slightly (but the system remains stable).

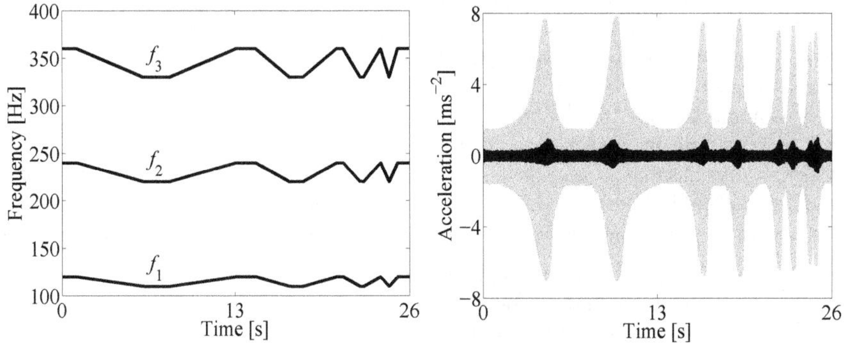

Figure 18. Results for a disturbance with time-varying frequencies. Variation of the frequencies (left) and measured acceleration (right) in open loop (gray) and closed loop (black)

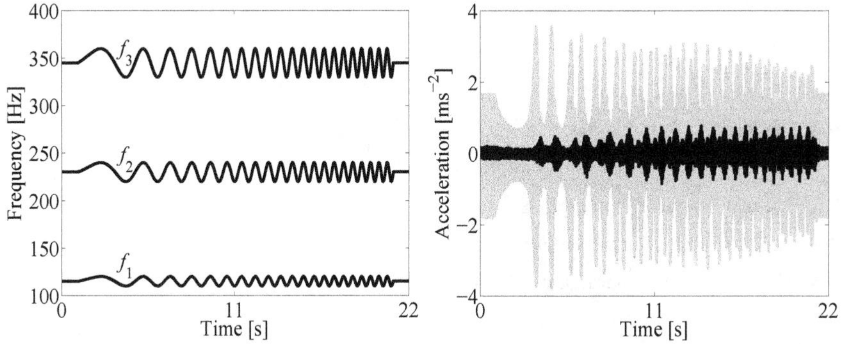

Figure 19. Results for a disturbance with time-varying frequencies. Variation of the frequencies (left) and measured acceleration (right) in open loop (gray) and closed loop (black)

7. Discussion and conclusion

Two discrete-time LPV observer-based state-feedback controllers for the rejection of harmonic disturbances with time-varying frequencies are presented. The control design methods are based on quadratic stability theory for pLPV systems. They guarantee stability of the closed-loop system also for arbitrarily fast changes of the disturbance frequencies. This is an advantage over other approaches such as adaptive filtering or heuristic gain scheduling. The experimental results show that an excellent disturbance rejection is achieved and that the proposed controllers can be applied in a real-time setup.

However, some degree of conservatism is present in these approaches. Using a single parameter independent Lyapunov function limits the range of admissible disturbance frequencies that can be covered with the resulting controller. Also, the polytope that contains the uncertain parameters could be chosen much smaller and with fewer vertices than the cuboid applied here, if information on the relations between the disturbance frequencies is given (as, for example, in the case of harmonically related frequencies). Thus, feasibility of the LMIs and the upper bound on the system performance could be improved as well as the computation time for the coordinates required for the on-line interpolation. This might be important for applications where many harmonics have to be cancelled, a wide frequency range has to be covered and the computational resources are limited, for example in automotive applications [7]. This aspect as well as a direct comparison of the proposed methods with conventional algorithms (such as the FxLMS) will be subject of future research.

To the best of the authors' knowledge, industrial applications of LPV controllers are rather limited. The results of this and the following chapter show that the implementation of even high-order LPV controllers can be quite straightforward.

Nomenclature

Acronyms

ANC	Active noise control.
AVC	Active vibration control.
FxLMS	Filtered-x least mean squares.
LFT	Linear fractional transformation.
LMI	Linear matrix inequality.
LPV	Linear parameter varying.
pLPV	Polytopic linear parameter varying.

Variables

(in order of appearance)

$\boldsymbol{A}_{\mathrm{p}}, \boldsymbol{B}_{\mathrm{p}}, \boldsymbol{C}_{\mathrm{p}}$	State-space matrices of the plant.
$\boldsymbol{x}_{\mathrm{p},k}$	State vector of the plant.
$y_{\mathrm{p},k}, u_{\mathrm{p},k}$	Output signal and control input of the plant.
$y_{\mathrm{d},k}$	Disturbance signal.
$\boldsymbol{A}_{\mathrm{M}}, \boldsymbol{B}_{\mathrm{M}}$	State-space matrices of the error filter.
$\boldsymbol{x}_{\mathrm{M},k}$	State-space vector of the error filter.
r_k, e_k	Reference and error signal.
$\boldsymbol{A}_{\mathrm{M},k}$	Time-varying system matrix of the error filter.

$\mathbf{0}$	Zero matrix.
$\mathbf{I}$	Identity matrix.
$\boldsymbol{L}_{\mathrm{p}}$	Observer gain for the plant.
$\widehat{\boldsymbol{x}}_{\mathrm{p,k}}$	Estimated state vector of the plant.
$\boldsymbol{K}_k$	Time-varying state-feedback gain for the overall system.
$\boldsymbol{K}_{\mathrm{M},k}, \boldsymbol{K}_{\mathrm{p},k}$	Time-varying state-feedback gains for the error filter and the plant.
$\widetilde{\boldsymbol{x}}_{\mathrm{p},k}$	Observer error of the plant.
$\boldsymbol{A}_{\mathrm{d},k}, \boldsymbol{C}_{\mathrm{d}}$	State-space matrices of the disturbance model.
$\boldsymbol{x}_{\mathrm{d},k}$	State vector of the disturbance model.
$\boldsymbol{A}_k, \boldsymbol{B}, \boldsymbol{C}$	State-space matrices of the overall system.
$\boldsymbol{L}_k$	Time-varying observer gain of the overall system.
$\boldsymbol{L}_{\mathrm{d},k}, \boldsymbol{L}_{\mathrm{p},k}$	Time-varying observer gains for the disturbance model and the plant.
$\widehat{\boldsymbol{x}}_{\mathrm{d},k}$	Estimated state vector of the disturbance model.
$\boldsymbol{K}_{\mathrm{d},k}, \boldsymbol{K}_{\mathrm{p},k}$	Time-varying state-feedback gains for the disturbance model and the plant.
$\boldsymbol{A}(\boldsymbol{\theta}), \boldsymbol{B}(\boldsymbol{\theta}),$ $\boldsymbol{C}(\boldsymbol{\theta}), \boldsymbol{D}(\boldsymbol{\theta})$	State-space matrices of an LPV system.
$\boldsymbol{\theta} = [\theta_1\,\theta_2\,\ldots\theta_N]^{\mathrm{T}}$	Parameter vector.
N	Size of the parameter vector.
$\boldsymbol{\Theta}$	Parameter polytope.
$\boldsymbol{\mathcal{A}}_i$	Constant matrices of a polytopic representation.
V	Set of vertices of the parameter polytope.
$\boldsymbol{v}_j$	Vertex j of the parameter polytope.
$\boldsymbol{\lambda} = [\lambda_1\,\lambda_2\,\ldots\lambda_M]^{\mathrm{T}}$	Coordinate vector.
M	Size of the coordinate vector.
$\boldsymbol{A}_{\mathrm{v},j}$	System matrices for the j-th vertex.
$\boldsymbol{P}$	Symmetric positive definite matrix.
$\boldsymbol{\theta}_k$	Vector of time-varying parameters.
$\boldsymbol{\lambda}_k = \left[\lambda_{1,k}\,\lambda_{2,k}\,\ldots\lambda_{M,k}\right]^{\mathrm{T}}$	Time-varying coordinate vector.
$\boldsymbol{K}_{\mathrm{v},j}$	State-feedback gain for the j-th vertex.

w_k, z_k, u_k	Process noise, performance output and control input.
Q, R	Weighting matrices.
G	Transfer function between w_k and z_k.
$\widetilde{A}, \widetilde{B}, \widetilde{C}$	State-space matrices of the transfer path from w_k to z_k.
$\widetilde{Q}, \widetilde{R}$	Matrices to build the LMIs.
W_c	Controllability gramian.
W_o	Observability gramian.
W	Symmetric positive definite matrix.
$Y_{v,j}$	Solution of the LMIs for the j-th vertex.
$L_{v,j}$	Observer gain for the j-th vertex.
$b_{i_{\min}}, b_{i_{\max}}$	Vectors for the computation of the coordinate vector.
$c_{i_{\min},k}, c_{i_{\max},k}$, $b_{i_{\min},j}, b_{i_{\max},j}$	Scalars for the computation of the coordinate vector.
$A_{M/d,k}$	System matrix of a general time-varying internal disturbance model.

Author details

Pablo Ballesteros, Christian Bohn, Wiebke Heins and Xinyu Shu
Institute of Electrical Information Technology, Clausthal University of Technology, Clausthal-Zellerfeld, Germany

8. References

[1] Amato, F. 2006. *Robust control of linear systems subject to uncertain time-varying parameters.* Berlin: Springer.

[2] Apkarian, P., P. Gahinet and G. Becker. 1995. Self-scheduled control of linear parameter-varying systems: A design example. *Automatica* 31:1251-61.

[3] Apkarian, P. 1997. On the discretization of LMI-synthesized linear parameter-varying controllers. *Automatica* 33:655-61.

[4] Ballesteros, P. and C. Bohn. 2011a. A frequency-tunable LPV controller for narrowband active noise and vibration control. *Proceedings of the American Control Conference.* San Francisco, June 2011. 1340-45.

[5] Ballesteros, P. and C. Bohn. 2011b. Disturbance rejection through LPV gain-scheduling control with application to active noise cancellation. *Proceedings of the IFAC World Congress.* Milan, August 2011. 7897-902.

[6] Balini, H. M. N. K., C. W. Scherer and J. Witte. 2011. Performance enhancement for AMB systems using unstable H_∞ controllers. *IEEE Transactions on Control Systems Technology* 19:1479-92.

[7] Bohn, C., A. Cortabarria, V. Härtel and K. Kowalczyk. 2003. Disturbance-observer-based active control of engine-induced vibrations in automotive vehicles. *Proceedings of the*

SPIE's 10th Annual International Symposium on Smart Structures and Materials. San Diego, March 2003. Paper No. 5049-68.

[8] Bohn, C., A. Cortabarria, V. Härtel and K. Kowalczyk. 2004. Active control of engine-induced vibrations in automotive vehicles using disturbance observer gain scheduling. *Control Engineering Practice* 12:1029-39.

[9] Daafouz, J., G. I. Bara, F. Kratz and J. Ragot. 2000: State observers for discrete-time LPV systems: an interpolation based approach. *Proceedings of the 39th IEEE Conference on Decision and Control* 5: 4571-72.

[10] Darengosse C. and P. Chevrel. 2000. Linear parameter-varying controller design for active power filters. *Proceedings of the IFAC Control Systems Design*. Bratislava, June 2000. 65-70.

[11] Du, H. and X. Shi. 2002. Gain-scheduled control for use in vibration suppression of system with harmonic excitation. *Proceedings of the American Control Conference*. Anchorage, May 2002. 4668-69.

[12] Du, H., L. Zhang and X. Shi. 2003. LPV technique for the rejection of sinusoidal disturbance with time-varying frequency. *IEE Proceedings on Control Theory and Applications* 150:132-38.

[13] Feintuch, P. L., N. J. Bershad and A. K. Lo. 1993. A frequency-domain model for filtered LMS algorithms - Stability analysis, design, and elimination of the training mode. *IEEE Transactions on Signal Processing* 41:1518-31.

[14] Francis, B. and W. Wonham. 1976. The internal model principle of control theory. *Automatica* 12:457-65.

[15] Gahinet, P., A. Nemirovskii, A. J. Laub and M. Chilali. 1995. LMI Control Toolbox. The Mathworks Inc.

[16] Heins, W., P. Ballesteros and C. Bohn. 2011. Gain-scheduled state-feedback control for active cancellation of multisine disturbances with time-varying frequencies. Presented at the *10th MARDiH Conference on Active Noise and Vibration Control Methods*. Krakow-Wojanow, Poland, June 2011.

[17] Heins, W., P. Ballesteros and C. Bohn. 2012. Experimental evaluation of an LPV-gain-scheduled observer for rejecting multisine disturbances with time-varying frequencies. *Proceedings of the American Control Conference*. Montreal, June 2012. Accepted for publication.

[18] Horn, R. A. and C. R. Johnson. 1985. *Matrix Analysis*. Cambridge: Cambridge UP.

[19] Kinney, C. E. and R. A. de Callafon. 2006a. Scheduling control for periodic disturbance attenuation. *Proceedings of the American Control Conference*. Minneapolis, June 2006. 4788-93.

[20] Kinney, C. E. and R. A. de Callafon. 2006b. An adaptive internal model-based controller for periodic disturbance rejection. *Proceedings of the 14th IFAC Symposium on System Identification*. Newcastle, Australia, March 2006. 273-78.

[21] Kinney, C. E. and R. A. de Callafon. 2007. A comparison of fixed point designs and time-varying observers for scheduling repetitive controllers. *Proceedings of the 46th IEEE Conference on Decision and Control*. New Orleans, December 2007. 2844-49.

[22] Köroğlu , H. and C. W. Scherer. 2008. Robust generalized asymptotic regulation against non-stationary sinusoidal disturbances. *Proceedings of the 47th IEEE Conference on Decision and Control*. Cancun, December 2008. 5426-31.

[23] Köroğlu, H. and C. W. Scherer. 2008. LPV control for robust attenuation of non-stationary sinusoidal disturbances with measurable frequencies. *Proceedings of the 17th IFAC World Congress*. Korea, July 2008. 4928-33.

[24] Kuo, S. M. and D. R. Morgan. 1996. *Active Noise Control Systems*. New York: Wiley.

[25] Morari, M. and E. Zafiriou 1989. *Robust Process Control*. Englewood Cliffs: Prentice Hall.

[26] Shu, X., P. Ballesteros and C. Bohn. 2011. Active vibration control for harmonic disturbances with time-varying frequencies through LPV gain scheduling. *Proceedings of the 23rd Chinese Control and Decision Conference*. Mianyang, China, May 2011. 728-33.

[27] Stilwell, D. J. and W. J. Rugh. 1998. Interpolation of observer state feedback controllers for gain scheduling. *Proceedings of the American Control Conference* 2:1215-19.

[28] Witte, J., H. M. N. K. Balini and C. W. Scherer. 2010. Experimental results with stable and unstable LPV controllers for active magnetic bearing systems. *Proceedings of the IEEE International Conference on Control Applications*. Yokohama, September 2010. 950-55.

Chapter 4

LPV Gain-Scheduled Output Feedback for Active Control of Harmonic Disturbances with Time-Varying Frequencies

Pablo Ballesteros, Xinyu Shu, Wiebke Heins and Christian Bohn

Additional information is available at the end of the chapter

1. Introduction

In this chapter, the same control problem as in the previous chapter is considered, which is the rejection of harmonic disturbances with time-varying frequencies for linear time-invariant (LTI) plants. In the previous chapter, gain-scheduled observer-based state-feedback controllers for this control problem were presented. In the present chapter, two methods for the design of general gain-scheduled output-feedback controllers are presented. As in the previous chapter, the control design is based on a description of the system in linear parameter-varying (LPV) form. One of the design methods presented is based on the polytopic linear parameter-varying (pLPV) system description (which has also been used in the previous chapter) and the other method is based on the description of an LPV system in linear fractional transformation (LPV-LFT) form. The basic idea is to use the well-established norm-optimal control framework based on the generalized plant setup shown in Fig. 1 with the generalized plant $\boldsymbol{G}$ and controller $\boldsymbol{K}$.

In this setup, $\boldsymbol{u}$ is the control signal and $\boldsymbol{y}$ consists of all signals that will be provided to the controller. The signal $\boldsymbol{w}$ is the performance input and the signal $\boldsymbol{q}$ is the performance output in the sense that the performance requirements are expressed in terms of the "overall gain" (usually measured by the H_∞ or the H_2 norm) of the transfer function from $\boldsymbol{w}$ to $\boldsymbol{q}$ in closed

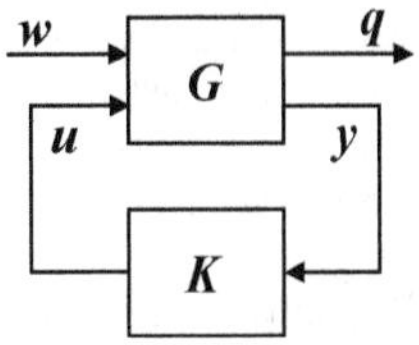

Figure 1. Generalized plant and controller

loop. In this setup, the aim of the controller design is to satisfy performance requirements expressed as upper bounds on the norm (in case of suboptimal control) or minimize the norm (in optimal control) of the transfer function from $\boldsymbol{w}$ to $\boldsymbol{q}$. Loosely speaking, a good controller should make the effect of $\boldsymbol{w}$ on $\boldsymbol{q}$ "small" (for suboptimal control) or "as small as possible" (for optimal control). The performance outputs usually consist of weighted versions of the controlled signal, the control error and the control effort. This is achieved by augmenting the original plant with output weighting functions. Good rejection of specific disturbances can be achieved in this framework by using a disturbance model as a weighting function in the transfer path from the performance input $\boldsymbol{w}$ to the performance output $\boldsymbol{q}$, that is, by modeling the disturbance to be rejected as a weighted version of the performance input. This forces the maximum singular value $\sigma_{\max}(\boldsymbol{G}_{qw}(\mathrm{j}\omega))$ or, in the single-input single-output case, the amplitude response $|G_{qw}(\mathrm{j}\omega)|$ of the open-loop transfer function to have a very high gain in the frequency regions specified by the disturbance model, or, loosely speaking, enlarges the effect of $\boldsymbol{w}$ on $\boldsymbol{q}$ in certain frequency regions. A reduction of the overall effect of $\boldsymbol{w}$ on $\boldsymbol{q}$ in closed loop will then be mostly achieved by reducing the effect in regions where it is large in open loop. From classical control arguments, it is intuitive that this requires a high loop gain in these frequency regions which in turn usually requires a high controller gain. A high loop gain will give a small sensitivity and in turn a good disturbance rejection (in specified frequency regions).

This control design setup is used in this chapter for the rejection of harmonic disturbances with time-varying frequencies. The control design problem is based on a generalized plant obtained through the introduction of a disturbance model that describes the harmonic disturbances and the addition of output weighting functions. Descriptions of the disturbance model in pLPV and in LPV-LFT form are used and lead to generalized plant descriptions that are also in pLPV or LPV-LFT form. Corresponding design methods are then employed to obtain controllers. For a plant in pLPV form, standard H_∞ design [11] is used to compute a set of controllers. The gain scheduling is then achieved by interpolation between these controllers. For a plant in LPV-LFT form, the design method of Apkarian & Gahinet [1] is used that directly yields a gain-scheduled controller also in LPV-LFT form.

LPV approaches for the rejection of harmonic disturbances have been used by Darengosse & Chevrel [7], Du & Shi [8], Du et al. [9], Bohn et al. [5, 6], Kinney & de Callafon [14, 15, 16], Köroğlu & Scherer [17], Witte et al. [19], Balini et al. [4], Heins et al. [12, 13], Ballesteros & Bohn [2, 3] and Shu et al. [18]. Darengosse & Chevrel [7], Du & Shi [8], Du et al. [9], Witte et al. [19], Balini et al. [4] suggested continuous-time LPV approaches. These approaches are tested for a single sinusoidal disturbance by Darengosse & Chevrel [7], Du et al. [9], Witte et al. [19] and Balini et al. [4]. Methods based on observer-based state-feedback controllers are presented by Bohn et al. [5, 6], Kinney & de Callafon [14, 15, 16] and Heins et al. [12, 13]. In the approach of Bohn et al. [5, 6], the observer gain is selected from a set of pre-computed gains by switching. In the other approaches of Kinney & de Callafon [16], Heins et al. [13] and in the previous chapter, the observer gain is calculated by interpolation. In the other approach presented in the previous chapter, which is also used by Kinney & de Callafon [14, 15] and Heins et al. [12], the state-feedback gain is scheduled using interpolation. A general output feedback LPV approach for the rejection of harmonic disturbances is suggested and applied in real time by Ballesteros & Bohn [2, 3] and Shu et al. [18].

The existing LPV approaches can be classified by the control design technique used to obtain the controller. Approaches based on pLPV control design are used by Heins et al. [12, 13],

Kinney & de Callafon [14], Du & Shi [8] and Du et al. [9]. An approach based on LPV-LFT control design is used by Ballesteros & Bohn [2, 3] and Shu et al. [18].

For a practical application, the resulting controller has to be implemented in discrete time. In applications of ANC/AVC, the plant model is often obtained through system identification. This usually gives a discrete-time plant model. If a continuous-time controller is computed, the controller has to be discretized. Since the controller is time varying, this discretization would have to be carried out at each sampling instant. An exact discretization involves the calculation of a matrix exponential, which is computationally too expensive and leads to a distortion of the frequency scale. Usually, this can be tolerated, but not for the suppression of harmonic disturbances. In this context, it is not surprising that the continuous-time design methods of Darengosse & Chevrel [7], Du et al. [9], Kinney & de Callafon [14] and Köroğlu & Scherer [17] are tested only in simulation studies with a very simple system as a plant and a single frequency in the disturbance signal. Exceptions are Witte et al. [19] and Balini et al. [4], who designed continuous-time controllers which then are approximately discretized. However, Witte et al. [19] use a very high sampling frequency of 40 kHz to reject a harmonic disturbance with a frequency up to 48 Hz (in fact, the authors state that they chose "the smallest [sampling time] available by the hardware") and Balini et al. [4] use a maximal sampling frequency of 50 kHz. The control design methods presented in this chapter are realized in discrete time.

The remainder of this chapter is organized as follows. In Sec. 2, pLPV systems and LPV-LFT systems are introduced and the control design for such systems is described. In Sec. 3, it is described how the control problem considered here can be transformed to a generalized plant setup. The required pLPV disturbance model for the harmonic disturbance is introduced in Sec. 3.1 and in Sec. 3.2, it is described how the generalized plant in pLPV form is obtained by combining the disturbance model, the plant and the weighting functions. In Sec. 4, the transformation of the control problem to a generalized plant in LPV-LFT form is treated in essentially the same way, by formulating an LPV-LFT disturbance model (Sec. 4.1) and building a generalized plant in LPV-LFT form (Sec. 4.2). The controller synthesis for both descriptions is described in Sec. 5. Experimental results are presented in Sec. 6 and the chapter finishes with a discussion and some conclusions in Sec. 7.

2. Control design setup

In this section, pLPV systems and LPV-LFT systems are introduced and the control design for such systems is described in Sec. 2.1 and 2.2, respectively.

2.1. Control design for pLPV systems

A pLPV system is of the form

$$\begin{bmatrix} x_{k+1} \\ \hline y_k \end{bmatrix} = \left[\begin{array}{c|c} A(\boldsymbol{\theta}) & B \\ \hline C & D \end{array}\right] \begin{bmatrix} x_k \\ \hline u_k \end{bmatrix}, \tag{1}$$

where the system matrix depends affinely on a parameter vector $\boldsymbol{\theta}$, that is

$$A(\boldsymbol{\theta}) = \mathcal{A}_0 + \theta_1 \mathcal{A}_1 + \theta_2 \mathcal{A}_2 + \cdots + \theta_N \mathcal{A}_N, \tag{2}$$

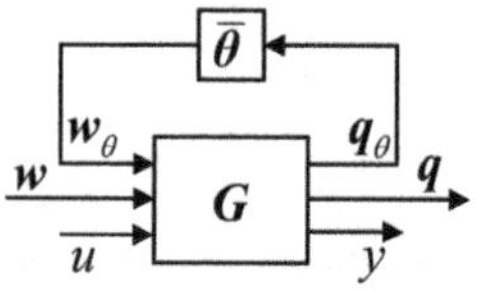

Figure 2. General LPV-LFT system

with constant matrices $\mathcal{A}_i$. The parameter vector $\boldsymbol{\theta}$ varies in a polytope Θ with M vertices $v_j \in \mathbb{R}^N$. A point $\boldsymbol{\theta} \in \Theta$ can be written as a convex combination of vertices, i.e. there exists a coordinate vector $\boldsymbol{\lambda} = [\lambda_1 \cdots \lambda_M]^{\mathrm{T}} \in \mathbb{R}^M$ such that $\boldsymbol{\theta}$ can be written as

$$\boldsymbol{\theta} = \sum_{j=1}^{M} \lambda_j v_j \tag{3}$$

with

$$\lambda_j \geq 0, \sum_{j=1}^{M} \lambda_j = 1. \tag{4}$$

Defining $\boldsymbol{A}_{\mathrm{v},j} = \boldsymbol{A}(v_j)$ for $j = 1, ..., M$, the system matrix $\boldsymbol{A}(\boldsymbol{\theta})$ can be represented as

$$\boldsymbol{A}(\boldsymbol{\theta}) = \boldsymbol{A}(\boldsymbol{\lambda}) = \lambda_1 \boldsymbol{A}_{\mathrm{v},1} + \lambda_2 \boldsymbol{A}_{\mathrm{v},2} + ... + \lambda_M \boldsymbol{A}_{\mathrm{v},M}. \tag{5}$$

The system matrix of a pLPV system $\boldsymbol{A}(\boldsymbol{\theta})$ can be calculated from the M vertices of the polytope Θ by finding the coordinate vector $\boldsymbol{\lambda}$ that fulfills the conditions of (3) and (4).

Once a representation of a system is obtained in pLPV form, it is possible to find a controller using H_∞ or H_2 techniques for each vertex of the polytope. The controller for a given $\boldsymbol{\theta} \in \Theta$ can be calculated through controllers for the vertex systems. The closed-loop stability is guaranteed even for arbitrarily fast changes of the scheduling parameters if a parameter-independent Lyapunov function is used (for the whole polytope) in the control design. This approach, however, is conservative because fast variations of the scheduling parameters are considered, which might not occur in a practical application. Parameter-dependent Lyapunov functions can be used to include bounds on the rate of change of the parameters, but are not considered here.

2.2. Control design for LPV-LFT systems

An LPV system in LFT form is shown in Fig. 2. It consists of a generalized plant $\boldsymbol{G}$ that includes input and output weighting functions and a parametric uncertainty block $\bar{\boldsymbol{\theta}}$ that has been "pulled out" of the system. For this general system, a gain-scheduling controller can be calculated following the method presented in Apkarian & Gahinet [1]. In this method, two sets of linear matrix inequalities (LMIs) are solved. The first set of LMIs determines the feasibility of the problem which means that a bound on the control system performance in the sense of the H_∞ norm can be satisfied. With the second set of LMIs, the controller matrices are calculated from the solution of the first set of LMIs.

As a result of applying this control design method, the gain-scheduling control structure of Fig. 3 is obtained. The time-varying plant parameters are directly used as the gain-scheduling

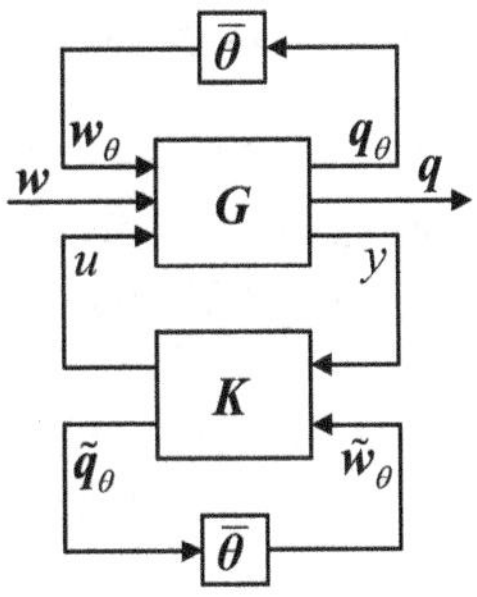

Figure 3. LPV-LFT gain-scheduling control structure

parameters of the controller. This control design method guarantees stability through the small gain theorem. It is often conservative, since the parameter ranges covered are usually larger than the ones that may occur in the real system.

3. Generalized plant in pLPV form

As stated in the previous section, to calculate the controller using the pLPV control design method, the generalized plant in pLPV form is needed. In this section, the steps to obtain the generalized plant in pLPV form are discussed. The disturbance model and a representation of the disturbance model in pLPV form are obtained in Sec. 3.1. In Sec. 3.2, the generalized plant is built by combining the plant, the disturbance model in pLPV form and the weighting functions.

3.1. Disturbance model

A general model for a harmonic disturbance with n_d fixed frequencies is described by

$$\left[\begin{array}{c|c} \boldsymbol{A}_\mathrm{d} & \boldsymbol{B}_\mathrm{d} \\ \hline \boldsymbol{C}_\mathrm{d} & 0 \end{array}\right] \tag{6}$$

with

$$\boldsymbol{A}_\mathrm{d} = \begin{bmatrix} \boldsymbol{A}_{\mathrm{d},1} & \cdots & \boldsymbol{0} \\ \vdots & \ddots & \vdots \\ \boldsymbol{0} & \cdots & \boldsymbol{A}_{\mathrm{d},n_\mathrm{d}} \end{bmatrix}, \quad \boldsymbol{A}_{\mathrm{d},i} = \begin{bmatrix} 0 & 1 \\ -1 & a_i \end{bmatrix}, \tag{7}$$

$$a_i = 2\cos(2\pi f_i T), \tag{8}$$

$$\boldsymbol{B}_\mathrm{d} = \begin{bmatrix} \boldsymbol{B}_{\mathrm{d},1} \\ \vdots \\ \boldsymbol{B}_{\mathrm{d},n_\mathrm{d}} \end{bmatrix}, \quad \boldsymbol{B}_{\mathrm{d},i} = \begin{bmatrix} 1 \\ 1 \end{bmatrix}, \tag{9}$$

$$\boldsymbol{C}_\mathrm{d} = \begin{bmatrix} \boldsymbol{C}_{\mathrm{d},1} \cdots \boldsymbol{C}_{\mathrm{d},n_\mathrm{d}} \end{bmatrix} \text{ and } \boldsymbol{C}_{\mathrm{d},i} = \begin{bmatrix} 1 & 0 \end{bmatrix}. \tag{10}$$

A harmonic disturbance can be modeled as the output of an unforced system with system matrix $\boldsymbol{A}_\mathrm{d}$ and output matrix $\boldsymbol{C}_\mathrm{d}$ given above in (7) and (10). An input matrix is not required. However, in the generalized plant setup, a performance input is required and the disturbance model acts as an input weighting function on the performance input. This is why the disturbance model above has been given with a nonzero input matrix $\boldsymbol{B}_{\mathrm{d}}$ in (9).

The frequency in (8) is fixed and denoted by f_i. As in Sec. 4 of the previous chapter, the pLPV disturbance model for n_{d} time-varying frequencies $f_{j,k} \in [f_{\min,j}, f_{\max,j}]$, $j = 1,2,\ldots,n_{\mathrm{d}}$, is defined as

$$\left[\begin{array}{c|c} \boldsymbol{A}_{\mathrm{d}}^{(\mathrm{pLPV})}(\boldsymbol{\theta}) & \boldsymbol{B}_{\mathrm{d}}^{(\mathrm{pLPV})} \\ \hline \boldsymbol{C}_{\mathrm{d}}^{(\mathrm{pLPV})} & 0 \end{array}\right] \tag{11}$$

with

$$\boldsymbol{A}_{\mathrm{d}}^{(\mathrm{pLPV})}(\boldsymbol{\theta}) = \boldsymbol{\mathcal{A}}_{\mathrm{d},0} + \theta_1 \boldsymbol{\mathcal{A}}_{\mathrm{d},1} + \cdots + \theta_{n_{\mathrm{d}}} \boldsymbol{\mathcal{A}}_{\mathrm{d},n_{\mathrm{d}}}. \tag{12}$$

As in Sec. 2.1, (12) can be written in the form of

$$\boldsymbol{A}_{\mathrm{d}}^{(\mathrm{pLPV})}(\boldsymbol{\theta}) = \boldsymbol{A}_{\mathrm{d}}^{(\mathrm{pLPV})}(\boldsymbol{\lambda}) = \lambda_1 \boldsymbol{A}_{\mathrm{v},1} + \cdots + \lambda_M \boldsymbol{A}_{\mathrm{v},M} = \sum_{i=1}^{M} \lambda_i \boldsymbol{A}_{\mathrm{v},i}, \tag{13}$$

where the matrices $\boldsymbol{A}_{\mathrm{v},i}$ are defined in the same way as $\boldsymbol{A}_{\mathrm{d}}^{(\mathrm{pLPV})}$ in (7) and (8), but with a_i evaluated for all the vertices of the polytope, with $j = 1,2,\ldots,n_{\mathrm{d}}$. The coordinate vector $\boldsymbol{\lambda}$ can be calculated using the method described in Sec. 4.4 of the previous chapter.

3.2. Generalized plant

A state-space representation of the plant is given by

$$G_{\mathrm{p}} = \left[\begin{array}{c|c} \boldsymbol{A}_{\mathrm{p}} & \boldsymbol{B}_{\mathrm{p}} \\ \hline \boldsymbol{C}_{\mathrm{p}} & \boldsymbol{D}_{\mathrm{p}} \end{array}\right] \tag{14}$$

and it is assumed that the disturbance is acting on the input of the plant.

The block diagram of the generalized plant with the disturbance, the plant and the weighting functions

$$W_y^{(\mathrm{pLPV})} = \left[\begin{array}{c|c} \boldsymbol{A}_{W_y}^{(\mathrm{pLPV})} & \boldsymbol{B}_{W_y}^{(\mathrm{pLPV})} \\ \hline \boldsymbol{C}_{W_y}^{(\mathrm{pLPV})} & D_{W_y}^{(\mathrm{pLPV})} \end{array}\right], \tag{15}$$

$$W_u^{(\mathrm{pLPV})} = \left[\begin{array}{c|c} \boldsymbol{A}_{W_u}^{(\mathrm{pLPV})} & \boldsymbol{B}_{W_u}^{(\mathrm{pLPV})} \\ \hline \boldsymbol{C}_{W_u}^{(\mathrm{pLPV})} & D_{W_u}^{(\mathrm{pLPV})} \end{array}\right] \tag{16}$$

is illustrated in Fig. 4.

For every vertex of the polytopic system, the generalized plant can be described by

$$\begin{bmatrix} \boldsymbol{x}_{k+1} \\ q_k \\ y_k \end{bmatrix} = \left[\begin{array}{c|cc} \boldsymbol{A}_i(\boldsymbol{\theta}) & \boldsymbol{B}_w^{(\mathrm{pLPV})} & \boldsymbol{B}_u^{(\mathrm{pLPV})} \\ \hline \boldsymbol{C}_q^{(\mathrm{pLPV})} & \boldsymbol{D}_{qw}^{(\mathrm{pLPV})} & \boldsymbol{D}_{qu}^{(\mathrm{pLPV})} \\ \boldsymbol{C}_y^{(\mathrm{pLPV})} & D_{yw}^{(\mathrm{pLPV})} & D_{yu}^{(\mathrm{pLPV})} \end{array}\right] \begin{bmatrix} \boldsymbol{x}_k \\ w_k \\ u_k \end{bmatrix} \tag{17}$$

where

$$\boldsymbol{x}_k = \left[\boldsymbol{x}_{\mathrm{p},k}^{\mathrm{T}}\ \boldsymbol{x}_{\mathrm{d},k}^{\mathrm{T}}\ \boldsymbol{x}_{W_y,k}^{\mathrm{T}}\ \boldsymbol{x}_{W_u,k}^{\mathrm{T}}\right]^{\mathrm{T}}, \tag{18}$$

$$A_i(\theta) = \begin{bmatrix} A_{\mathrm{p}} & B_{\mathrm{p}} C_{\mathrm{d}}^{(\mathrm{pLPV})} & 0 & 0 \\ 0 & A_{\mathrm{v},i} & 0 & 0 \\ B_{W_y}^{(\mathrm{pLPV})} C_{\mathrm{p}} & 0 & A_{W_y}^{(\mathrm{pLPV})} & 0 \\ 0 & 0 & 0 & A_{W_u}^{(\mathrm{pLPV})} \end{bmatrix}, \tag{19}$$

$$\begin{bmatrix} B_w^{(\mathrm{pLPV})} & B_u^{(\mathrm{pLPV})} \end{bmatrix} = \begin{bmatrix} 0 & B_{\mathrm{p}} \\ B_{\mathrm{d}}^{(\mathrm{pLPV})} & 0 \\ 0 & 0 \\ 0 & B_{W_u}^{(\mathrm{pLPV})} \end{bmatrix}, \tag{20}$$

$$\begin{bmatrix} C_q^{(\mathrm{pLPV})} \\ C_y^{(\mathrm{pLPV})} \end{bmatrix} = \left[\begin{array}{cccc} D_{W_y}^{(\mathrm{pLPV})} C_{\mathrm{p}} & 0 & C_{W_y}^{(\mathrm{pLPV})} & 0 \\ 0 & 0 & 0 & C_{W_u}^{(\mathrm{pLPV})} \\ \hline C_{\mathrm{p}} & 0 & 0 & 0 \end{array}\right] \tag{21}$$

and

$$\begin{bmatrix} D_{qw}^{(\mathrm{pLPV})} & D_{qu}^{(\mathrm{pLPV})} \\ D_{yw}^{(\mathrm{pLPV})} & D_{yu}^{(\mathrm{pLPV})} \end{bmatrix} = \left[\begin{array}{cc} 0 & 0 \\ 0 & D_{W_u}^{(\mathrm{pLPV})} \\ \hline 0 & 0 \end{array}\right]. \tag{22}$$

Once the generalized plant is obtained, the controller can be calculated using the algorithms in the following section.

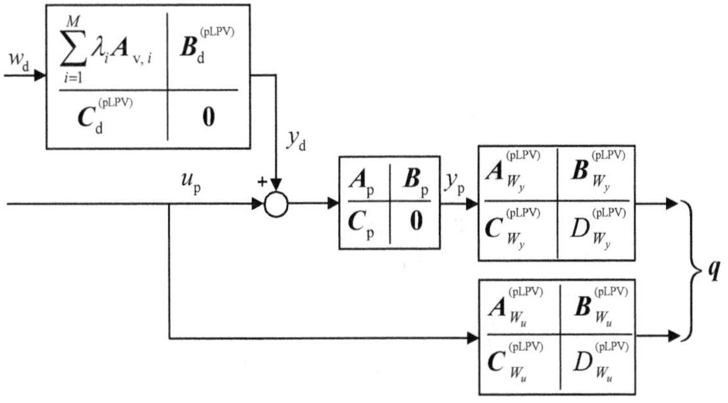

Figure 4. Plant with pLPV disturbance model and weighting functions

4. Generalized plant in LPV-LFT form

The same steps as in the previous section are carried out, but in this section the generalized plant in LPV-LFT form is obtained such that the control design method of Apkarian & Gahinet [1] can be used. The model of the harmonic disturbance and the generalized plant in LFT form are obtained in Sec. 4.1 and 4.2, respectively. The generalized plant is the result of combining plant, harmonic disturbance and weighting functions.

4.1. Disturbance model

The state-space representation of a harmonic disturbance for n_{d} fixed frequencies was given by (6-10). If the frequencies of a harmonic disturbance change between minimal values $f_{i,\min}$ and maximal values $f_{i,\max}$, a representation for the variations of the frequencies is given by

$$a_i(f_i) = 2\cos(2\pi f_i T) = \bar{a}_i + p_i \bar{\theta}_{i,k}(f_i) \tag{23}$$

with

$$\bar{a}_i = \cos(2\pi f_{i,\max} T) + \cos(2\pi f_{i,\min} T), \tag{24}$$

$$p_i = \cos(2\pi f_{i,\max} T) - \cos(2\pi f_{i,\min} T) \tag{25}$$

and

$$\bar{\theta}_{i,k} \in [-1, 1]. \tag{26}$$

An LPV-LFT model of the disturbance can be written as

$$\boldsymbol{x}_{\mathrm{d},k+1} = \boldsymbol{A}_{\mathrm{d}} \boldsymbol{x}_{\mathrm{d},k} + \boldsymbol{B}_{\mathrm{d},\theta} \boldsymbol{w}_{\theta,k} + \boldsymbol{B}_{\mathrm{d},w} w_{\mathrm{d},k}, \tag{27}$$

$$\boldsymbol{q}_{\theta,k} = \boldsymbol{C}_{\mathrm{d},\theta} \boldsymbol{x}_{\mathrm{d},k}, \tag{28}$$

$$y_{\mathrm{d},k} = \boldsymbol{C}_{\mathrm{d},y} \boldsymbol{x}_{\mathrm{d},k}, \tag{29}$$

$$\boldsymbol{w}_{\theta,k} = \bar{\boldsymbol{\theta}}_k \boldsymbol{q}_{\theta,k} \tag{30}$$

with

$$\boldsymbol{A}_{\mathrm{d}} = \begin{bmatrix} \boldsymbol{A}_{\mathrm{d},1} & \cdots & \boldsymbol{0} \\ \vdots & \ddots & \vdots \\ \boldsymbol{0} & \cdots & \boldsymbol{A}_{\mathrm{d},n_{\mathrm{d}}} \end{bmatrix}, \ \boldsymbol{A}_{\mathrm{d},i} = \begin{bmatrix} 0 & 1 \\ -1 & \bar{a}_i \end{bmatrix}, \tag{31}$$

$$\boldsymbol{B}_{\mathrm{d},\theta} = \begin{bmatrix} \boldsymbol{B}_{\mathrm{d},\theta,1} & \cdots & \boldsymbol{0} \\ \vdots & \ddots & \vdots \\ \boldsymbol{0} & \cdots & \boldsymbol{B}_{\mathrm{d},\theta,n_{\mathrm{d}}} \end{bmatrix}, \ \boldsymbol{B}_{\mathrm{d},\theta,i} = \begin{bmatrix} 0 \\ p_i \end{bmatrix}, \tag{32}$$

$$\boldsymbol{B}_{\mathrm{d},w} = \begin{bmatrix} \boldsymbol{B}_{\mathrm{d},w,1} \\ \vdots \\ \boldsymbol{B}_{\mathrm{d},w,n_{\mathrm{d}}} \end{bmatrix}, \ \boldsymbol{B}_{\mathrm{d},w,i} = \begin{bmatrix} 1 \\ 1 \end{bmatrix}, \tag{33}$$

$$\boldsymbol{C}_{\mathrm{d},\theta} = \begin{bmatrix} \boldsymbol{C}_{\mathrm{d},\theta,1} & \cdots & \boldsymbol{0} \\ \vdots & \ddots & \vdots \\ \boldsymbol{0} & \cdots & \boldsymbol{C}_{\mathrm{d},\theta,n_{\mathrm{d}}} \end{bmatrix}, \ \boldsymbol{C}_{\mathrm{d},\theta,i} = \begin{bmatrix} 0 & 1 \end{bmatrix}, \tag{34}$$

$$\boldsymbol{C}_{\mathrm{d},y} = \begin{bmatrix} \boldsymbol{C}_{\mathrm{d},y,1} & \cdots & \boldsymbol{C}_{\mathrm{d},y,n_{\mathrm{d}}} \end{bmatrix}, \ \boldsymbol{C}_{\mathrm{d},y,i} = \begin{bmatrix} 1 & 0 \end{bmatrix} \tag{35}$$

and

$$\bar{\boldsymbol{\theta}}_k = \begin{bmatrix} \bar{\theta}_{1,k} & \cdots & 0 \\ \vdots & \ddots & \vdots \\ 0 & \cdots & \bar{\theta}_{n_{\mathrm{d}},k} \end{bmatrix}. \tag{36}$$

4.2. Generalized plant

The generalized plant is the result of combining the plant, the harmonic disturbance and the weighting functions and it is shown in Fig. 5. The weighting functions are defined the same way as in (15) and (16). A representation of the generalized plant in LFT form is given by

$$\begin{bmatrix} x_{k+1} \\ q_{\theta,k} \\ q_k \\ y_k \end{bmatrix} = \left[\begin{array}{c|ccc} A & B_\theta & B_w^{(\mathrm{LFT})} & B_u^{(\mathrm{LFT})} \\ \hline C_\theta & D_{\theta\theta} & D_{\theta w} & D_{\theta u} \\ C_q^{(\mathrm{LFT})} & D_{q\theta} & D_{qw}^{(\mathrm{LFT})} & D_{qu}^{(\mathrm{LFT})} \\ C_y^{(\mathrm{LFT})} & D_{y\theta} & D_{yw}^{(\mathrm{LFT})} & D_{yu}^{(\mathrm{LFT})} \end{array}\right] \begin{bmatrix} x_k \\ w_{\theta,k} \\ w_k \\ u_k \end{bmatrix} \tag{37}$$

with

$$x_k = \begin{bmatrix} x_{\mathrm{p},k}^{\mathrm{T}} & x_{\mathrm{d},k}^{\mathrm{T}} & x_{\mathrm{W}_y,k}^{\mathrm{T}} & x_{\mathrm{W}_u,k}^{\mathrm{T}} \end{bmatrix}^{\mathrm{T}}, \tag{38}$$

$$A = \begin{bmatrix} A_{\mathrm{p}} & B_{\mathrm{p}} C_{\mathrm{d},y} & 0 & 0 \\ 0 & A_{\mathrm{d}} & 0 & 0 \\ B_{\mathrm{W}_y}^{(\mathrm{LFT})} C_{\mathrm{p}} & 0 & A_{\mathrm{W}_y}^{(\mathrm{LFT})} & 0 \\ 0 & 0 & 0 & A_{\mathrm{W}_u}^{(\mathrm{LFT})} \end{bmatrix}, \tag{39}$$

$$\begin{bmatrix} B_\theta & B_w^{(\mathrm{LFT})} & B_u^{(\mathrm{LFT})} \end{bmatrix} = \begin{bmatrix} 0 & 0 & B_{\mathrm{p}} \\ B_{\mathrm{d},\theta} & B_{\mathrm{d},w} & 0 \\ 0 & 0 & 0 \\ 0 & 0 & B_{\mathrm{W}_u}^{(\mathrm{LFT})} \end{bmatrix}, \tag{40}$$

$$\begin{bmatrix} C_\theta \\ C_q^{(\mathrm{LFT})} \\ C_y^{(\mathrm{LFT})} \end{bmatrix} = \left[\begin{array}{cccc} 0 & C_{\mathrm{d},\theta} & 0 & 0 \\ \hline D_{\mathrm{W}_y}^{(\mathrm{LFT})} C_{\mathrm{p}} & 0 & C_{\mathrm{W}_y}^{(\mathrm{LFT})} & 0 \\ 0 & 0 & 0 & C_{\mathrm{W}_u}^{(\mathrm{LFT})} \\ \hline C_{\mathrm{p}} & 0 & 0 & 0 \end{array}\right] \tag{41}$$

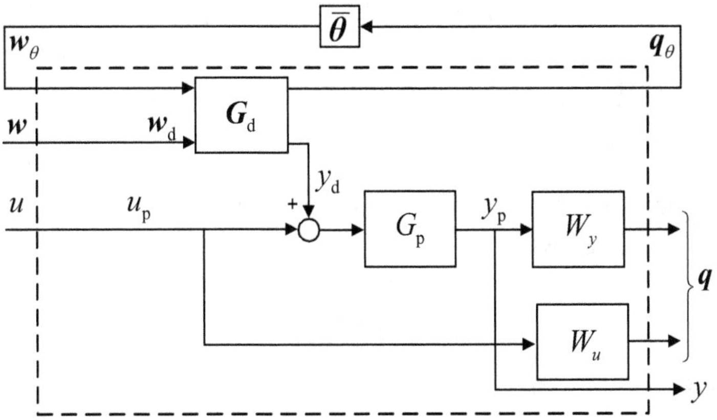

Figure 5. Plant with LPV-LFT disturbance model and weighting functions

and

$$\begin{bmatrix} D_{\theta\theta} & D_{\theta w} & D_{\theta u} \\ D_{q\theta} & D_{qw}^{(\mathrm{LFT})} & D_{qu}^{(\mathrm{LFT})} \\ D_{y\theta} & D_{yw}^{(\mathrm{LFT})} & D_{yu}^{(\mathrm{LFT})} \end{bmatrix} = \begin{bmatrix} 0 & 0 & 0 \\ \hline 0 & 0 & 0 \\ 0 & 0 & D_{W_u}^{(\mathrm{LFT})} \\ \hline 0 & 0 & 0 \end{bmatrix}. \quad (42)$$

5. Controller synthesis and implementation for LPV systems

In this section, algorithms for the calculation of the pLPV and LPV-LFT gain-scheduling controllers are explained in detail. Suboptimal controllers using H_∞ techniques are obtained.

5.1. Controller synthesis and implementation for pLPV systems

With the generalized plant in pLPV form, an H_∞-suboptimal controller for each vertex of the polytope can be calculated using standard H_∞ techniques [11]. The steps to obtain them are explained here in detail.

First, two outer factors

$$N_X = \mathrm{null}\left[C_y^{(\mathrm{pLPV})} \; D_{yw}^{(\mathrm{pLPV})} \; 0\right] \quad (43)$$

and

$$N_Y = \mathrm{null}\left[(B_u^{(\mathrm{pLPV})})^{\mathrm{T}} \; (D_{qu}^{(\mathrm{pLPV})})^{\mathrm{T}} \; 0\right] \quad (44)$$

are defined, where null$[\cdot]$ denotes the basis of the null space of a matrix.

Then, the LMIs

$$N_X^{\mathrm{T}} \begin{bmatrix} A_i^{\mathrm{T}} X_1 A_i - X_1 & A_i^{\mathrm{T}} X_1 B_w^{(\mathrm{pLPV})} & (C_q^{(\mathrm{pLPV})})^{\mathrm{T}} \\ (B_w^{(\mathrm{pLPV})})^{\mathrm{T}} X_1 A_i & -\gamma + (B_w^{(\mathrm{pLPV})})^{\mathrm{T}} X_1 B_w^{(\mathrm{pLPV})} & (D_{qu}^{(\mathrm{pLPV})})^{\mathrm{T}} \\ C_q^{(\mathrm{pLPV})} & D_{qu}^{(\mathrm{pLPV})} & -\gamma \mathbf{I} \end{bmatrix} N_X < 0, \quad (45)$$

$$N_Y^{\mathrm{T}} \begin{bmatrix} A_i Y_1 A_i^{\mathrm{T}} - Y_1 & A_i Y_1 (C_q^{(\mathrm{pLPV})})^{\mathrm{T}} & B_w^{(\mathrm{pLPV})} \\ C_q^{(\mathrm{pLPV})} Y_1 A_i^{\mathrm{T}} & -\gamma \mathbf{I} + C_q^{(\mathrm{pLPV})} Y_1 (C_q^{(\mathrm{pLPV})})^{\mathrm{T}} & D_{qu}^{(\mathrm{pLPV})} \\ (B_w^{(\mathrm{pLPV})})^{\mathrm{T}} & (D_{qu}^{(\mathrm{pLPV})})^{\mathrm{T}} & -\gamma \end{bmatrix} N_Y < 0, \quad (46)$$

$$\begin{bmatrix} X_1 & \mathbf{I} \\ \mathbf{I} & Y_1 \end{bmatrix} \geq 0 \quad (47)$$

for feasibility and optimality are solved for X_1 and Y_1 for every $A_i = A_i(\theta)$.

With X_1 and Y_1, the matrices

$$X_1 - Y_1^{-1} = X_2^{\mathrm{T}} X_2, \quad (48)$$

$$X^{(\mathrm{pLPV})} = \begin{bmatrix} X_1 & X_2 \\ X_2^{\mathrm{T}} & \mathbf{I} \end{bmatrix} \quad (49)$$

are calculated.

With

$$\overline{A}_i = \begin{bmatrix} A_i & 0 \\ 0 & 0 \end{bmatrix}, \quad (50)$$

$$\overline{B} = \begin{bmatrix} B_w^{(\mathrm{pLPV})} \\ 0 \end{bmatrix}, \tag{51}$$

$$\overline{C} = \begin{bmatrix} C_q^{(\mathrm{pLPV})} & 0 \end{bmatrix}, \tag{52}$$

the matrix

$$\psi_i = \begin{bmatrix} -(X^{(\mathrm{pLPV})})^{-1} & \overline{A}_i & \overline{B} & 0 \\ \overline{A}_i^{\mathrm{T}} & -X^{(\mathrm{pLPV})} & 0 & \overline{C}^{\mathrm{T}} \\ \overline{B}^{\mathrm{T}} & 0 & -\gamma & (D_{qw}^{(\mathrm{pLPV})})^{\mathrm{T}} \\ 0 & \overline{C} & D_{qw}^{(\mathrm{pLPV})} & -\gamma \mathrm{I} \end{bmatrix}. \tag{53}$$

is calculated. The matrices

$$P^{(\mathrm{pLPV})} = \begin{bmatrix} \underline{B}^{\mathrm{T}} & 0 & 0 & \underline{D}_{qu}^{\mathrm{T}} \end{bmatrix} \tag{54}$$

and

$$Q^{(\mathrm{pLPV})} = \begin{bmatrix} 0 & \underline{C} & \underline{D}_{yw} & 0 \end{bmatrix} \tag{55}$$

are composed with

$$\underline{B} = \begin{bmatrix} 0 & B_u^{(\mathrm{pLPV})} \\ \mathrm{I} & 0 \end{bmatrix}, \tag{56}$$

$$\underline{C} = \begin{bmatrix} 0 & \mathrm{I} \\ C_y^{(\mathrm{pLPV})} & 0 \end{bmatrix}, \tag{57}$$

$$\underline{D}_{qu} = \begin{bmatrix} 0 & D_{qu}^{(\mathrm{pLPV})} \end{bmatrix}, \tag{58}$$

$$\underline{D}_{yw} = \begin{bmatrix} 0 \\ D_{yw}^{(\mathrm{pLPV})} \end{bmatrix}. \tag{59}$$

Finally, the basic LMIs

$$\psi_i + (P^{(\mathrm{pLPV})})^{\mathrm{T}} \Omega_i Q^{(\mathrm{pLPV})} + (Q^{(\mathrm{pLPV})})^{\mathrm{T}} \Omega_i P^{(\mathrm{pLPV})} < 0 \tag{60}$$

are solved for Ω_i for every i.

The state-spaces matrices of the controllers for each vertex can be extracted from

$$\Omega_i = \begin{bmatrix} A_{K_i} & B_{K_i} \\ C_{K_i} & D_{K_i} \end{bmatrix}. \tag{61}$$

The implemented controller is interpolated using the coordinate vector λ in

$$\Omega^{(\mathrm{pLPV})} = \Sigma_{i=1}^{m} \lambda_i \Omega_i. \tag{62}$$

5.2. Controller synthesis and implementation for LPV-LFT systems

In this section, the algorithm for the calculation of the H_∞-suboptimal gain-scheduling controller from [1] is explained in detail.

From the state-space representation of the generalized plant the outer factors for the LMIs that have to be solved in the design can be calculated as

$$N_R = \mathrm{null} \begin{bmatrix} (B_u^{(\mathrm{LFT})})^{\mathrm{T}} & D_{\theta u}^{(\mathrm{LFT})} & (D_{qu}^{(\mathrm{LFT})})^{\mathrm{T}} & 0 \end{bmatrix} \tag{63}$$

and

$$N_S = \text{null}\left[C_y^{(\text{LFT})}\ D_{y\theta}\ D_{yw}^{(\text{LFT})}\ 0\right]. \tag{64}$$

With the outer factors, a first set of LMIs corresponding to the feasibility and optimality condition is given as

$$N_R^{\mathrm{T}} \begin{bmatrix} ARA^{\mathrm{T}} - R & ARC_\theta^{\mathrm{T}} & AR(C^{(\text{LFT})})_q^{\mathrm{T}} & B_\theta & B_w \\ C_\theta RA^{\mathrm{T}} & -\gamma J_3 + C_\theta RC_\theta^{\mathrm{T}} & C_\theta R(C^{(\text{LFT})})_q^{\mathrm{T}} & D_{\theta\theta} & D_{\theta w} \\ C_q^{(\text{LFT})} RA^{\mathrm{T}} & C_q^{(\text{LFT})} RC_\theta^{\mathrm{T}} & C_q^{(\text{LFT})} R(C^{(\text{LFT})})_q^{\mathrm{T}} - \gamma \mathbf{I} & D_{q\theta} & D_{qw}^{(\text{LFT})} \\ B_\theta^{\mathrm{T}} & D_{\theta\theta}^{\mathrm{T}} & D_{q\theta}^{\mathrm{T}} & -\gamma L_3 & \mathbf{0} \\ (B_w^{(\text{LFT})})^{\mathrm{T}} & D_{\theta w}^{\mathrm{T}} & (D_{qw}^{(\text{LFT})})^{\mathrm{T}} & \mathbf{0} & -\gamma \end{bmatrix} N_R < 0, \tag{65}$$

$$N_S^{\mathrm{T}} \begin{bmatrix} A^{\mathrm{T}}SA - S & A^{\mathrm{T}}SB_\theta & A^{\mathrm{T}}SB_w^{(\text{LFT})} & C_\theta^{\mathrm{T}} & (C_q^{(\text{LFT})})^{\mathrm{T}} \\ B_\theta^{\mathrm{T}}SA & -\gamma L_3 + B_\theta^{\mathrm{T}}SB_\theta & B_\theta^{\mathrm{T}}SB_w^{(\text{LFT})} & D_{\theta\theta}^{\mathrm{T}} & D_{q\theta}^{\mathrm{T}} \\ (B_w^{(\text{LFT})})^{\mathrm{T}}SA & (B_w^{(\text{LFT})})^{\mathrm{T}}SB_\theta & (B_w^{(\text{LFT})})^{\mathrm{T}}S(B_w^{(\text{LFT})}) - \gamma & D_{\theta w}^{\mathrm{T}} & (D_{qw}^{(\text{LFT})})^{\mathrm{T}} \\ C_\theta & D_{\theta\theta} & D_{\theta w} & -\gamma J_3 & \mathbf{0} \\ C_q^{(\text{LFT})} & D_{q\theta} & D_{qw}^{(\text{LFT})} & \mathbf{0} & -\gamma \mathbf{I} \end{bmatrix} N_S < 0, \tag{66}$$

$$\begin{bmatrix} R & \mathbf{I} \\ \mathbf{I} & S \end{bmatrix} \geq 0, \tag{67}$$

$$\begin{bmatrix} L_3 & \mathbf{I} \\ \mathbf{I} & J_3 \end{bmatrix} \geq 0. \tag{68}$$

The scalar γ is an upper bound of the maximum singular value, which is given as a constraint. This set of LMIs is solved for R, S, J_3 and L_3.

The matrices L_1 and L_2 are calculated through

$$L_3 - J_3^{-1} = L_2^{\mathrm{T}} L_1^{-1} L_2, \tag{69}$$

and the matrix $X^{(\text{LFT})}$ is computed as

$$X^{(\text{LFT})} = \begin{bmatrix} S & \mathbf{I} \\ N^{\mathrm{T}} & \mathbf{0} \end{bmatrix} \begin{bmatrix} \mathbf{I} & R \\ \mathbf{0} & M^{\mathrm{T}} \end{bmatrix}, \tag{70}$$

with M and N satisfying

$$MN^{\mathrm{T}} = \mathbf{I} - RS. \tag{71}$$

Then, the basic LMI

$$\psi + (Q^{(\text{LFT})})^{\mathrm{T}} (\Omega^{(\text{LFT})})^{\mathrm{T}} P^{(\text{LFT})} + (P^{(\text{LFT})})^{\mathrm{T}} \Omega^{(\text{LFT})} Q^{(\text{LFT})} < 0, \tag{72}$$

where

$$\psi = \begin{bmatrix} -X^{-1} & A_0 & B_0 & \mathbf{0} \\ A_0^{\mathrm{T}} & -X & \mathbf{0} & C_0^{\mathrm{T}} \\ B_0^{\mathrm{T}} & \mathbf{0} & -\gamma L_0 & D_0^{\mathrm{T}} \\ \mathbf{0} & C_0 & D_0 & -\gamma J_0 \end{bmatrix}, \tag{73}$$

$$P^{(\text{LFT})} = \left[\tilde{B}^{\mathrm{T}}\ \mathbf{0}\ \mathbf{0}\ \tilde{D}_{12}^{\mathrm{T}}\right], \tag{74}$$

$$Q^{(\mathrm{LFT})} = \begin{bmatrix} 0 \ \tilde{C} \ \tilde{D}_{21} \ 0 \end{bmatrix}, \tag{75}$$

$$A_0 = \begin{bmatrix} A & 0 \\ 0 & 0 \end{bmatrix}, \ B_0 = \begin{bmatrix} 0 & B_\theta & B_w^{(\mathrm{LFT})} \\ 0 & 0 & 0 \end{bmatrix}, \ \tilde{B} = \begin{bmatrix} 0 & B_u^{(\mathrm{LFT})} & 0 \\ \mathrm{I} & 0 & 0 \end{bmatrix}, \tag{76}$$

$$C_0 = \begin{bmatrix} 0 & 0 \\ C_\theta & 0 \\ C_q^{(\mathrm{LFT})} & 0 \end{bmatrix}, \ \tilde{C} = \begin{bmatrix} 0 & \mathrm{I} \\ C_y^{(\mathrm{LFT})} & 0 \\ 0 & 0 \end{bmatrix}, \ D_0 = \begin{bmatrix} 0 & 0 & 0 \\ 0 & D_{\theta\theta} & D_{\theta w} \\ 0 & D_{q\theta} & D_{qw}^{(\mathrm{LFT})} \end{bmatrix}, \tag{77}$$

$$\tilde{D}_{12} = \begin{bmatrix} 0 & 0 & \mathrm{I} \\ 0 & D_{\theta u} & 0 \\ 0 & D_{qu}^{(\mathrm{LFT})} & 0 \end{bmatrix}, \ \tilde{D}_{21} = \begin{bmatrix} 0 & 0 & 0 \\ 0 & D_{y\theta} & D_{yw}^{(\mathrm{LFT})} \\ \mathrm{I} & 0 & 0 \end{bmatrix}, \tag{78}$$

and

$$L = \begin{bmatrix} L_1 & L_2 \\ L_2^{\mathrm{T}} & L_3 \end{bmatrix}, \ L_0 = \begin{bmatrix} L & 0 \\ 0 & 1 \end{bmatrix}, \ J = L^{-1}, \ J_0 = \begin{bmatrix} J & 0 \\ 0 & \mathrm{I} \end{bmatrix}, \tag{79}$$

is solved for the controller matrix $\Omega^{(\mathrm{LFT})}$. In the last step, the state-space matrices of the controller are extracted from

$$\Omega^{(\mathrm{LFT})} = \begin{bmatrix} A_K^{(\mathrm{LFT})} & B_K^{(\mathrm{LFT})} \\ C_K^{(\mathrm{LFT})} & D_K^{(\mathrm{LFT})} \end{bmatrix}. \tag{80}$$

6. Experimental results

The gain-scheduled output-feedback controllers obtained through the design procedures presented in this chapter are validated with experimental results. Both controllers have been tested on the ANC and AVC systems. Results are presented for the pLPV gain-scheduled controller on the ANC system in Sec. 6.1 and for the LPV-LFT controller on the AVC test bed in Sec. 6.2. Identical hardware setup and sampling frequency as in the previous chapter are used.

6.1. Experimental results for the pLPV gain-scheduled controller

The pLPV gain-scheduled controller is validated with experimental results on the ANC headset. The controller is designed to reject a disturbance signal which contains four harmonically related sine signals with fundamental frequency between 80 and 90 Hz. The controller obtained is of 21st order.

Amplitude frequency responses and pressure measured when the fundamental frequency rises suddenly from 80 to 90 Hz are shown in Figs. 6 and 7. An excellent disturbance rejection is achieved even for unrealistically fast variations of the disturbance frequencies. In Fig. 8, results for time-varying frequencies are shown. The performance for fast variations of the fundamental frequency is further studied in Fig. 9. As in the previous chapter, with fast changes of the fundamental frequency the disturbance attenuation performance decreases but the system remains stable.

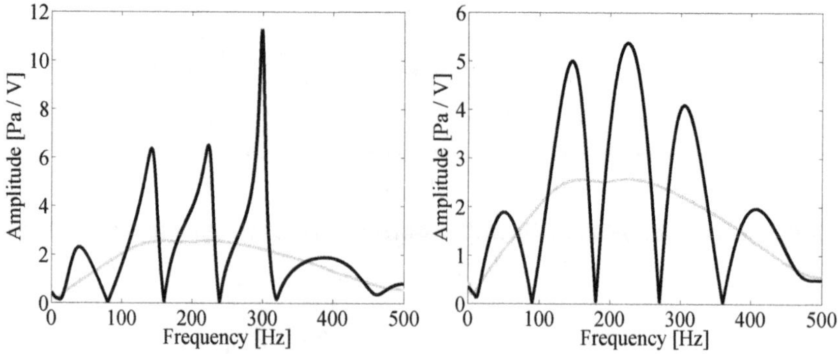

Figure 6. Open-loop (gray) and closed-loop (black) amplitude frequency responses for fixed disturbance frequencies of 80, 160, 240 and 320 Hz (left) and of 90, 180, 270 and 360 Hz (right)

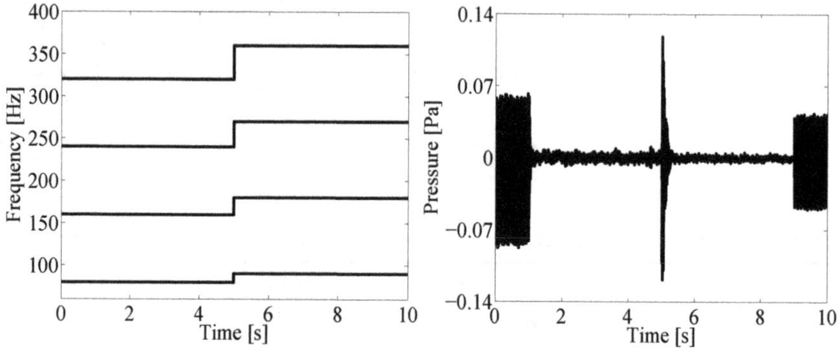

Figure 7. Results for a disturbance with time-varying frequencies. Variation of the frequencies (left) and measured sound pressure (right). The control sequence is off/on/off

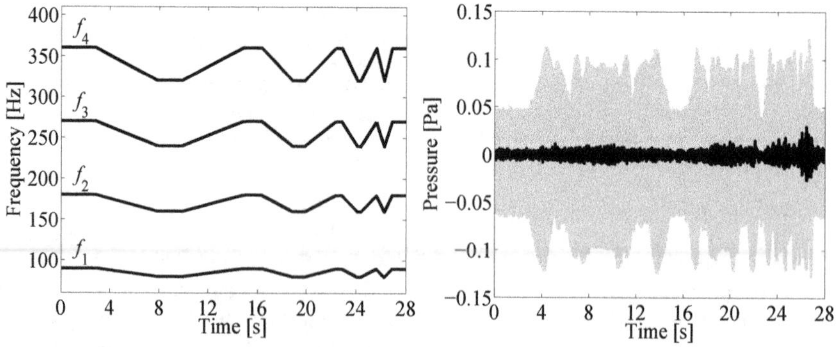

Figure 8. Results for a disturbance with time-varying frequencies. Variation of the frequencies (left) and measured sound pressure (right) in open loop (gray) and closed loop (black)

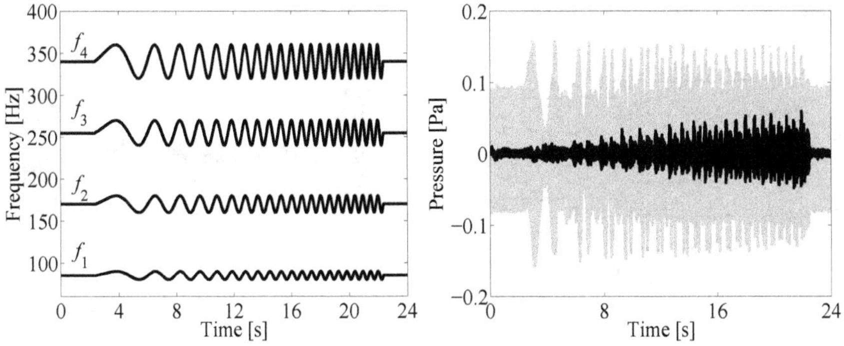

Figure 9. Results for a disturbance with time-varying frequencies. Variation of the frequencies (left) and measured sound pressure (right) in open loop (gray) and closed loop (black)

6.2. Experimental results for the LFT gain-scheduled controller

The AVC test bed is used to test the LFT gain-scheduled controller experimentally. The controller is designed to reject a disturbance with eight harmonic components which are selected to be uniformly distributed from 80 to 380 Hz in intervals of 20 Hz. The resulting controller is of 27th order.

Amplitude frequency responses are shown in Fig. 10 and results for an experiment where the frequencies change drastically as a step function in Fig. 11. Results from experiments with time-varying frequencies are shown in Figs. 12 and 13. Excellent disturbance rejection is achieved.

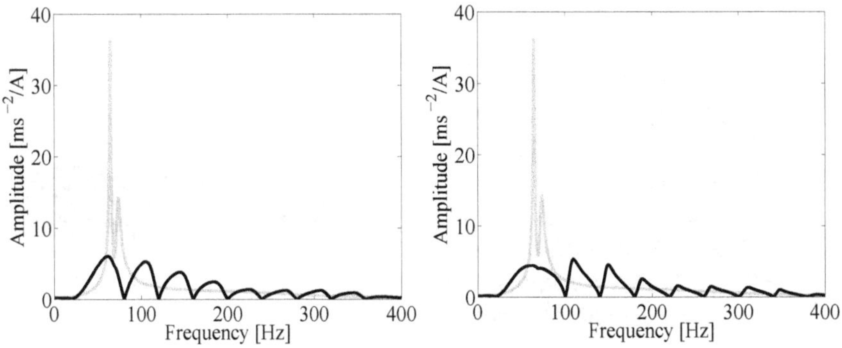

Figure 10. Open-loop (gray) and closed-loop (black) amplitude frequency responses for fixed disturbance frequencies of 80, 120, 160, 200, 240, 280, 320 and 360 Hz (left) and 100, 140, 180, 220, 260, 300, 340 and 380 Hz (right)

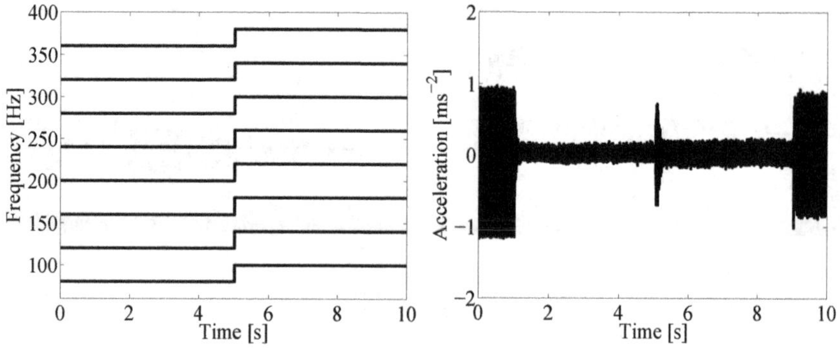

Figure 11. Results for a disturbance with time-varying frequencies. Variation of the frequencies (left) and measured acceleration (right). The control sequence is off/on/off

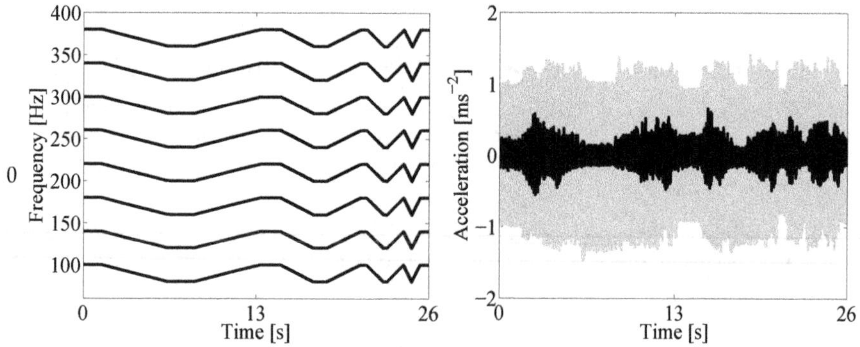

Figure 12. Results for a disturbance with time-varying frequencies. Variation of the frequencies (left) and measured acceleration (right) in open loop (gray) and closed loop (black)

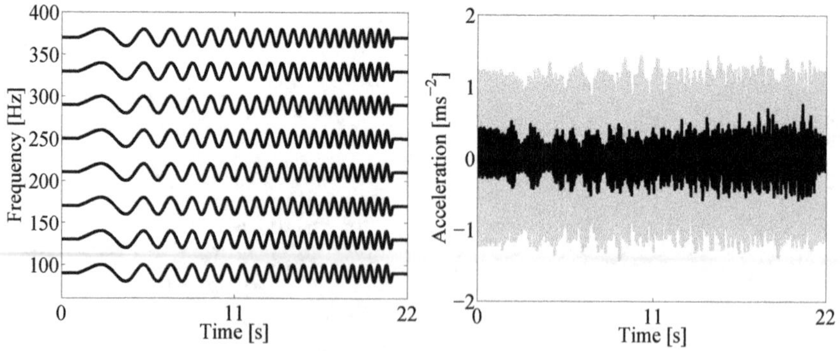

Figure 13. Results for a disturbance with time-varying frequencies. Variation of the frequencies (left) and measured acceleration (right) in open loop (gray) and closed loop (black)

7. Discussion and conclusion

Two discrete-time control design methods have been presented in this chapter for the rejection of time-varying frequencies. The output-feedback controllers are obtained through pLPV and LPV-LFT gain-scheduling techniques. The controllers obtained are validated experimentally on an ANC and AVC system. The experimental results show an excellent disturbance rejection even for the case of eight frequency components of the disturbance.

The control design guarantees stability even for arbitrarily fast changes of the disturbance frequencies. This is an advantage over heuristic interpolation methods or adaptive filtering, for which none or only "approximate stability results" are available [10].

To the best of the authors' knowledge, industrial applications of LPV controllers are rather limited. The results of this chapter show that the implementation of even high-order LPV controllers can be quite straightforward.

Nomenclature

Acronyms

ANC Active noise control.

AVC Active vibration control.

LFT Linear fractional transformation.

LMI Linear matrix inequality.

LPV Linear parameter varying.

LTI Linear time invariant.

pLPV Polytopic linear parameter varying.

Variables

(in order of appearance)

$\boldsymbol{G}$	Generalized plant.
$\boldsymbol{K}$	Controller.
$\boldsymbol{u}, \boldsymbol{y}$	Control input, output signal.
$\boldsymbol{w}, \boldsymbol{q}$	Performance input, performance output.
$\sigma_{\max}$	Maximum singular value.
G_{qw}	Transfer path between performance input and performance output.
$\boldsymbol{A}(\boldsymbol{\theta}), \boldsymbol{B}, \boldsymbol{C}, \boldsymbol{D}$	State-space matrices of a pLPV system.
$\boldsymbol{x}_k, \boldsymbol{y}_k, \boldsymbol{u}_k$	State vector, output and input.

$\mathcal{A}_i$	Constant matrices of the polytopic representation of $\boldsymbol{A}(\boldsymbol{\theta})$.
$\boldsymbol{\theta}$	Parameter vector.
θ_i	The i-th element of the parameter vector.
$\boldsymbol{\Theta}$	Parameter polytope.
$\boldsymbol{v}_j$	Vertices of the polytope.
M	Number of vertices of the polytope.
N	Number of parameters.
$\boldsymbol{\lambda}$	Coordinate vector.
λ_j	The j-th element of the coordinate vector.
$\boldsymbol{A}_{\mathrm{v},j}, \boldsymbol{A}(v_j)$	System matrix for the j-th vertex.
$\overline{\boldsymbol{\theta}}$	Parametric uncertainty block.
$\boldsymbol{w}_\theta, \boldsymbol{q}_\theta$	Output and input of the parameter block for the plant in LFT form.
$\tilde{\boldsymbol{w}}_\theta, \tilde{\boldsymbol{q}}_\theta$	Output and input of the parameter block for the controller in LFT form.
n_{d}	Number of frequencies of the disturbance.
$\boldsymbol{A}_{\mathrm{d}}^{(2n_{\mathrm{d}}\times 2n_{\mathrm{d}})}$, $\boldsymbol{B}_{\mathrm{d}}^{(2n_{\mathrm{d}}\times 1)}$, $\boldsymbol{C}_{\mathrm{d}}^{(1\times 2n_{\mathrm{d}})}$	State-space matrices of the disturbance model for fixed frequencies.
$\boldsymbol{A}_{\mathrm{d},i}, \boldsymbol{B}_{\mathrm{d},i}, \boldsymbol{C}_{\mathrm{d},i}$	Block matrices of $\boldsymbol{A}_{\mathrm{d}}$, $\boldsymbol{B}_{\mathrm{d}}$ and $\boldsymbol{C}_{\mathrm{d}}$.
a_i	Scalar parameter for the disturbance model.
T, f_i	Sampling time and the i-th frequency.
$\boldsymbol{A}_{\mathrm{d}}(\boldsymbol{\theta})^{(2n_{\mathrm{d}}\times 2n_{\mathrm{d}})}$, $\boldsymbol{B}_{\mathrm{d}}^{(2n_{\mathrm{d}}\times 1)}$, $\boldsymbol{C}_{\mathrm{d}}^{(1\times 2n_{\mathrm{d}})}$	State-space matrices of the pLPV disturbance model.
$\mathcal{A}_{\mathrm{d},i}$	Constant matrices of the polytopic representation of $\boldsymbol{A}_{\mathrm{d}}(\boldsymbol{\theta})$.
n_{p}	Order of the plant.
G_{p}	System representation of the plant.
$\boldsymbol{A}_{\mathrm{p}}^{(n_{\mathrm{p}}\times n_{\mathrm{p}})}$, $\boldsymbol{B}_{\mathrm{p}}^{(n_{\mathrm{p}}\times 1)}$, $\boldsymbol{C}_{\mathrm{p}}^{(1\times n_{\mathrm{p}})}$, $D_{\mathrm{p}}^{(1\times 1)}$	State-space matrices of the plant.
n_{W_y}	Order of the weighting function for y.

W_y, W_u	System representations of the weighting functions.
$\boldsymbol{A}_{W_y}^{(n_{W_y}\times n_{W_y})}, \boldsymbol{B}_{W_y}^{(n_{W_y}\times 1)}, \boldsymbol{C}_{W_y}^{(1\times n_{W_y})}, D_{W_y}^{(1\times 1)}$	State-space matrices of the weighting function for y.
n_{W_u}	Order of the weighting function for u.
$\boldsymbol{A}_{W_u}^{(n_{W_u}\times n_{W_u})}, \boldsymbol{B}_{W_u}^{(n_{W_u}\times 1)}, \boldsymbol{C}_{W_u}^{(1\times n_{W_u})}, D_{W_u}^{(1\times 1)}$	State-space matrices of the weighting function for u.
$\boldsymbol{x}_{\mathrm{p},k}, \boldsymbol{x}_{\mathrm{d},k},$	State vectors of plant, disturbance and
$\boldsymbol{x}_{W_y,k}, \boldsymbol{x}_{W_u,k}$	weighting functions.
$\boldsymbol{A}_i(\boldsymbol{\theta}), \boldsymbol{B}_w^{(\mathrm{pLPV})}, \boldsymbol{B}_u^{(\mathrm{pLPV})}, \boldsymbol{C}_q^{(\mathrm{pLPV})}, \boldsymbol{D}_{qw}^{(\mathrm{pLPV})}, \boldsymbol{D}_{qu}^{(\mathrm{pLPV})}$ $\boldsymbol{C}_y^{(\mathrm{pLPV})}, D_{yw}^{(\mathrm{pLPV})}, D_{qw}^{(\mathrm{pLPV})}$	State-space matrices of the pLPV generalized plant.
$\mathbf{0}$	Zero matrix.
$\boldsymbol{A}_{\mathrm{d}}, \boldsymbol{B}_{\mathrm{d},\theta}, \boldsymbol{B}_{\mathrm{d},w}, \boldsymbol{C}_{\mathrm{d},\theta}, \boldsymbol{C}_{\mathrm{d},y}$	State-space matrices of the LFT disturbance model.
$w_{\mathrm{d}}, y_{\mathrm{d}}$	Input and output of the disturbance model.
$u_{\mathrm{p}}, y_{\mathrm{p}}$	Input and output of the plant.
$\bar{a}_i, p_i$	Scalar parameters for the disturbance model.
$\boldsymbol{A}, \boldsymbol{B}_\theta, \boldsymbol{B}_w^{(\mathrm{LFT})}, \boldsymbol{B}_u^{(\mathrm{LFT})},$ $\boldsymbol{C}_\theta, \boldsymbol{D}_{\theta\theta}, \boldsymbol{D}_{\theta w}, \boldsymbol{D}_{\theta u},$ $\boldsymbol{C}_q^{(\mathrm{LFT})}, \boldsymbol{D}_{q\theta}, \boldsymbol{D}_{qw}^{(\mathrm{LFT})}, \boldsymbol{D}_{qu}^{(\mathrm{LFT})},$ $\boldsymbol{C}_y^{(\mathrm{LFT})}, \boldsymbol{D}_{y\theta}, \boldsymbol{D}_{yw}^{(\mathrm{LFT})}, \boldsymbol{D}_{yu}^{(\mathrm{LFT})}$	State-space matrices of the LFT generalized plant.
$\boldsymbol{N}_{\boldsymbol{X}}^{((n+3)\times(n+2))}, \boldsymbol{N}_{\boldsymbol{Y}}^{((n+3)\times(n+2))}$	Outer factors to build the LMIs.
$\boldsymbol{X}_1^{(n\times n)}, \boldsymbol{Y}_1^{(n\times n)}$	Solutions of the first set of LMIs.
$\mathbf{I}$	Identitiy matrix.
$n = n_{\mathrm{p}} + 2n_{\mathrm{d}} + n_{W_y} + n_{W_u}$	Order of matrices $\boldsymbol{X}_1$ and $\boldsymbol{Y}_1$.
$\boldsymbol{\psi}_i^{(4n+3)\times(4n+3)}$	Matrix to build the basic LMI.
$\boldsymbol{X}^{(2n\times 2n)}, \overline{\boldsymbol{A}}_i^{(2n\times 2n)}, \overline{\boldsymbol{B}}^{(2n\times 1)}, \overline{\boldsymbol{C}}^{(2\times 2n)}$	Matrices to build matrix $\boldsymbol{\psi}_i$.
$\boldsymbol{P}^{((n+1)\times(4n+3))}, \boldsymbol{Q}^{((n+1)\times(4n+3))}$	Matrices to build the basic LMI.

$\underline{\boldsymbol{B}}^{(2n\times(n+1))}, \underline{\boldsymbol{C}}^{((n+1)\times 2n)}, \underline{\boldsymbol{D}}_{qu}^{(2\times(n+1))}, \underline{\boldsymbol{D}}_{yw}^{((n+1)\times 1)}$	Matrices to obtain $\boldsymbol{P}^{(\mathrm{pLPV})}$ and $\boldsymbol{Q}^{(\mathrm{pLPV})}$.
$\boldsymbol{\Omega}_i^{((n+1)\times(n+1))}$	Solution of the basic LMI for the i-th vertex.
$\boldsymbol{A}_{K_i}^{(n\times n)}, \boldsymbol{B}_{K_i}^{(n\times 1)}, \boldsymbol{C}_{K_i}^{(1\times n)}, D_{K_i}^{(1\times 1)}$	State-space matrices of the controller for the i-th vertex.
$\boldsymbol{N}_{\boldsymbol{R}}^{((n+2n_\mathrm{d}+3)\times(n+2n_\mathrm{d}+2))}, \boldsymbol{N}_{\boldsymbol{S}}^{((n+2n_\mathrm{d}+3)\times(n+2n_\mathrm{d}+2))}$	Outer factors to build the LMIs.
$\boldsymbol{R}^{(n\times n)}, \boldsymbol{S}^{(n\times n)}, \boldsymbol{J}_3^{(n_\mathrm{d}\times n_\mathrm{d})}, \boldsymbol{L}_3^{(n_\mathrm{d}\times n_\mathrm{d})}$	Solutions of the first set of LMIs.
γ	Upper bound of the maximum singular value.
$\boldsymbol{M}^{(n\times n)}, \boldsymbol{N}^{(n\times n)}$	Matrices calculated from $\boldsymbol{R}$ and $\boldsymbol{S}$.
$\boldsymbol{L}_1^{(n_\mathrm{d}\times n_\mathrm{d})}, \boldsymbol{L}_2^{(n_\mathrm{d}\times n_\mathrm{d})}$	Matrices to build $\boldsymbol{L}$.
$\boldsymbol{\psi}^{((4n+4n_\mathrm{d}+3)\times(4n+4n_\mathrm{d}+3))}$	Matrix to build the basic LMI.
$\boldsymbol{X}^{(2n\times 2n)}, \boldsymbol{A}_0^{(2n\times 2n)}, \boldsymbol{B}_0^{(2n\times(2n_\mathrm{d}+1))}, \boldsymbol{C}_0^{((2n_\mathrm{d}+2)\times 2n)}, \boldsymbol{D}_0^{((2n_\mathrm{d}+2)\times(2n_\mathrm{d}+1))}, \boldsymbol{J}_0^{((2n_\mathrm{d}+2)\times(2n_\mathrm{d}+2))}, \boldsymbol{L}_0^{((2n_\mathrm{d}+1)\times(2n_\mathrm{d}+1))}, \boldsymbol{J}^{(2n_\mathrm{d}\times 2n_\mathrm{d})}, \boldsymbol{L}^{(2n_\mathrm{d}\times 2n_\mathrm{d})}$	Matrices needed to build $\boldsymbol{\psi}$.
$\boldsymbol{P}^{((n+n_\mathrm{d}+1)\times(4n+4n_\mathrm{d}+3))}, \boldsymbol{Q}^{((n+n_\mathrm{d}+1)\times(4n+4n_\mathrm{d}+3))}$	Matrices to build the basic LMI.
$\tilde{\boldsymbol{B}}^{(2n\times(n+n_\mathrm{d}+1))}, \tilde{\boldsymbol{C}}^{((n+n_\mathrm{d}+1)\times 2n)}, \tilde{\boldsymbol{D}}_{12}^{((2n_\mathrm{d}+2)\times(n+n_\mathrm{d}+1))}, \tilde{\boldsymbol{D}}_{21}^{((n+n_\mathrm{d}+1)\times(2n_\mathrm{d}+1))}$	Matrices to obtain $\boldsymbol{P}^{(\mathrm{LFT})}$ and $\boldsymbol{Q}^{(\mathrm{LFT})}$.
$\boldsymbol{\Omega}^{((n+n_\mathrm{d}+1)\times(n+n_\mathrm{d}+1))}$	Controller matrix.
$\boldsymbol{A}_K^{(n\times n)}, \boldsymbol{B}_K^{(n\times(n_\mathrm{d}+1))}, \boldsymbol{C}_K^{((n_\mathrm{d}+1)\times n)}, \boldsymbol{D}_K^{((n_\mathrm{d}+1)\times(n_\mathrm{d}+1))}$	State-space matrices of the controller.

Author details

Pablo Ballesteros, Xinyu Shu, Wiebke Heins and Christian Bohn
Institute of Electrical Information Technology, Clausthal University of Technology, Clausthal-Zellerfeld, Germany

8. References

[1] Apkarian, P. and P. Gahinet. 1995. A convex charachterization of gain-scheduled H_∞ controllers. *IEEE Transactions on Automatic Control* 40:853-64.

[2] Ballesteros, P. and C. Bohn. 2011a. A frequency-tunable LPV controller for narrowband active noise and vibration control. *Proceedings of the American Control Conference*. San Francisco, June 2011. 1340-45.

[3] Ballesteros, P. and C. Bohn. 2011b. Disturbance rejection through LPV gain-scheduling control with application to active noise cancellation. *Proceedings of the IFAC World Congress*. Milan, August 2011. 7897-902.

[4] Balini, H. M. N. K., C. W. Scherer and J. Witte. 2011. Performance enhancement for AMB systems using unstable H_∞ controllers. *IEEE Transactions on Control Systems Technology* 19:1479-92.

[5] Bohn, C., A. Cortabarria, V. Härtel and K. Kowalczyk. 2003. Disturbance-observer-based active control of engine-induced vibrations in automotive vehicles. *Proceedings of the SPIE's 10th Annual International Symposium on Smart Structures and Materials*. San Diego, March 2003. Paper No. 5049-68.

[6] Bohn, C., A. Cortabarria, V. Härtel and K. Kowalczyk. 2004. Active control of engine-induced vibrations in automotive vehicles using disturbance observer gain scheduling. *Control Engineering Practice* 12:1029-39.

[7] Darengosse, C. and P. Chevrel. 2000. Linear parameter-varying controller design for active power filters. *Proceedings of the IFAC Control Systems Design*. Bratislava, June 2000. 65-70.

[8] Du, H. and X. Shi. 2002. Gain-scheduled control for use in vibration suppression of system with harmonic excitation. *Proceedings of the American Control Conference*. Anchorage, May 2002. 4668-69.

[9] Du, H., L. Zhang and X. Shi. 2003. LPV technique for the rejection of sinusoidal disturbance with time-varying frequency. *IEE Proceedings on Control Theory and Applications* 150:132-38.

[10] Feintuch, P. L., N. J. Bershad and A. K. Lo. 1993. A frequency-domain model for filtered LMS algorithms - Stability analysis, design, and elimination of the training mode. *IEEE Transactions on Signal Processing* 41:1518-31.

[11] Gahinet, P. and P. Apkarian. 1994. A linear matrix inequality approach to H_∞ control. *International Journal of Robust and Nonlinear Control* 4:421-48.

[12] Heins, W., P. Ballesteros and C. Bohn. 2011. Gain-scheduled state-feedback control for active cancellation of multisine disturbances with time-varying frequencies. Presented at the *10th MARDiH Conference on Active Noise and Vibration Control Methods*. Krakow-Wojanow, Poland, June 2011.

[13] Heins, W., P. Ballesteros and C. Bohn. 2012. Experimental evaluation of an LPV-gain-scheduled observer for rejecting multisine disturbances with time-varying frequencies. *Proceedings of the American Control Conference*. Montreal, June 2012. Accepted for publication.

[14] Kinney, C. E. and R. A. de Callafon. 2006a. Scheduling control for periodic disturbance attenuation. *Proceedings of the American Control Conference*. Minneapolis, June 2006. 4788-93.

[15] Kinney, C. E. and R. A. de Callafon. 2006b. An adaptive internal model-based controller for periodic disturbance rejection. *Proceedings of the 14th IFAC Symposium on System Identification*. Newcastle, Australia, March 2006. 273-78.

[16] Kinney, C. E. and R. A. de Callafon. 2007. A comparison of fixed point designs and time-varying observers for scheduling repetitive controllers. *Proceedings of the 46th IEEE Conference on Decision and Control*. New Orleans, December 2007. 2844-49.
[17] Köroğlu, H. and C. W. Scherer. 2008. LPV control for robust attenuation of non-stationary sinusoidal disturbances with measurable frequencies. *Proceedings of the 17th IFAC World Congress*. Korea, July 2008. 4928-33.
[18] Shu, X., P. Ballesteros and C. Bohn. 2011. Active vibration control for harmonic disturbances with time-varying frequencies through LPV gain scheduling. *Proceedings of the 23rd Chinese Control and Decision Conference*. Mianyang, China, May 2011. 728-33.
[19] Witte, J., H. M. N. K. Balini and C. W. Scherer. 2010. Experimental results with stable and unstable LPV controllers for active magnetic bearing systems. *Proceedings of the IEEE International Conference on Control Applications*. Yokohama, September 2010. 950-55.

Chapter 5

The Active Suspension of a Cab in a Heavy Machine

Grzegorz Tora

Additional information is available at the end of the chapter

1. Introduction

The work of operators in heavy machinery requires constant attention to gather information about the machine's surroundings, its current status and the operations performed. Operators have to analyse the received information on the continuous basis and make decisions accordingly, to have them implemented via the control system and to perform the scheduled tasks in the optimal manner. The more powerful the machine, the more serious the consequence of errors committed by operators. The typical frequency range of vibration of machines and their equipment is determined based on testing done on heavy machines used in Europe [1] and is found to be 0.5- 80 Hz.

Machine vibrations are induced by the drives' action, movements of the equipment, variable loading and machine ride. The ride of heavy machines, tractors, forestry vehicles over a rough terrain lead to cyclic tilting of the machines, which can be regarded as low-frequency (up to several Hz) and high-amplitude (about 10 degrees) vibration of the machine. The angular motions of the frame are transmitted onto the cab, and the higher the cab position, the larger the amplitude range of linear vibration of the point SIP (about 70 cm). Vibrations negatively impact on the machine structure, control processes, performance quality and the operator's comfort. Growing ergonomic concerns and competition on the market have prompted the design of machines ensuring the better comfort for the operator.

Cab suspensions are now incorporated in the machine structure as a new solution. The active suspension is a system whose components are based on existing vibration reduction solutions. Early vehicles were also provided with suspension systems to suppress vibrations due to the ride in the rough terrain. At first these were passive suspension systems, in which the characteristics of the components i.e. elastic and damping elements are fixed. Suspensions incorporating semiactive elements perform better as vibration isolation systems since their characteristics can be varied according to the adopted control strategy. Active suspensions lend a new quality to control of low-frequency vibrations in vehicles. The operating ranges of passive, semiactive and active suspensions, given as velocity-force characteristics, are shown in Fig 1.

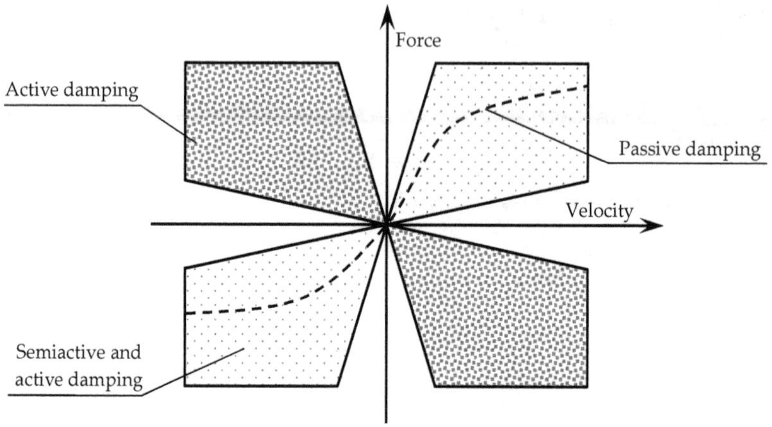

Figure 1. The range of velocity-force characteristics of passive, semiactive and active suspensions [2]

Passive suspensions are described with a damper characteristics with fixed parameters (broke line). These suspensions typically comprise elastic elements featuring a linear or nonlinear elasticity and damping elements with nonlinear characteristics. On one hand, stability of parameters of a passive suspension system is considered an advantage as its construction can be made simple, but on the other hand the vehicle suspension will not perform optimally in response to inputs other than average. Steel and rubber connectors used as joint components in vehicle suspensions play a major role in damping higher-frequency vibrations.

The areas in the first and third quadrant have relevance to the family of semiactive characteristics. A semiactive element comprises a damping element whose damping ratio can be varied through the real-time control process. A semiactive suspension does not generate any active force, hence the power demand remains on a low level. The damping ratio can be varied using throttling valves or through application of rheological fluids whose stiffness and viscosity depend on electric or magnetic field intensity. The work [3] investigates the potential applications of dry friction in semiactive dampers. Another solution uses a lever system wherein the attachment point of the spring can be varied and its elasticity controllable [4]. The upper frequency limit for the effective performance of semiactive suspension systems is about 100 Hz. Because of their low power demand, semiactive suspensions systems are being vigorously researched and widely implemented in vehicles.

The operation of active suspension systems is revealed in characteristics in all four quadrants, whereas the second and fourth quadrant capture the conditions where the actuator requires an external energy source, otherwise it acts as a passive element, dissipating energy. Active suspension systems incorporate a force actuator, either independent, or connected in parallel to a damper or a spring. Active suspensions prove most effective in the case of low-frequency vibrations [5], about 5 Hz and are often added to passive systems which well handle high-frequency vibrations. High efficiency of active systems, however, comes at the cost of high power demand. Active suspension systems typically utilise small hydraulic cylinders to achieve high accelerations of masses subjected to considerable loads. Pneumatic cylinders are capable of delivering higher velocity than hydraulic ones, yet the acceleration is strongly dependent on the external load applied. There are also hydro-pneumatic cylinders, displaying the advantages of the two previous types. Active suspension systems in truck cabs utilise electric cylinders DC and AC-servo.

The new developments of the operator's seats are in line with the advancements in vehicle suspensions. In the simplest solution where the seat is fixed rigid to the machine frame or to the cab floor, vibrations are transmitted from the cab attachment point onto the operator. The mobile seat support allows it to be moved only in the vertical so the passive, semiactive or active vibration reduction system can be added along this direction only. Typically, a shearing mechanism is used as seat support [6], though other solutions with a greater number of DOFs are reported as well [7]. On account of small mass of the seat together with the operator, the upper frequency limit is reached and the active support becomes more effective than a vehicle suspension. Active suspension systems are now incorporated in agricultural and forestry machinery, where typical excitations are in the form of low-frequency and high-amplitude vibrations [8].

The correct control strategy is a key element in active and semiactive vibration reduction systems, resulting in a good compromise between numerous and sometimes mutually excluding requirements. Optimal control techniques that handle this problem include the linear quadratic regulator LQR [9] and the linear Gaussian regulator LQG [10]. In simpler cases the PID control can be applied. Active control systems may also use regulators based on neural networks [11] and fuzzy logics [12]. The control schemes that are commonly used for semiactive suspension systems include the 'sky-hook' control strategy [13] where the damping force is related to the absolute velocity of the vehicle body. The $\mathcal{H}_\infty$ control is insensitive to uncertain input quantities, encumbered with major errors (for instant: time constant, reduced mass, damping force, acceleration, velocity, damping ratio for the tire [14]. The $\mathcal{H}_\infty$ control scheme concurrently executes the mutlicriterial optimisation and in active and semiactive suspension systems it takes into account the acceleration of sprung mass, peak accelerations, jerks of the front and rear suspension, road holding, forces acting upon the relevant masses, deflection of the tire and of the suspension [15]. The work [16] focused on $\mathcal{H}_\infty$ control in an active suspension of vehicle investigates the influence of the time delay on stability of the control process. It is shown that the delay time (i.e. time before the cylinder in the active system is activated), the reduced mass and power ratings are major determinants of the frequency limit for effective operation of these type of suspensions. The stability condition is formulated for the predetermined range of time delay and system parameters. Extensive expertise prompts the use of filters, beside robust control for the purpose of estimation. The Kalman filter, reported in literature on the subject [17], is now widely used.

To improve the operator's comfort, an active suspension of a cab can be incorporated in the machine structure (Fig 2), to reduce the cab's vibration. The active suspension system comprises several sub-systems:

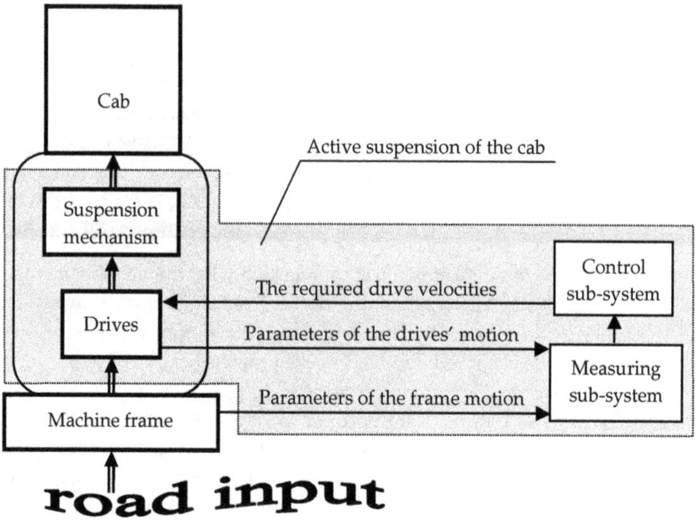

Figure 2. Block diagram of the active suspension of a cab

1. The actuator mechanism, connected to the machine frame and the cab is placed in between. The main element (link) of the mechanism is a mobile platform to which the cab is attached. The platform is suspended or supported on the frame and depending on the mechanism's mobility, it can move with respect to the frame in the selected DOFs.
2. The drives set in motion the passive links in the active suspension mechanism. On account of the stiffness requirements and availability of the given type of energy, and to ensure fast response the control signals the hydraulic drives are going to be used. The drives are provided with actuators to capture the instantaneous velocities, derived in the control sub-system.
3. Measuring sub-system. Displacement and velocity are chosen as control quantities for the active suspension system. Directly measured data yield the error signal to be used in the control process. The machine frame subjected to kinematic excitations executes a spatial movement, measured with a set of sensors.
4. Control sub-system. Errors of drive positions and their derivatives are going to be used in the feedback control of the active suspension system. Basing on the frame motion measurements, the control sub-system performs the real time calculation of the anticipated loads and the required drive velocities.

2. Actuator mechanism in the active suspension system

The active suspension mechanism, shown schematically in Fig 3 has been engineered specifically for the purpose of modelling and simulations and its design involves a certain trade-off between functionality and simplicity. The presented active suspension mechanism is capable of reducing the amplitudes of the cab's linear vibrations in the direction y_r and its angular vibrations around the axes x_r, y_r. The active suspension mechanism comprises just three passive links, set in motion by two linear drives. The separate seat suspension mechanism reduces the vibrations along the axis z_r. The main function of the active suspension system is to stabilise the cab such that the correct control of the drives 1 and 4 should enable its vertical movement in the direction of the gravity force. The active suspension system comprises a platform *p* suspended on three limbs with spherical pairs having the centres B_1, B_2, B_3. The two limbs are rocker arms 2 and 3, connected to the machine frame *r* via a revolving pair. The third limb is the actuator 1 with the length s_1. The cylinder in the actuator 1*c* is connected to the frame *r* via a cross pair with the point A_1. The piston in the actuator 4*t* is connected to the rocker arm via a spherical joint with the point C_4. The length of the actuator 4 equals s_4. The part of the active suspension system comprising the rocker arms 2 and 3, a actuator and the platform *p* along the line segment B_2B_3 can be treated as a planar mechanism where the points A_4, C_4, A_2, B_2 and A_3 are on the plane y_rz_r and the axes of joints A_2 and A_3 are parallel to x_r. The structure of the mechanism is such that the actuators 1 and 4, when in their middle position, do not carry the cab's gravity load and when in their extreme positions, the load due to the gravity force is carried mostly by the joints A_2, A_3. Besides, the performance of the mechanism is affected by manufacturing imprecision, though this influence is found to be negligible.

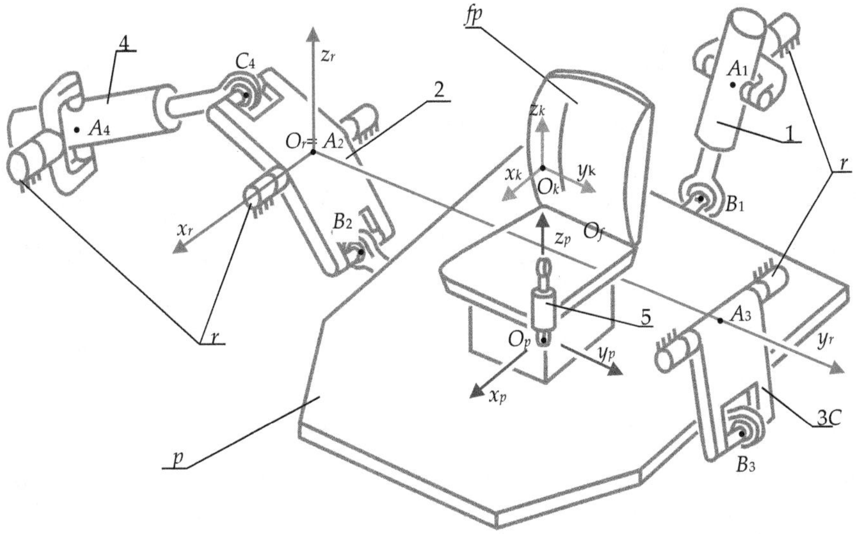

Figure 3. Platform mechanism stabilising the cab in the vertical: *r* - machine frame, *p* - platform, 2,3 - rocker arms,4*w* - forks in cross joints, 1*c*, 4*c* - cylinders in actuators, 1*t*, 4*t* - pistons in cylinders

The cab is rigidly attached to the platform *p*. The centre of gravity (c.o.g) of the cab is at the point Q_k. The reference systems associated with the platform $\{Q_px_py_pz_p\}$ and with the cab $\{Q_kx_ky_kz_k\}$ are parallel and immobile with respect to one another. Inside the cab there is a movable operator's seat *f*, which can be moved with respect to the cab, along its vertical axis z_k. The seat suspension mechanism with the operator is not the subject matter of the present study. The centre of gravity of the seat and the operator is at the point Q_f, whose vertical coordinate in the reference system associated with the platform is controlled by the drive 5, implementing seat elevation.

3. Model of the kinematic excitation of the machine frame motion

To implement the control, we need to know the angles of deflection of the vertical frame axis in the direction of the gravity force (Fig 4). The first measured angle defines the frame rotation around the longitudinal axis α_x, the other angle defines the frame rotation round the lateral axis α_y.

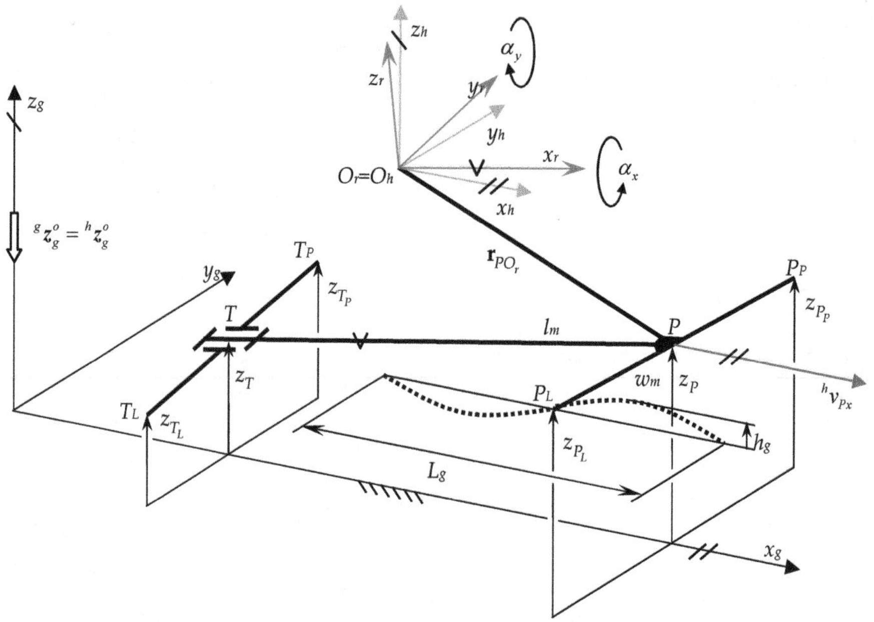

Figure 4. Model of the kinematic excitation inducing the motion of the machine frame

The model of the active suspension mechanism uses several reference systems. The immobile system $\{Q_g x_g y_g z_g\}$ is associated with the road travelled by the machine. This system is used to define the function of road profile on which the machine travels. The mobile system $\{Q_r x_r y_r z_r\}$ is associated with the machine frame (Fig 3, Fig 4). Its origin $O_r=A_2$ is one of the attachments points of the active suspension mechanism to the frame. This system is recalled to define all kinematic and dynamic quantities (no superscript on the left). The system $\{Q_h x_h y_h z_h\}$ (Fig 4) is intermediate between the inertial system associated with the road and the mobile system associated with the frame. The origins of the reference systems $\{Q_h x_h y_h z_h\}$and $\{Q_r x_r y_r z_r\}$ will coincide: $O_h=O_r$. The axes z_h and z_g are parallel and the plane determined by $x_h z_h$ contains the axis x_r. In the model of the active suspension, the directional versor of the gravity force ${}^{h}\mathbf{z}_{g}^{o}=[0,0,-1]^{T}$ given in the reference system $\{Q_h x_h y_h z_h\}$ should be expressed in the system $\{Q_r x_r y_r z_r\}$:

$$\mathbf{z}_{g}^{o}={}_{h}^{r}\mathbf{R}\,{}^{h}\mathbf{z}_{g}^{o} \tag{1}$$

where: ${}_{h}^{r}\mathbf{R}(\alpha_x,\alpha_y)=\begin{bmatrix}\cos\alpha_y & 0 & \sin\alpha_y\\ \sin\alpha_x\sin\alpha_y & \cos\alpha_x & -\sin\alpha_x\cos\alpha_y\\ -\cos\alpha_x\sin\alpha_y & \sin\alpha_x & \cos\alpha_x\cos\alpha_y\end{bmatrix}$ - the transition matrix from the system $\{Q_h x_h y_h z_h\}$ to $\{Q_r x_r y_r z_r\}$, derived basing on the frame deflection angles α_x, α_y measured in the measuring sub-system.

Underlying the simulation procedure is the model of the kinematic excitation applied to the machine frame, shown in Fig 4.

The machine frame is represented by a front bridge (P_P, P_T), a longitudinal frame (P, T) and a joint at the point O_r, where the active suspension mechanism is connected to the frame. The rear bridge (T_L, T_P) is connected to the longitudinal frame via a revolving pair T. It is assumed that:

- the longitudinal axis of the frame TP moves in the plane $x_g z_g$,
- velocity components of the extreme points of the front and rear bridge along the axis x_g are identical and equal to ${}^{h}v_{Px}$, whilst their velocity components in the direction of y_g are negligible,
- velocity components in the direction z_g are associated with vertical displacements of the wheels during the ride in rough terrain.

The vertical displacement of the centre of the front right wheel is governed by a harmonic function:

$$z_{P_P}=h_o+h_g\left[1-\cos\left(2\pi\frac{{}^{h}v_{Px}t}{L_g}\right)\right] \tag{2}$$

When $t<t_\varphi$, then vertical displacement of the centre of the front left wheel is taken to be $z_{P_L}=h_o$. When $t\ge t_\varphi$, vertical displacement of the centre of the front left wheel is expressed as the harmonic:

$$z_{P_L}=h_o+h_g\left[1-\cos\left(2\pi\frac{{}^{h}v_{Px}(t-t_\varphi)}{L_g}\right)\right] \tag{3}$$

When $t<t_{l_m}$, the vertical displacement of the centre of the rear right wheel is taken to be $z_{T_P}=h_o$. When $t\ge t_{l_m}$, the vertical displacement is expressed as harmonic:

$$z_{T_P}=h_o+h_g\left[1-\cos\left(2\pi\frac{{}^{h}v_{Px}(t-t_{l_m})}{L_g}\right)\right] \tag{4}$$

When $t < t_{l_m} + t_\varphi$, the vertical displacement of the centre of the rear left wheel is taken to be $z_{T_L} = h_o$. When $t \geq t_{l_m} + t_\varphi$, the vertical displacement is expressed as harmonic:

$$z_{T_L} = h_o + h_g \left[1 - \cos\left(2\pi \frac{{}^{h}v_{Px}(t - t_{l_m} - t_\varphi)}{L_g} \right) \right] \tag{5}$$

where: $t_\varphi = \dfrac{\varphi L_g}{2\pi\, {}^{h}v_{Px}}$ - phase shift time between the left and right hand side of the machine,

$t_{l_m} = \dfrac{l_m}{v_{Px}}$ - phase shift time between the front and rear part of the machine,

L_g - distance corresponding to the full wave, h_g- amplitude, w_m - width of the front and rear bridge,

l_m -distance between the front and rear bridge, φ - phase shift angle of the road profile between the left and right-hand side of the machine.

4. Kinematic model of the active suspension mechanism

To determine the influence of the active suspension system on the cab motion, the kinematic model is developed based on vector calculus. Versors used to define the positions of the active suspension mechanism links are shown in Fig 5.

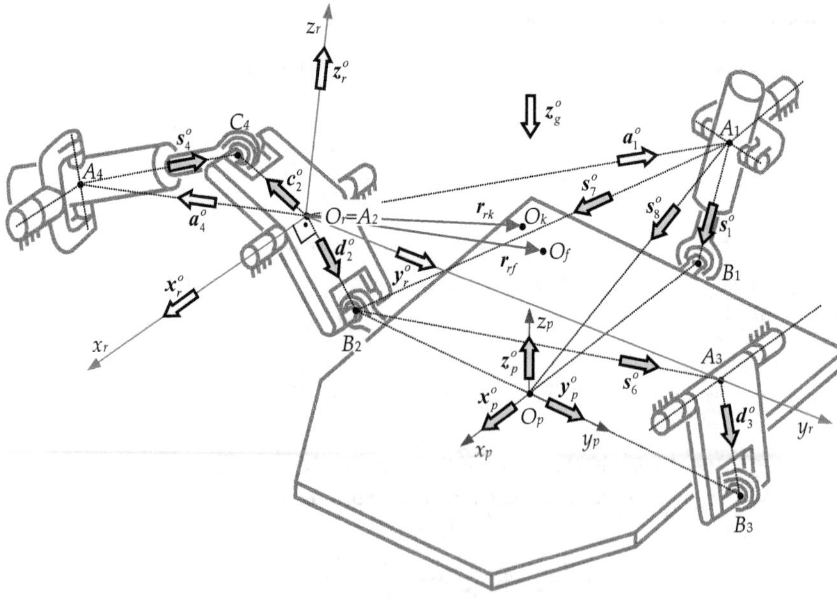

Figure 5. Versors in the kinematic model of cab stabilisation in the vertical

4.1. Direct kinematics problem of links position

Solving the direct problem consists in finding the cab orientation and position of its centre of gravity Q_k and of the point Q_f - the centre of gravity of the seat-operator system with respect to the reference system associated with the machine frame. The cab orientation is determined by directional versors in the reference system associated with the platform $\mathbf{x}_p^o$, $\mathbf{y}_p^o$, $\mathbf{z}_p^o$. The cab position (and of the platform and the seat) and $\mathbf{r}_{rk}$ - the radius vector of the of the cab's c.o.g with respect to the reference system associated with the frame depend on variable lengths of actuators s_1, s_4 and the known constant dimensions of links in the active suspension mechanism. The radius vector of the c.o.g of the seat-operator system; $\mathbf{r}_{rf}$ is controlled by the lengths of three actuators s_1, s_4, s_5 and the fixed dimensions of links in the active suspension mechanism. The solution of the simple problem involving the link position is explicit and consists in determining versors on the basis of two already known or already established ones.

The versor $\mathbf{c}_2^o$ is derived basing on two known versors $\mathbf{a}_4^o = [0, a_{4y}^o, a_{4z}^o]^T$ and $\mathbf{x}_r^o = [1,0,0]^T$:

$$\mathbf{c}_2^o = \mathbf{a}_4^o \frac{c_2^2 - s_4^2 + a_4^2}{2a_4 c_2} + (\mathbf{a}_4^o \times \mathbf{x}_r^o)\sqrt{1 - \left(\frac{c_2^2 - s_4^2 + a_4^2}{2a_4 c_2}\right)^2} \tag{6}$$

where: : $s_4 = A_4C_4$, $c_2 = O_rC_4$, $a_4 = O_rA_4$.

The versor $\mathbf{s}_4^o$ is obtained basing on $\mathbf{a}_4^o$ and $\mathbf{c}_2^o$:

$$\mathbf{s}_4^o = -\frac{a_4}{s_4}\mathbf{a}_4^o + \frac{c_2}{s_4}\mathbf{c}_2^o \tag{7}$$

The versor $\mathbf{d}_2^o$ is obtained basing on $\mathbf{x}_r^o$ and $\mathbf{c}_2^o$:

$$\mathbf{d}_2^o = c_{22}\mathbf{c}_2^o + \sqrt{1 - c_{22}^2}(\mathbf{x}_r^o \times \mathbf{c}_2^o) \tag{8}$$

where: $c_{22} = \mathbf{c}_2^o \cdot \mathbf{d}_2^o = \cos\measuredangle(\mathbf{c}_2^o, \mathbf{d}_2^o)$, $\measuredangle(\mathbf{c}_2^o, \mathbf{d}_2^o)$ - a known fixed angle.

The versor $\mathbf{s}_6^o$ is obtained basing on $\mathbf{d}_2^o$ and $\mathbf{y}_r^o = [0,1,0]^T$:

$$\mathbf{s}_6^o = -\frac{d_2}{s_6}\mathbf{d}_2^o + \frac{a_3}{s_6}\mathbf{y}_r^o \tag{9}$$

where: $s_6 = \sqrt{d_2^2 + a_3^2 - 2d_2 a_3 (\mathbf{d}_2^o \cdot \mathbf{a}_3^o)}$, $d_2 = O_rB_2$, $a_3 = O_rA_3$.

The versor $\mathbf{d}_3^o$ is obtained basing on $\mathbf{s}_6^o$ i and $\mathbf{x}_r^o$:

$$\mathbf{d}_3^o = \mathbf{s}_6^o \frac{d_3^2 - b_{23}^2 + s_6^2}{2s_6 d_3} - \mathbf{s}_6^o \times \mathbf{x}_r^o \sqrt{1 - \left(\frac{d_3^2 - b_{23}^2 + s_6^2}{2s_6 d_3}\right)^2} \tag{10}$$

where: $d_3 = A_3B_3$, $b_{23} = B_2B_3$.

The versor $\mathbf{y}_p^o$ is obtained basing on $\mathbf{s}_6^o$ and $\mathbf{d}_3^o$:

$$\mathbf{y}_p^o = \frac{s_6}{b_{23}}\mathbf{s}_6^o + \frac{d_3}{b_{23}}\mathbf{d}_3^o \tag{11}$$

The versor $\mathbf{s}_7^o$ is obtained basing on $\mathbf{d}_2^o$ and $\mathbf{a}_1^o = [a_{1x}^o, a_{1y}^o, a_{1z}^o]^T$:

$$\mathbf{s}_7^o = -\frac{d_2}{s_7}\mathbf{d}_2^o + \frac{a_1}{s_7}\mathbf{a}_1^o \tag{12}$$

where: $s_7 = \sqrt{d_2^2 + a_1^2 - 2d_2a_1\,(\mathbf{d}_2^o \cdot \mathbf{a}_1^o)}$, $a_1 = O_rA_1$.

The versor $\mathbf{s}_8^o$ is obtained basing on $\mathbf{s}_7^o$ and $\mathbf{y}_p^o$:

$$\mathbf{s}_8^o = \frac{s_7}{s_8}\mathbf{s}_7^o + \frac{b_2}{s_8}\mathbf{y}_p^o \tag{13}$$

where: $s_8 = \sqrt{s_7^2 + b_2^2 - 2s_7b_2\,(\mathbf{s}_7^o \cdot \mathbf{y}_p^o)}$, $b_2 = O_pB_2$.

The versor $\mathbf{s}_1^o$ is obtained basing on $\mathbf{s}_7^o$ and $\mathbf{s}_8^o$:

$$\mathbf{s}_1^o = \frac{c_{71} - c_{78}c_{81}}{1 - c_{78}^2}\mathbf{s}_7^o + \frac{c_{81} - c_{78}c_{71}}{1 - c_{78}^2}\mathbf{s}_8^o - \frac{\sqrt{1 - c_{78}^2 - c_{71}^2 - c_{81}^2 + 2c_{78}c_{71}c_{81}}}{1 - c_{78}^2}(\mathbf{s}_7^o \times \mathbf{s}_8^o) \tag{14}$$

where: $c_{71} = \dfrac{s_7^2 + s_1^2 - b_{12}^2}{2s_1s_7}$, $c_{78} = \dfrac{s_7^2 + s_8^2 - b_2^2}{2s_7s_8}$, $c_{81} = \dfrac{s_8^2 + s_1^2 - b_1^2}{2s_1s_8}$, $b_1 = O_pB_1$, $b_{12} = B_1B_2$.

The versor $\mathbf{x}_p^o$ is obtained basing on $\mathbf{s}_1^o$ and $\mathbf{s}_8^o$:

$$\mathbf{x}_p^o = -\frac{s_1}{b_1}\mathbf{s}_1^o + \frac{s_8}{b_1}\mathbf{s}_8^o \tag{15}$$

where: $b_1 = O_pB_1$.

The versor $\mathbf{z}_p^o$ is obtained basing on $\mathbf{x}_p^o$ and $\mathbf{y}_p^o$:

$$\mathbf{z}_p^o = \mathbf{x}_p^o \times \mathbf{y}_p^o \tag{16}$$

Eq (11), (15) and (16) yield the versors of the platform p, and the matrix ${}_p^r\mathbf{R}$ - the direction matrix of the reference system associated with the platform with respect to the system associated with the frame:

$$ {}^{r}_{p}\mathbf{R} = \begin{bmatrix} \mathbf{x}_p^o \cdot \mathbf{x}_r^o & \mathbf{y}_p^o \cdot \mathbf{x}_r^o & \mathbf{z}_p^o \cdot \mathbf{x}_r^o \\ \mathbf{x}_p^o \cdot \mathbf{y}_r^o & \mathbf{y}_p^o \cdot \mathbf{y}_r^o & \mathbf{z}_p^o \cdot \mathbf{y}_r^o \\ \mathbf{x}_p^o \cdot \mathbf{z}_r^o & \mathbf{y}_p^o \cdot \mathbf{z}_r^o & \mathbf{z}_p^o \cdot \mathbf{z}_r^o \end{bmatrix} \tag{17} $$

The solution to the simple problem involving the link position is complete when the positions of points O_k. and O_f are found in relation to s_1, s_4, s_5. The radius vector from the origin of the reference system associated with the frame O_r to the point O_f becomes (Fig 5):

$$ \mathbf{r}_{rf} = d_2\mathbf{d}_2^o + b_2\mathbf{y}_p^o + {}^{r}_{p}\mathbf{R}\,{}^{p}\mathbf{r}_{pf} \tag{18} $$

where: ${}^{p}(\overline{O_pO_f}) = {}^{p}\mathbf{r}_{pf} = [{}^{p}r_{(pf)x} \quad {}^{p}r_{(pf)y} \quad {}^{p}r_{(pf)z}]^T$ - radius vector of the point O_r in the system associated with the platform, ${}^{p}r_{(pf)z} = df_z + s_5$ - variable vertical coordinate of the seat controlled by the seat elevating drive 5. ${}^{p}r_{(pf)x}$, ${}^{p}r_{(pf)y}$, df_z - fixed coordinates.

The radius vector from the origin of the reference system associated with the frame O_r to the point O_k becomes:

$$ \mathbf{r}_{rk} = d_2\mathbf{d}_2^o + b_2\mathbf{y}_p^o + {}^{r}_{p}\mathbf{R}\,{}^{p}\mathbf{r}_{pk} \tag{19} $$

where: ${}^{p}(\overline{O_pO_k}) = {}^{p}\mathbf{r}_{pk} = [{}^{p}r_{(pk)x} \quad {}^{p}r_{(pk)y} \quad {}^{p}r_{(pk)z}]^T$ - radius vector of the point O_k in the system associated with the platform. ${}^{p}r_{(pk)x}$, ${}^{p}r_{(pk)y}$, ${}^{p}r_{(pk)z}$ - fixed coordinates

4.2. Inverse kinematics problem of link position

The inverse problem handled in the coordinate system associated with the machine frame involves the orientation of the platform p. The platform should be stabilised in the vertical whilst the active suspension system is in use. The platform position is related to the gravity force versor $\mathbf{z}_g^o$ (Fig 4, 5, 6), which can be expressed in the coordinate system associated with the frame according to the formula (1).

In order to solve the inverse problem it is required that the lengths of the actuators s_{1o}, s_{4o} should be established, corresponding to the predetermined and expected platform position with respect to the system associated with the frame and expressed by versor coordinates: $\mathbf{x}_{po}^o$, $\mathbf{y}_{po}^o$ and $\mathbf{z}_{po}^o$. Actually, the cab will reach the position nearing the expected one. The anticipated values (indicated with a subscript "o") obtained from solving the inverse problem will be used to derive the error signal required for the control process. The direction of the cab's vertical axis versor should be opposite to that of the gravity force versor $\mathbf{z}_{po}^o = -\mathbf{z}_g^o$.

The versor $\mathbf{y}_{po}^o$ is obtained basing on $\mathbf{z}_{po}^o$ and $\mathbf{x}_r^o$:

$$ \mathbf{y}_{po}^o = \mathbf{z}_{po}^o \times \mathbf{x}_r^o \tag{20} $$

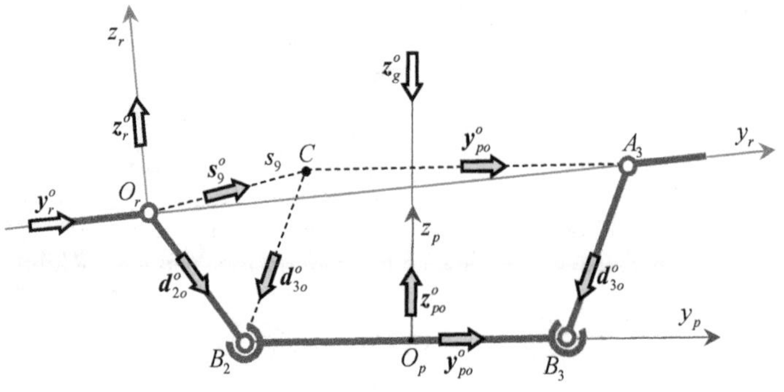

Figure 6. Solving the inverse problem- schematic diagram

The versor $\mathbf{x}_{po}^o$ is obtained basing on $\mathbf{y}_{po}^o$ and $\mathbf{z}_{po}^o$:

$$\mathbf{x}_{po}^o = \mathbf{y}_{po}^o \times \mathbf{z}_{po}^o \tag{21}$$

The modulus of the vector between points O_r and C and its versor are computed using the triangle O_rA_3C (Fig 6):

$$s_9 = \sqrt{a_{23}^2 + b_{23}^2 - 2a_{23}b_{23}(\mathbf{y}_{po}^o \cdot \mathbf{y}_r^o)} \tag{22}$$

$$\mathbf{s}_9^o = \frac{a_{23}\mathbf{y}_r^o - b_{23}\mathbf{y}_{po}^o}{s_9} \tag{23}$$

The versor $\mathbf{d}_{2o}^o$ is obtained basing on $\mathbf{s}_9^o$, s_9, d_2 and d_3:

$$\mathbf{d}_{2o}^o = -\sqrt{1 - \left(\frac{d_2^2 - d_3^2 + s_9^2}{2s_9d_2}\right)^2}\,\mathbf{x}_r^o \times \mathbf{s}_9^o + \frac{d_2^2 - d_3^2 + s_9^2}{2s_9d_2}\mathbf{s}_9^o \tag{24}$$

The versor $\mathbf{c}_{2o}^o$ is obtained basing on $\mathbf{x}_r^o$ and $\mathbf{d}_{2o}^o$:

$$\mathbf{c}_{2o}^o = c_{22}\mathbf{d}_{2o}^o - \sqrt{1 - c_{22}^2}\,(\mathbf{x}_r^o \times \mathbf{d}_{2o}^o) \tag{25}$$

The expected length of the drive 4 and its versor are obtained from the triangle $A_4A_2C_4$:

$$s_{4o} = \sqrt{(\mathbf{s}_{4o} \cdot \mathbf{x}_r^o)^2 + (\mathbf{s}_{4o} \cdot \mathbf{y}_r^o)^2 + (\mathbf{s}_{4o} \cdot \mathbf{z}_r^o)^2}\,, \qquad \mathbf{s}_{4o}^o = \frac{\mathbf{s}_{4o}}{s_{4o}} \tag{26}$$

where: $\mathbf{s}_{4o} = c_2\mathbf{c}_{2o}^o - a_4\mathbf{a}_4^o$ - vector of the drive 4.

The expected length of the actuator and its axis versor are obtained on the basis of a polygon $A_1A_2B_2O_pB_1$:

$$s_{1o} = \sqrt{(\mathbf{s}_{1o} \cdot \mathbf{x}_r^o)^2 + (\mathbf{s}_{1o} \cdot \mathbf{y}_r^o)^2 + (\mathbf{s}_{1o} \cdot \mathbf{z}_r^o)^2}, \qquad \mathbf{s}_{1o}^o = \frac{\mathbf{s}_{1o}}{s_{1o}} \tag{27}$$

where: $\mathbf{s}_{1o} = d_2\mathbf{d}_{2o}^o + b_2\mathbf{y}_{po}^o - a_1\mathbf{a}_1^o - b_1\mathbf{x}_{po}^o$ - vector of the actuator 1.

4.3. Direct kinematics problem of links velocity

In order to solve the simple problem to derive velocity of the active suspension system in the coordinate system associated with the frame it is required that the following vectors have to be determined:

$\boldsymbol{\omega}_{p,r} = \boldsymbol{\omega}_{k,r} = \boldsymbol{\omega}_{f,r}$ - identical angular velocity of the platform p, of the cab k and the operator seat f,

$\mathbf{v}_{O_k,r}$ - linear velocity of the cab's c.o.g Q_k,

$\mathbf{v}_{O_f,r}$ - linear velocity of the c.o.g. in the seat-operator system O_f as functions of cylinders' velocity.

The kinematic chain determined by points $A_4C_4A_2$ (Fig 5) satisfies the closing condition:

$$a_4\mathbf{a}_4^o + s_4\mathbf{s}_4^o = c_2\mathbf{c}_2^o \tag{28}$$

Differentiating Eq (28) over time yields:

$$\dot{s}_4\mathbf{s}_4^o + s_4(\boldsymbol{\omega}_{4,r} \times \mathbf{s}_4^o) = c_2(\boldsymbol{\omega}_{2,r} \times \mathbf{c}_2^o) \tag{29}$$

The versor $\mathbf{x}_r^o$ determines the direction of the angular velocity vector of the rocker arm 2:

$$\boldsymbol{\omega}_{2,r} = \omega_{2,r}\mathbf{x}_r^o \tag{30}$$

Scalar-multiplying Eq (29) by $\mathbf{s}_4^o$ yields:

$$\omega_{2,r} = \frac{\dot{s}_4}{r_2} \tag{31}$$

where: $r_2 = c_2[(\mathbf{x}_r^o \times \mathbf{c}_2^o) \cdot \mathbf{s}_4^o]$.

The kinematic chain determined by points $A_2B_2B_3A_3$ (Fig 5.) satisfies the closing condition:

$$d_2\mathbf{d}_2^o + b_{23}\mathbf{y}_p^o = a_{23}\mathbf{y}_r^o + d_3\mathbf{d}_3^o \tag{32}$$

Differentiating Eq (32) over time yields:

$$d_2(\boldsymbol{\omega}_{2,r} \times \mathbf{d}_2^o) + b_{23}(\boldsymbol{\omega}_{p,r} \times \mathbf{y}_p^o) = d_3(\boldsymbol{\omega}_{3,r} \times \mathbf{d}_3^o) \tag{33}$$

The versor $\mathbf{x}_r^o$ determines the direction of the angular velocity vector of the rocker arm 3:

$$\boldsymbol{\omega}_{3,r} = \omega_{3,r}\mathbf{x}_r^o \tag{34}$$

Substituting (30), (31) and (34) into Eq (33) and scalar-multiplying all vectors by the versor $\mathbf{y}_p^o$, yields the modulus of angular velocity vector $\boldsymbol{\omega}_{3,r}$:

$$\omega_{3,r} = \frac{\dot{s}_4}{r_3} \tag{35}$$

where: $r_3 = \frac{c_2 d_3}{d_2} \frac{[(\mathbf{x}_r^o \times \mathbf{d}_3^o)\cdot \mathbf{y}_p^o][(\mathbf{x}_r^o \times \mathbf{c}_2^o)\cdot \mathbf{s}_4^o]}{(\mathbf{x}_r^o \times \mathbf{d}_2^o)\cdot \mathbf{y}_p^o}$.

The angular velocity vector of the platform p can be expressed as the sum of three components whose axis directions are determined by versors $\mathbf{x}_p^o$, $\mathbf{y}_p^o$ and $\mathbf{z}_p^o$:

$$\boldsymbol{\omega}_{p,r} = \omega_{(p,r)x_p}\mathbf{x}_p^o + \omega_{(p,r)y_p}\mathbf{y}_p^o + \omega_{(p,r)z_p}\mathbf{z}_p^o \tag{36}$$

Recalling Eq (30), (31), (34), (35), (36), Eq (33) can be rewritten as:

$$d_2(\frac{\dot{s}_4}{r_2}\mathbf{x}_r^o \times \mathbf{d}_2^o) + b_{23}\omega_{(p,r)x_p}\mathbf{z}_p^o - b_{23}\omega_{(p,r)z_p}\mathbf{x}_p^o = d_3(\frac{\dot{s}_4}{r_3}\mathbf{x}_r^o \times \mathbf{d}_3^o) \tag{37}$$

Scalar-multiplying Eq (37) by $\mathbf{x}_p^o$ yields:

$$\omega_{(p,r)z_p} = \frac{\dot{s}_4}{r_{pz_p}} \tag{38}$$

where: $r_{pz_p} = \frac{b_{23}}{(\mathbf{x}_p^o \times \mathbf{x}_r^o)\cdot\left(\frac{d_2}{r_2}\mathbf{d}_2^o - \frac{d_3}{r_3}\mathbf{d}_3^o\right)}$.

Scalar-multiplying Eq (37) by $\mathbf{z}_p^o$ yields:

$$\omega_{(p,r)x_p} = \frac{\dot{s}_4}{r_{px_p}} \tag{39}$$

where: $r_{px_p} = \frac{b_{23}}{(\mathbf{z}_p^o \times \mathbf{x}_r^o)\cdot\left(\frac{d_3}{r_3}\mathbf{d}_3^o - \frac{d_2}{r_2}\mathbf{d}_2^o\right)}$.

The closed kinematic chain comprising a actuator s_1 and represented as a pentagon $A_2A_1B_1O_pB_2$ in Fig 5 satisfies the closing condition:

$$a_1\mathbf{a}_1^o + s_1\mathbf{s}_1^o + b_1\mathbf{x}_p^o = d_2\mathbf{d}_2^o + b_2\mathbf{y}_p^o \tag{40}$$

Differentiating Eq (40) over time yields:

$$\dot{s}_1\mathbf{s}_1^o + s_1(\boldsymbol{\omega}_{1,r} \times \mathbf{s}_1^o) + b_1(\boldsymbol{\omega}_{p,r} \times \mathbf{x}_p^o) = d_2(\boldsymbol{\omega}_{2,r} \times \mathbf{d}_2^o) + b_2(\boldsymbol{\omega}_{p,r} \times \mathbf{y}_p^o) \tag{41}$$

Recalling the relationships expressing the angular velocity vector of the platform (Eq 36) and angular velocity vector of the rocker arm 2 (Eq 30), Eq (41) can be rearranged accordingly:

$$\dot{s}_1\mathbf{s}_1^o + s_1(\boldsymbol{\omega}_{1,r} \times \mathbf{s}_1^o) + b_1(\omega_{(p,r)z_p}\mathbf{y}_p^o - \omega_{(p,r)y_p}\mathbf{z}_p^o) = d_2(\frac{\dot{s}_4}{r_2}\mathbf{x}_r^o \times \mathbf{d}_2^o) + b_2(\omega_{(p,r)x_p}\mathbf{z}_p^o - \omega_{(p,r)z_p}\mathbf{x}_p^o) \tag{42}$$

Scalar multiplying vectors present in Eq (42) by $\mathbf{s}_1^o$ yields the coordinate of the angular velocity of the platform in the direction determined by the versor $\mathbf{y}_p^o$:

$$\omega_{(p,r)y_p} = \frac{\dot{s}_1}{r_{py_p1}} + \frac{\dot{s}_4}{r_{py_p4}} \tag{43}$$

where: $r_{py_p1} = b_1(\mathbf{z}_p^o \cdot \mathbf{s}_1^o)$, $r_{py_p4} = \dfrac{b_1\mathbf{z}_p^o \cdot \mathbf{s}_1^o}{\left[\dfrac{b_1}{r_{pz_p}}\mathbf{y}_p^o - \dfrac{d_2}{r_2}(\mathbf{x}_r^o \times \mathbf{d}_2^o) - \dfrac{b_2}{r_{px_p}}\mathbf{z}_p^o + \dfrac{b_2}{r_{pz_p}}\mathbf{x}_p^o\right] \cdot \mathbf{s}_1^o}$.

The angular velocity vector of the platform, based on (36), (39), (43), (38), will become:

$$\boldsymbol{\omega}_{p,r} = \frac{\dot{s}_4}{r_{px_p}}\mathbf{x}_p^o + (\frac{\dot{s}_1}{r_{py_p1}} + \frac{\dot{s}_4}{r_{py_p4}})\mathbf{y}_p^o + \frac{\dot{s}_4}{r_{pz_p}}\mathbf{z}_p^o \tag{44}$$

Finally, the angular velocity vector of the platform, cab and the seat is linearly related to the velocity of actuators s_1 and s_4:

$$\boldsymbol{\omega}_{p,r} = \boldsymbol{\omega}_{k,r} = \boldsymbol{\omega}_{f,r} = \dot{s}_1\mathbf{h}_1 + \dot{s}_4\mathbf{h}_4 \tag{45}$$

where: $\mathbf{h}_1 = \dfrac{\mathbf{y}_p^o}{r_{py_p1}}$, $\mathbf{h}_4 = \dfrac{\mathbf{x}_p^o}{r_{px_p}} + \dfrac{\mathbf{y}_p^o}{r_{py_p4}} + \dfrac{\mathbf{z}_p^o}{r_{p\mathbf{z}_p}}$.

Eq (45) can be written in the matrix format:

$$\boldsymbol{\omega}_{p,r} = \mathbf{J}_\omega\,\dot{\mathbf{s}} \tag{46}$$

where: $\mathbf{J}_\omega = \begin{bmatrix}\mathbf{h}_1 & \mathbf{h}_4 & \mathbf{0}\end{bmatrix}$, $\dot{\mathbf{s}} = \begin{bmatrix}\dot{s}_1 & \dot{s}_4 & \dot{s}_5\end{bmatrix}^T$.

The radius vector of the point O_k - the origin of the reference system associated with the frame (Fig 5), is expressed as:

$$\mathbf{r}_{rk} = d_2\mathbf{d}_2^o + b_2\mathbf{y}_p^o + \mathbf{r}_{pk} \tag{47}$$

where: $\mathbf{r}_{pk} = \mathbf{x}_p^o\, {}^p r_{(pk)x} + \mathbf{y}_p^o\, {}^p r_{(pk)y} + \mathbf{z}_p^o\, {}^p r_{(pk)z}$.

Differentiating Eq (47) over time yields the linear velocity vector of the point O_k:

$$\mathbf{v}_{O_k,r} = d_2(\boldsymbol{\omega}_{2,r} \times \mathbf{d}_2^o) + \boldsymbol{\omega}_{p,r} \times (\mathbf{y}_p^o b_2 + \mathbf{r}_{pk}) \tag{48}$$

Recalling Eq (30), (31), (45), Eq (48) can be rewritten as:

$$\mathbf{v}_{O_k,r} = \dot{s}_1 \mathbf{k}_{1O_k} + \dot{s}_4 \mathbf{k}_{4O_k} \tag{49}$$

where: $\mathbf{k}_{1O_k} = \mathbf{h}_1 \times (b_2 \mathbf{y}_p^o + \mathbf{r}_{pk})$, $\mathbf{k}_{4O_k} = \frac{d_2}{r_2}(\mathbf{x}_r^o \times \mathbf{d}_2^o) + \mathbf{h}_4 \times (b_2 \mathbf{y}_p^o + \mathbf{r}_{pk})$.

Eq (49) can be expressed in the matrix format:

$$\mathbf{v}_{O_k,r} = \mathbf{J}_{vk}\dot{\mathbf{s}} \tag{50}$$

where: $\mathbf{J}_{vk} = \begin{bmatrix} \mathbf{k}_{1O_k} & \mathbf{k}_{4O_k} & \mathbf{0} \end{bmatrix}$.

The radius vector of the c.o.g of the seat-operator system O_f becomes:

$$\mathbf{r}_{rf} = d_2 \mathbf{d}_2^o + b_2 \mathbf{y}_p^o + \mathbf{r}_{pf} \tag{51}$$

where: $\mathbf{r}_{pf} = \mathbf{x}_p^o\, {}^p r_{(pf)x} + \mathbf{y}_p^o\, {}^p r_{(pf)y} + \mathbf{z}_p^o (s_5 + df_z)$.

Differentiating Eq (51) over time yields the linear velocity of the point O_f:

$$\mathbf{v}_{O_f,r} = d_2 \boldsymbol{\omega}_{2,r} \times \mathbf{d}_2^o + \boldsymbol{\omega}_{p,r} \times (\mathbf{y}_p^o b_2 + \mathbf{r}_{pf}) + \dot{s}_5 \mathbf{z}_p^o \tag{52}$$

Recalling Eq (30), (31), (45), Eq (52) can be rewritten as:

$$\mathbf{v}_{O_f,r} = \dot{s}_1 \mathbf{k}_{1O_f} + \dot{s}_4 \mathbf{k}_{4O_f} + \dot{s}_5 \mathbf{k}_{5O_f} \tag{53}$$

where: : $\mathbf{k}_{1O_f} = \mathbf{h}_1 \times (b_2 \mathbf{y}_p^o + \mathbf{r}_{pf})$, $\mathbf{k}_{4O_f} = \frac{d_2}{r_2}(\mathbf{x}_r^o \times \mathbf{d}_2^o) + \mathbf{h}_4 \times (\mathbf{y}_p^o b_2 + \mathbf{r}_{pf})$, $\mathbf{k}_{5O_f} = \mathbf{z}_p^o$.

Eq (53) can be expressed in the matrix format:

$$\mathbf{v}_{O_f,r} = \mathbf{J}_{vf}\dot{\mathbf{s}} \tag{54}$$

where: $\mathbf{J}_{vf} = \begin{bmatrix} \mathbf{k}_{1O_f} & \mathbf{k}_{4O_f} & \mathbf{k}_{5O_f} \end{bmatrix}$.

To define the operating conditions of the drives in the active suspension mechanism, velocity vectors related to the road system are of key importance. The absolute angular velocity of the cab in the reference system associated with the frame becomes:

$$\boldsymbol{\omega}_{p,g} = \boldsymbol{\omega}_{p,r} + \boldsymbol{\omega}_{r,g} = \mathbf{J}_\omega \dot{\mathbf{s}} + \boldsymbol{\omega}_{r,g} \tag{55}$$

where: $\boldsymbol{\omega}_{r,g}$ - angular velocity of the frame with respect to road, based on measurement data.

Absolute linear velocities of points O_k and O_f expressed in the reference system associated with the frame are:

$$\mathbf{v}_{O_k,g} = \mathbf{v}_{O_m,g} + \boldsymbol{\omega}_{r,g} \times (\mathbf{r}_{mr} + \mathbf{r}_{rk}) + \mathbf{v}_{O_k,r} \tag{56}$$

$$\mathbf{v}_{O_f,g} = \mathbf{v}_{O_m,g} + \boldsymbol{\omega}_{r,g} \times (\mathbf{r}_{mr} + \mathbf{r}_{rf}) + \mathbf{v}_{O_f,r} \tag{57}$$

where: $\mathbf{v}_{O_m,g}$ - measured linear velocity of the control point O_m associated with the frame with respect to the road; $\mathbf{r}_{mr} = [r_{(mr)x}, r_{(mr)y}, r_{(mr)z}]^T$ - vector between the points O_m and O_r expressed in the reference system associated with the frame.

4.4. Inverse kinematics problem of links velocity

The inverse problem involves finding the drive velocities for the predetermined cab velocity with respect to the road. As the active suspension mechanism displays three degrees of freedom (DOFs), three constraints can be imposed upon the cab velocity. The function of the active suspension system is to stabilise the cab in the vertical direction, hence the condition is adopted prohibiting the absolute rotating motion of the platform around its two axes $\mathbf{x}_{po}^{o}$ and $\mathbf{y}_{po}^{o}$. The third condition implicates that the absolute value of linear velocity of the point O_f in the direction of the gravity force should be zero:

$$\boldsymbol{\omega}_{(p,g)o} \cdot \mathbf{x}_{po}^{o} = 0 \tag{58}$$

$$\boldsymbol{\omega}_{(p,g)o} \cdot \mathbf{y}_{po}^{o} = 0 \tag{59}$$

$$\mathbf{v}_{(O_f,g)o} \cdot \mathbf{z}_{g}^{o} = 0 \tag{60}$$

Recalling Eq (55) and conditions (58), (59), we get the formulas expressing the expected velocities of actuators 1 and 4:

$$\left.\begin{aligned} \dot{s}_{1o}\mathbf{h}_{1o} \cdot \mathbf{x}_{po}^{o} + \dot{s}_{4o}\mathbf{h}_{4o} \cdot \mathbf{x}_{po}^{o} + \boldsymbol{\omega}_{r,g} \cdot \mathbf{x}_{po}^{o} = 0 \\ \dot{s}_{1o}\mathbf{h}_{1o} \cdot \mathbf{y}_{po}^{o} + \dot{s}_{4o}\mathbf{h}_{4o} \cdot \mathbf{y}_{po}^{o} + \boldsymbol{\omega}_{r,g} \cdot \mathbf{y}_{po}^{o} = 0 \end{aligned}\right\} \tag{61}$$

When the actuators 1 and 4 should move at velocities governed by Eq (61), the cab will perform a slight rotating motion around in the direction $\mathbf{z}_{po}^{o}$ only.

The third condition (60) in relation to Eq (54), (57) gives:

$$\dot{s}_{1o}(\mathbf{k}_{1O_f} \cdot \mathbf{z}_g^o) + \dot{s}_{4o}(\mathbf{k}_{4O_f} \cdot \mathbf{z}_g^o) + \dot{s}_{5o}(\mathbf{k}_{5O_f} \cdot \mathbf{z}_g^o) + \mathbf{v}_{O_m,g} \cdot \mathbf{z}_g^o + [\boldsymbol{\omega}_{r,g} \times (\mathbf{r}_{mr} + \mathbf{r}_{(rf)o})] \cdot \mathbf{z}_g^o = 0 \tag{62}$$

The solution to linear system of equations (61) and (62) can be written as a matrix equation:

$$\dot{\mathbf{s}}_o = \mathbf{J}\mathbf{v} \tag{63}$$

$$\text{where: } \dot{\mathbf{s}}_o = \begin{bmatrix} \dot{s}_{1o} \\ \dot{s}_{4o} \\ \dot{s}_{5o} \end{bmatrix}, \ \mathbf{J} = -\begin{bmatrix} \mathbf{h}_{1o} \cdot \mathbf{x}_{po}^o & \mathbf{h}_{4o} \cdot \mathbf{x}_{po}^o & 0 \\ \mathbf{h}_{1o} \cdot \mathbf{y}_{po}^o & \mathbf{h}_{4o} \cdot \mathbf{y}_{po}^o & 0 \\ \mathbf{k}_{1O_f} \cdot \mathbf{z}_g^o & \mathbf{k}_{4O_f} \cdot \mathbf{z}_g^o & \mathbf{k}_{5O_f} \cdot \mathbf{z}_g^o \end{bmatrix}^{-1} \begin{bmatrix} \mathbf{0}_{1\times3} & \mathbf{x}_{po}^o \\ \mathbf{0}_{1\times3} & \mathbf{y}_{po}^o \\ \mathbf{z}_g^o & (\mathbf{r}_{mr} + \mathbf{r}_{rf}) \times \mathbf{z}_g^o \end{bmatrix}, \ \mathbf{v} = \begin{bmatrix} \mathbf{v}_{O_m,g} \\ \boldsymbol{\omega}_{r,g} \end{bmatrix}.$$

4.5. Constraining the motion of the drive responsible for the seat movement in the vertical direction

The machine, when in service or during the ride, may change its position in the vertical direction such that in order to stabilise the seat position in this direction the operating range of the actuator 5 should be exceeded. To solve this problem, it is suggested that a penalty function should be introduced, its argument being the instantaneous length of the actuator s_5:

$$\dot{s}_{5k} = K_5 \dot{s}_{5k\max} \left[\frac{s_{5śr} - s_5}{0.5(s_{5\max} - s_{5\min})} \right]^{2n-1} \tag{64}$$

where: $s_{5śr} = 0.5(s_{5\max} + s_{5\min})$, K_5 - amplification factor penalty function $\dot{s}_{5k\max}$ - maximal velocity of the actuator, $n \in N$.

The final expected velocity of the actuator 5 should involve a term responsible for the seat "drifting" towards the middle position:

$$\dot{s}_{5o}^{*} = \dot{s}_{5o} + \dot{s}_{5k} \tag{65}$$

where: $\dot{s}_{5o}$ - expected velocity in the actuator 5, derived from formula (63).

The second term in (65) represents the seat movement towards the middle position being superimposed on its relative movement. The assumed penalty function (64) guarantees that the relative velocity during the seat's return movement (Fig 7) to the middle position should be significant at extreme points of the actuator's displacement range.

4.6. Cab and seat acceleration

In order to solve the simple problem involving acceleration of the active suspension mechanism in the reference system associated with the frame, it is required that certain quantities should be determined in the function of length, velocity and acceleration of actuators 1,4,5. These include: $\boldsymbol{\varepsilon}_{p,r}$ - angular acceleration of the platform, cab and seat, $\mathbf{a}_{O_k,r}$ - linear acceleration of the cab's c.o.g, $\mathbf{a}_{O_f,r}$ - linear acceleration of the c.o.g of the seat-operator system. Differentiating Eq (46), (50), (54) with respect to time yields the angular acceleration of the platform, cab and seat and linear acceleration at points O_k and O_f:

$$\boldsymbol{\varepsilon}_{p,r} = \dot{\mathbf{J}}_{\omega} \dot{\mathbf{s}} + \mathbf{J}_{\omega} \ddot{\mathbf{s}} \tag{66}$$

$$\mathbf{a}_{O_k,r} = \dot{\mathbf{J}}_{vk} \dot{\mathbf{s}} + \mathbf{J}_{vk} \ddot{\mathbf{s}} \tag{67}$$

$$\mathbf{a}_{O_f,r} = \dot{\mathbf{J}}_{vf}\,\dot{\mathbf{s}} + \mathbf{J}_{vf}\ddot{\mathbf{s}} \tag{68}$$

where: $\ddot{\mathbf{s}} = \left[\ddot{\mathbf{s}}_1 \;\; \ddot{\mathbf{s}}_4 \;\; \ddot{\mathbf{s}}_5\right]^T$.

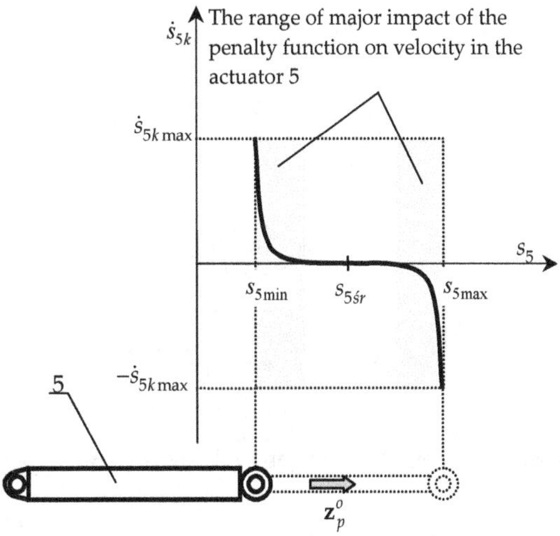

Figure 7. Penalty function

The inertia loads are determined basing on absolute acceleration values related to the inertial reference system $\{O_g x_g y_g z_g\}$. In accordance with Eq (55), the absolute angular accelerations of the platform, cab and seat expressed in the mobile reference system associated with the frame become:

$$\boldsymbol{\varepsilon}_{p,g} = \boldsymbol{\varepsilon}_{k,g} = \boldsymbol{\varepsilon}_{f,g} = \boldsymbol{\varepsilon}_{r,g} + \boldsymbol{\varepsilon}_{p,r} + \boldsymbol{\omega}_{r,g} \times \boldsymbol{\omega}_{p,r} \tag{69}$$

Recalling Eq (56), (57), the absolute linear acceleration of the points O_r, O_k and O_f in the mobile reference system associated with the frame become:

$$\mathbf{a}_{O_r,g} = \mathbf{a}_{O_m,g} + \boldsymbol{\varepsilon}_{r,g} \times \mathbf{r}_{mr} + \boldsymbol{\omega}_{r,g} \times (\boldsymbol{\omega}_{r,g} \times \mathbf{r}_{mr}) \tag{70}$$

$$\mathbf{a}_{O_k,g} = \mathbf{a}_{O_r,g} + \boldsymbol{\varepsilon}_{r,g} \times \mathbf{r}_{rk} + 2\boldsymbol{\omega}_{r,g} \times \mathbf{v}_{O_k,r} + \boldsymbol{\omega}_{r,g} \times (\boldsymbol{\omega}_{r,g} \times \mathbf{r}_{rk}) + \mathbf{a}_{O_k,r} \tag{71}$$

$$\mathbf{a}_{O_f,g} = \mathbf{a}_{O_k,g} + \boldsymbol{\varepsilon}_{k,g} \times \mathbf{r}_{kf} + 2\boldsymbol{\omega}_{k,g} \times \mathbf{v}_{O_f,k} + \boldsymbol{\omega}_{k,g} \times (\boldsymbol{\omega}_{k,g} \times \mathbf{r}_{kf}) + \mathbf{a}_{O_f,k} \tag{72}$$

where: $\mathbf{a}_{O_m,g}$ - measured linear acceleration of a point on the frame Q_m with respect to the inertial reference system $\{O_g x_g y_g z_g\}$, $\boldsymbol{\omega}_{r,g}$, $\boldsymbol{\varepsilon}_{r,g}$ - measured angular velocity and acceleration of the machine frame with respect to the inertial reference system $\{O_g x_g y_g z_g\}$, $\boldsymbol{\omega}_{k,g} = \boldsymbol{\omega}_{p,g} = \boldsymbol{\omega}_{r,g} + \boldsymbol{\omega}_{p,r}$, $\mathbf{r}_{kf} = \overrightarrow{O_k O_f}$, $\mathbf{r}_{rk} = \overrightarrow{O_r O_k}$, $\mathbf{r}_{mr} = \overrightarrow{O_m O_r}$.

5. Inverse problem of dynamics

The external loads acting on the active suspension mechanism involve the gravity forces, inertia forces and moments of inertial force of the platform together with the cab, the seat and the operator. These are governed by the Newton-Euler equations, referenced in [18]:

$$\mathbf{P}_{bk} = -m_k \mathbf{a}_{O_k,g} \tag{73}$$

$$\mathbf{M}_{bk} = -\boldsymbol{\varepsilon}_{k,g} \mathbf{I}_k - \tilde{\boldsymbol{\omega}}_{k,g} \mathbf{I}_k \boldsymbol{\omega}_{k,g} \tag{74}$$

$$\mathbf{P}_{bf} = -m_f \mathbf{a}_{O_f,g} \tag{75}$$

$$\mathbf{M}_{bf} = -\boldsymbol{\varepsilon}_{f,g} \mathbf{I}_f - \tilde{\boldsymbol{\omega}}_{f,g} \mathbf{I}_f \boldsymbol{\omega}_{f,g} \tag{76}$$

where: $\tilde{\boldsymbol{\omega}} = \begin{bmatrix} 0 & -\omega_z & \omega_y \\ \omega_z & 0 & -\omega_x \\ -\omega_y & \omega_x & 0 \end{bmatrix}$, $\mathbf{I}_k = {}^r_p\mathbf{R}\, {}^k\mathbf{I}_k\, {}^r_p\mathbf{R}^T$, $\mathbf{I}_f = {}^r_p\mathbf{R}\, {}^f\mathbf{I}_f\, {}^r_p\mathbf{R}^T$,

${}^k\mathbf{I}_k$, ${}^f\mathbf{I}_f$ - mass moments of inertia of the cab and the seat with operator in their own reference systems.

The sum total of instantaneous power applied by the active suspension mechanism and power of the gravity and inertia forces are brought down to zero:

$$\begin{aligned} &\dot{\mathbf{s}}^T \mathbf{F}_s + m_k \mathbf{v}^T_{O_k,r}(\mathbf{g} - \mathbf{a}_{O_k,g}) + \boldsymbol{\omega}^T_{k,r}(-\mathbf{I}_k \boldsymbol{\varepsilon}_{k,g} - \tilde{\boldsymbol{\omega}}_{k,g} \mathbf{I}_k \boldsymbol{\omega}_{k,g}) + \\ &+ m_f \mathbf{v}^T_{O_f,r}(\mathbf{g} - \mathbf{a}_{O_f,g}) + \boldsymbol{\omega}^T_{f,r}(-\mathbf{I}_f \boldsymbol{\varepsilon}_{f,g} - \tilde{\boldsymbol{\omega}}_{f,g} \mathbf{I}_f \boldsymbol{\omega}_{f,g}) = 0 \end{aligned} \tag{77}$$

where: $\mathbf{F}_s = [F_1, F_4, F_5]^T$ - forces developed by the drives.

Recalling Jacobean matrices (46), (50), (54), Eq (77) can be rewritten as:

$$\begin{aligned} &\dot{\mathbf{s}}^T [\mathbf{F}_s + m_k \mathbf{J}^T_{vk}(\mathbf{g} - \mathbf{a}_{O_k,g}) + \mathbf{J}^T_\omega(-\mathbf{I}_k \boldsymbol{\varepsilon}_{k,g} - \tilde{\boldsymbol{\omega}}_{k,g} \mathbf{I}_k \boldsymbol{\omega}_{k,g}) + \\ &+ m_f \mathbf{J}^T_{vf}(\mathbf{g} - \mathbf{a}_{O_f,g}) + \mathbf{J}^T_\omega(-\mathbf{I}_f \boldsymbol{\varepsilon}_{f,g} - \tilde{\boldsymbol{\omega}}_{f,g} \mathbf{I}_f \boldsymbol{\omega}_{f,g})] = 0 \end{aligned} \tag{78}$$

Knowing the loads due to gravity and inertia, Eq (78) yields the forces acting in the drives:

$$\mathbf{F}_s = m_k \mathbf{J}^T_{vk}(\mathbf{a}_{O_k,g} - \mathbf{g}) + m_f \mathbf{J}^T_{vf}(\mathbf{a}_{O_f,g} - \mathbf{g}) + \mathbf{J}^T_\omega\{(\mathbf{I}_k + \mathbf{I}_f)\boldsymbol{\varepsilon}_{k,g} + [\tilde{\boldsymbol{\omega}}_{k,g}(\mathbf{I}_f + \mathbf{I}_k)\boldsymbol{\omega}_{k,g}]\} \tag{79}$$

6. Simulation of the active suspension system

The operation of the active suspension system is investigated using two mutually supportive programmes. MSC visualNastran 4D is used to develop the model of the input inducing the machine motion, of the machine suspension, the active suspension mechanism for the cab and the seat. All these modelled elements are simplified (Fig 8).

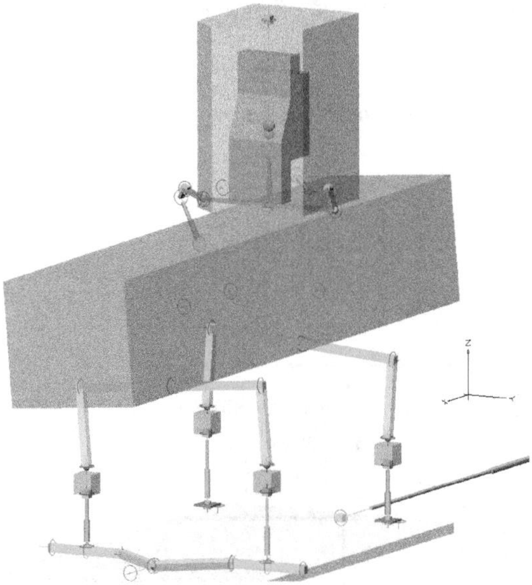

Figure 8. Model of the road input, machine suspension mechanism, active suspension of the cab and seat developed in MSC visualNASTRAN 4D

The programme enables the measurements of the actuator length, the angle of frame tilting α_x and pitching α_y, velocity and acceleration O_m and of velocity and acceleration of the cab's angular motion. These are shown in the block diagram "Measurements of the machine frame movements". During the simulation procedure, these quantities are sent to be further handled by Matlab/Simulink (Fig 9).

The proposed control strategy to be applied to the active suspension of the cab uses the feedback control system with compensation for the measured disturbances in the form of the machine frame movements. The expected states of the cab motion, determined in the block "Preset cab motion" involve the requirement whereby the cab is to be stabilised in the vertical direction and the seat must not be displaced along the cab's vertical axis, at the same time the operating range of the actuator 5 should be duly taken into account. Once frame movements are known from measurements and assumptions as to the anticipated cab movements being taken into account, an unambiguous procedure is applied to compute drive movements in the active suspension mechanism. On the output from the block "Inverse problem of kinematics of the active suspension mechanism" we get the expected velocities and accelerations of three drives, represented by vectors $\mathbf{s}_o = [s_{1o} \quad s_{4o} \quad s_{5o}]^T$, $\dot{\mathbf{s}}_o = [\dot{s}_{1o} \quad \dot{s}_{4o} \quad \dot{s}_{5o}]^T$. Actuators should be equipped with sensors for measuring their actual lengths $\mathbf{s} = [s_1 \quad s_4 \quad s_5]^T$ in order to determine the control error

$\mathbf{e} = \begin{bmatrix} s_{1o} - s_1 & s_{4o} - s_4 & s_{5o} - s_5 \end{bmatrix}^T$. The control error should tend to zero if the velocities implemented in actuators are in accordance with the formula:

$$\dot{\mathbf{s}}_w = \dot{\mathbf{s}}_o + \mathbf{K}_P \mathbf{e} \tag{80}$$

where: $\mathbf{K}_P = \begin{bmatrix} k_{p1} & 0 & 0 \\ 0 & k_{p4} & 0 \\ 0 & 0 & k_{p5} \end{bmatrix}$ - gain matrix in the position path.

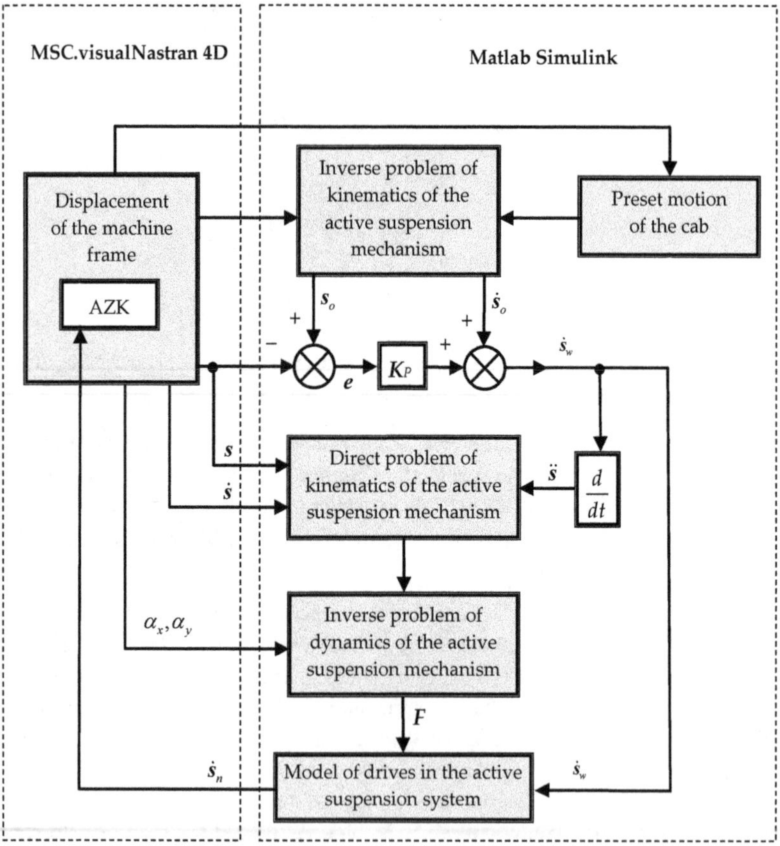

Figure 9. Model of drives control in the active suspension system - schematic diagram

Computed accelerations $\ddot{\mathbf{s}}$ and measured velocities $\dot{\mathbf{s}}$ and displacements $\mathbf{s}$ of the drives become the inputs to the block "Direct problem of kinematics of the active suspension mechanism problem of kinematics of the active suspension", which calculates the anticipated movement of the active suspension mechanism of the cab and of the cab itself. Basing on anticipated cab movements, inertia interactions are found which, alongside the gravity forces, become the major loads applied to the cab. The inverse problem of the active suspension dynamics involves the calculation of the driving forces in the form of a vector $\mathbf{F} = [F_1 \quad F_4 \quad F_5]^T$, counterbalancing the external loads. The contribution of gravity forces to the load of particular drives depends on the frame tilt angles: α_x, α_y. Basing on computed loads F and the required instantaneous velocities $\dot{\mathbf{s}}_w$, the block " Model of active suspension drives" generates the realisable instantaneous velocities of actuators $\dot{\mathbf{s}}_n = \begin{bmatrix} \dot{s}_{n1} & \dot{s}_{n4} & \dot{s}_{n5} \end{bmatrix}^T$. Velocity values $\dot{\mathbf{s}}_n$ are sent to be further processed by MSCvisualNASTRAN 4D. This work does not include the analysis of the drive model. It is assumed in simulations $\dot{\mathbf{s}}_n = \dot{\mathbf{s}}_w$.

The control of the active suspension system gives rise to certain errors ***e***, and in consequence the constraints imposed on the angular velocity of the cab and linear velocity of the cab and the seat cannot be accurately reproduced. These errors are attributable to inaccurate measurements of the frame movements, the time delay involved in implementation of the drive velocity or the drives' failure to implement the required velocity (moving beyond the limits of their typical operating range).

7. Dependings of link dimensions of the active suspension cab mechanism

It is demonstrated in [19] that dimensions of key parts of the mechanism $A_2A_3B_3B_2$ can be chosen such that the instantaneous centre of the platform rotation with respect to the machine frame C_{pr} should be included in the road unevenness path when the active suspension system is on. When this condition is satisfied, the actuator 4, controlled in accordance with the cab vertical stabilisation requirement, will at the same time reduce the absolute movement of the platform in the direction transverse to the ride.

Assuming the central position of the cab on the machine frame, the points A_2 and A_3 should be arranged symmetrically with respect to the frame's longitudinal axis and the lengths of the rocker arms 2 and 3 should be identical $d_2 = d_3 = d$.

The road unevenness range and a typical crescent - shaped field of instantaneous centres of the platform rotation with respect to the frame C_{pr} are shown in Fig 10. It is assumed that when the machine frame is in a horizontal position, C_{pr} is found on the line of wheel-ground contact. The distance of joints in the rocker arm connections A_2 and A_3 from the ground is h_m. The relevant dimensions are related as follows:

$$h^2 + \left(\frac{a_3 - b_{23}}{2}\right)^2 = d^2, \qquad \frac{h_m}{a_3} = \frac{h_m - h}{b_{23}} \tag{81}$$

where: h - distance between the platform and the points of joints A_2 and A_3 for the frame in the horizontal position.

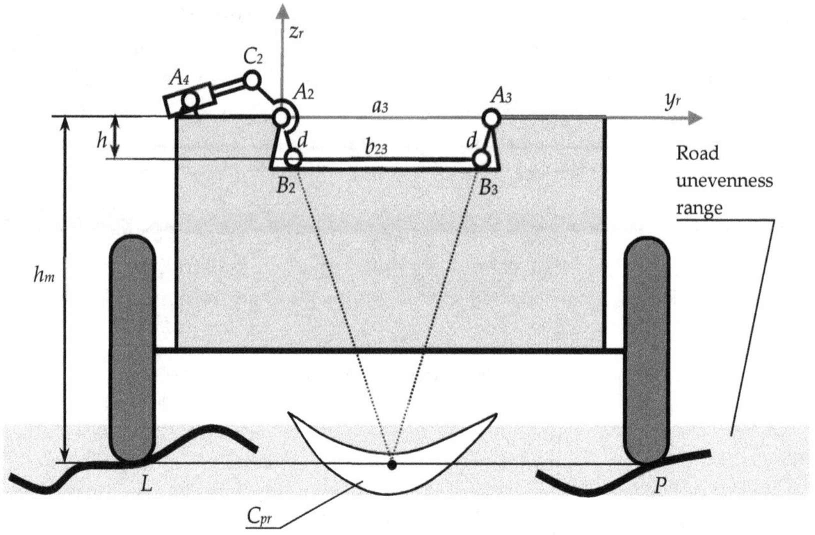

Figure 10. Distribution of the field made of points C_{pr}

Eliminating h from Eq (81) yields the relationship between the dimension of the four bar linkage $A_2A_3B_3B_2$ and h_m. When this condition is satisfied, C_{pr} is found in the road unevenness path:

$$h_m\left(1-\frac{b_{23}}{a_3}\right)=\sqrt{d^2-\left(\frac{a_3-b_{23}}{2}\right)^2} \tag{82}$$

Displacements of the four bar linkage in the active suspension mechanism are constrained by the occurrence of singular positions. The mechanism should not come near the singular position, when controllability of the system deteriorates and the loads acting upon the drives and mobile connections tend to increase. For the predetermined maximal height of the road unevenness range $2h_g$ and for the machine wheel spacing w_m, the maximal angle $\alpha_{\max}$ a of the machine tilting with respect to the axis y_g should be such that the four bar linkage should not assume a singular position (Fig 9):

$$\alpha_{x\max}<\varphi \;\rightarrow\; \arcsin\frac{2h_g}{w_m}<\arccos\frac{a_3^2+(b_{23}+d)^2-d^2}{2a_3(b_{23}+d)} \tag{83}$$

When the dimensions of links in the four bar linkage $A_2A_3B_3B_2$ as well as h_g satisfy the condition (83), the mechanism is able to operate in a single configuration.

Another geometric condition stems from the assumption that the cab can move freely without colliding with the joints A_2, A_3, when the machine assumes its extreme position due to tilting by the angle $\alpha_{x\max}$ (Fig 9):

$$a_3 \cos\alpha_{x\max} > \frac{b_{23}}{2} + \frac{w_k}{2} + \delta_w + d \tag{84}$$

where: w_k - cab width, $w_k < b_{23}$, δ_w - allowable distance between cab walls and the point of the joint A_2 or A_3.

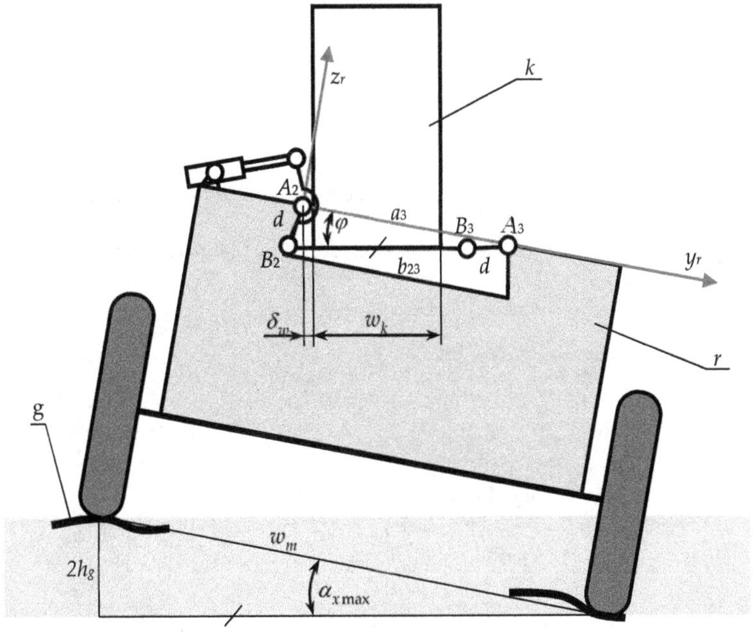

Figure 11. Frame tilt angle in relation to the singular position of the four bar linkage

Satisfying the inequality (84) quarantines a fail-safe operation of the four bar linkage $A_2A_3B_3B_2$ and of the cab. Conditions (82), (83), (84) yield the dimensions: a_3, b_2, d. When the machine is operated in uneven terrain where $2h_g$ exceeds the predetermined value, the active suspension mechanism can reach the limits of its working field and the cab will momentarily deviate from the vertical direction.

8. Simulation data of the active suspension system

Machine specification data used in simulations:

$l_m = 2.810\,[m]$ - distance between the front and rear axle of the machine frame, $w_m = 1.980\,[m]$ - wheel spacing, w_r =1.4 [m] - frame width, $\mathbf{r}_{PO_r} = [-2.1, -0.818, 1.6]\,[m]$ - position vector of the point O_r, $w_k = 1.200\,[m]$ - cab width, ${}^p r_{(pk)x} = 0.000\,[m]$, ${}^p r_{(pk)y} = 0.000\,[m]$, ${}^p r_{(pk)z} = 0.685\,[m]$

${}^{p}r_{(pf)x} = 0{,}000\,[m]$, ${}^{p}r_{(pf)y} = 0{,}000\,[m]$, $df_z = 0.335\,[m]$, m_k=480 $[kg]$ - cab mass, m_f=160 $[kg]$ - mass of the seat with an operator, ${}^{k}J_k = \begin{bmatrix} 180 & 0 & 0 \\ 0 & 208 & 0 \\ 0 & 0 & 133 \end{bmatrix} [kg\,m^2]$ - inertia matrix of the cab in the reference system associated with the cab, ${}^{f}J_f = \begin{bmatrix} 23.8 & 0 & 0 \\ 0 & 24.7 & 0 \\ 0 & 0 & 13.2 \end{bmatrix} [kg\,m^2]$ - inertia matrix of the seat and operator in the system associated with the seat,

Road profile:

$2h_g$= 0.250 $[m]$ - height of the unevenness range, L_g= 2$[m]$ - wave length of the road unevenness, $\varphi = \pi / 2\,[rad]$ - the phase shift angle between the left-and right-hand side of the machine, ${}^{h}v_{Px,\max} = \frac{L_g}{2\pi}\sqrt{\frac{g}{h_g}} = 2.82\,[m/s]$ maximal speed of the machine ride computed for the free wheel in contact with the road surface.

Active suspension mechanism for the cab:

$h_m = 2.420\,[m]$ - distance of joints in the rocker arm connections A_2 and A_3 from the ground, $\delta_w = 0.05\,[m]$ - admissible distance between the cab's side wall from the joint axis A_2 or A_3, $a_3 = 1.636\,[m]$, $d = d_3 = d = 0.227\,[m]$, $b_{23} = 1.490\,[m]$, $c_2 = 0.099\,[m]$, $\measuredangle(\mathbf{c}_2^o, \mathbf{d}_2^o) = 4.5606\,[rad]$ $a_4 = 0.541\,[m]$, $\mathbf{a}_4^o = [0.0000, 0.2181, -0.9759]\,[m]$, $a_1 = 1.182\,[m]$, $\mathbf{a}_1^o = [-0.7218, 0.6921, 0.0000]\,[m]$, $b_2 = b_3 = 0.5\,b_{23}$, $b_1 = 0.850\,[m]$, $s_{5\min} = 0.415\,[m]$, $s_{5\max} = 0.715\,[m]$, $\dot{s}_{5k\max} = 1\,[m/s]$, $K_5 = 50\,[-]$.

$\mathbf{K}_P = \begin{bmatrix} 20 & 0 & 0 \\ 0 & 20 & 0 \\ 0 & 0 & 0 \end{bmatrix} [1/s]$ - matrix of gain in the position path. The distance covered during the simulation - 10$[m]$. The time step in the simulation procedure - 0.005 $[s]$.

Simulation data relating to the cab's and machine frame angular motion are given in Fig 12 and 13, each showing two plots of one angular velocity component in the function of time.

Simulation data relevant to the linear movement of the point O_f on the seat are given in Fig 14, showing the plots of vibration reduction factors for the three components of the rms acceleration derived from the formula:

$$\mu_l = \sqrt{\int_0^{T_s} a_{f,off,l}^2 dt \Big/ \int_0^{T_s} a_{f,on,l}^2 dt} \tag{85}$$

where: l = (x_g, y_g, z_g), $a_{f,off,l}$ - linear acceleration in the direction l, the active suspension system being off, $a_{f,on,l}$ - linear acceleration in the direction l, the active suspension system being on; T_s - simulation time associated with the ride velocity.

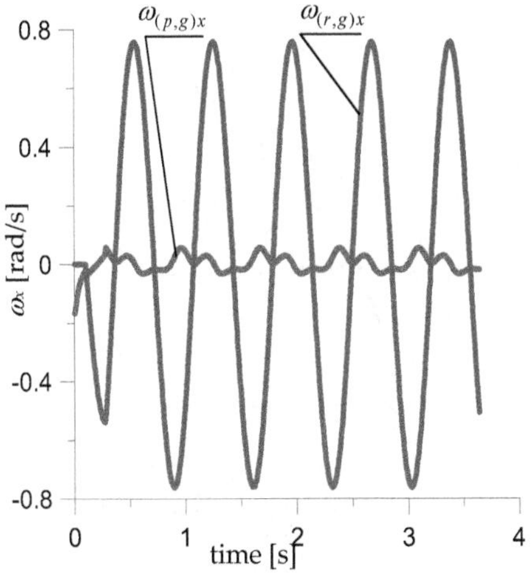

Figure 12. Coordinate x_r of angular velocity of the frame and platform in the function of time, expressed in the system associated with the frame.

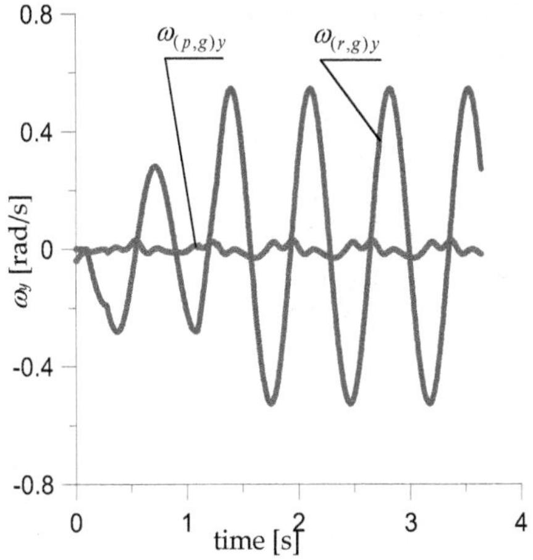

Figure 13. Coordinate y_r of angular velocity of the frame and platform in the function of time, expressed in the system associated with the frame.

The mean power expended by the drives (shown in Fig 15) is derived from the formula:

$$N_{med,j} = \frac{\int_0^{T_s} N_j^+ dt}{T_s} \tag{86}$$

where: $j = (1, 4, 5)$, N_j^+ - instantaneous positive power expended by the drive j.

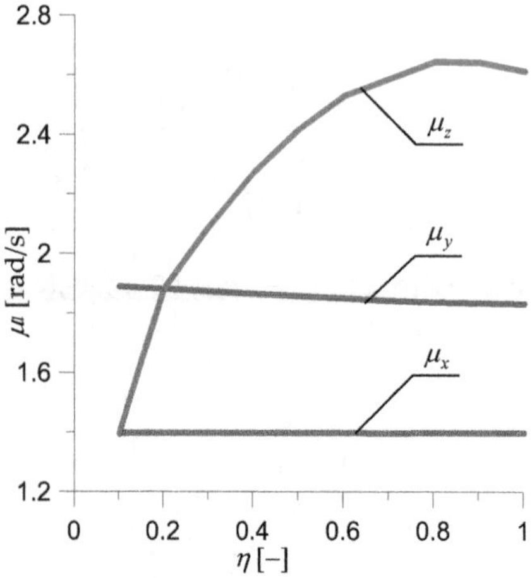

Figure 14. Vibration reduction of the point O_f on the operator seat, in the directions x_r, y_r, z_r

Plots in Fig 14 and 15 show the relevant parameters in the function of the coefficient η linearly related to the machine ride velocity, whilst for ${}^h v_{Px} = {}^h v_{Px,\max}$ the value of η becomes 1.

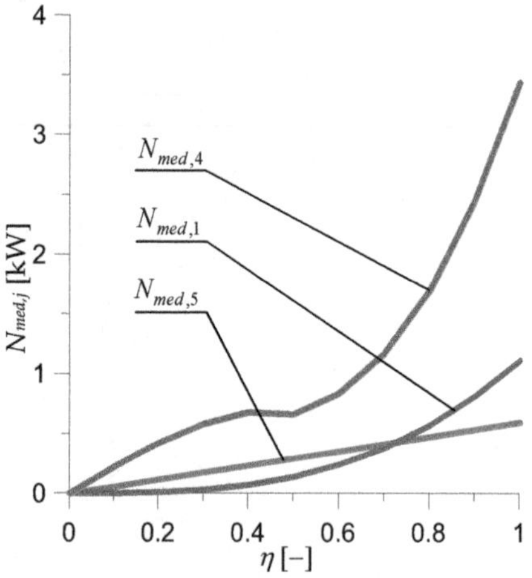

Figure 15. Mean power ratings of the drives 1, 4, 5

9. Conclusions

Simulations of the active suspension system performance have proved its adequacy in vibration reduction of angular vibrations of the cab around the longitudinal axis of the machine x_r and around the transverse axis y_r. Seat vibrations along the vertical axis z_r are successfully controlled, too. The applied procedure of dimension synthesis of the active suspension mechanism links leads to significant reduction of the cab and seat vibration in the direction y_r. Reduction of angular cab vibration around y_r leads to reduction of linear seat vibrations in the direction x_r.

The operation of the active suspension system involves the real-time measurements of mechanical quantities which can be accurately measured with state-of-the-art sensors: angular velocity and acceleration of the frame, linear acceleration of a selected point on the machine frame, two angles of the frame tilting from the direction of the gravity forces and the length and velocities implemented by actuators.

Underlying the simulation procedure is that assumption that each computed drive velocity will be implemented without any time delay (provided that is allowed by collaborating programmes). Results therefore can be utilised when selecting drives which, when in extreme conditions, may not be able to perform the required movements. Besides, the overall time constant, taking into account the response time of the measurement system, the controls and drives becomes another limiting factor, particularly at higher frequencies of road input.

The actuator 4 handles two DOFs (i.e. the cab rotation around the x_r axis and its translation along the y_r-axis) and induces slight movements of the cab in the direction of the z_r-axis, hence its power demand is higher than in drive 1.

Author details

Grzegorz Tora
Institute of Machine Design, Faculty of Mechanical Engineering, Cracow University of Technology, Cracow, Poland

Acknowledgement

I am particularly indebted to the Institute of Machine Design at Cracow Polytechnic for financial support needed to prepare this chapter and have it published in the book Vibration Control.

10. References

[1] Achen A, Toscano J, Marjoram R, StClair K, McMahon B, Goelz A, Shutto S (2008) Semiactive vehicle cab suspension using magnetorheological (mr) technology. Proc. of the 7th JFPS Int. Symp. on Fluid Power, TOYAMA, 561-564.

[2] Jonasson M, Roos F (2008) Design and evaluation of an active electromechanical wheel suspension system, Mechatronics, 18: 218–230.

[3] Al Sayed B, Chatelet E, Baguet S, Jacquet-Richardet G (2011) Dissipated energy and boundary condition effects associated to dry friction on the dynamics of vibrating structures, Mechanism and Machine Theory 46: 479–491.

[4] Sampaio J V R (2009) Design of a Low Power Active Truck Cab Suspension, Eindhoven University of Technology, DCT № 119.

[5] Du H, Lam J, Sze K Y (2003) Non-fragile output feedback H_∞ vehicle suspension control using genetic algorithm, Engineering Applications of Artificial Intelligence 16: 667–680.

[6] Duke M, Goss G, (2007) Investigation of Tractor Driver Seat Performance with Non-linear Stiffness and On–off Damper, Biosystems Engineering 96 (4): 477–486.

[7] Jarviluoma M, Nevala K (1997) An Active Vibration Damping System of a Driver's Seat for Off-Road Vehicles, IEEE, 3: 38-43.

[8] Ruotsalainen P,Nevala K, Marjanen Y (2006) Design of an adjustable hydropneumatic damper for cab suspension, The XIII International Congress on Sound and Vibration, Vienna, July.

[9] Savkoor A, Manders S, Riva P (2001) Design of actively controlled aerodynamic devices for reducing pitch and heave of truck cabins, JSAE Review 22: 421–434.

[10] He Y, McPhee J (2005) Multidisciplinary design optimization of mechatronic vehicles with active suspensions, Journal of Sound and Vibration 283: 217–241.

[11] Yildirim S (2004) Vibration control of suspension systems using a proposed neural Network, Journal of Sound and Vibration 277: 1059–1069.

[12] Liu H, Nonami K, Hagiwara T (2008) Active following fuzzy output feedback sliding mode control of real-vehicle semi-active suspensions, Journal of Sound and Vibration 314: 39–52.

[13] Graf Ch, Maas J, Pflug H-Ch (2009) Concept for an Active Cabin Suspension, Proc. of the 2009 IEEE International Conference on Mechatronics. Malaga, Spain, April.

[14] Akcay H, Turkay S (2009) Influence of tire damping on mixed H_2 / H_∞ synthesis of half-car active suspensions, Journal of Sound and Vibration 322: 15–28.

[15] Koch G, Fritsch O, Lohmann B (2010) Potential of low bandwidth active suspension control with continuously variable damper, Control Engineering Practice 18: 1251–1262.

[16] Du H, Zhang N, Lam J (2008) Parameter-dependent input-delayed control of uncertain vehicle suspension, Journal of Sound and Vibration 317: 537–556.

[17] Marzbanrad J, Ahmadi G, Zohoor H, Hojjat Y (2004) Stochastic optimal previewcontrol of a vehicle suspension, Journal of Sound and Vibration 275, 973–990.

[18] Dasgupta B, Choudhury P (1999) A general strategy based on the Newton-Euler approach for the dynamic formulation of parallel manipulators, Mechanism and Machine Theory 34: 801-824.

[19] Tora G (2008) Kinematyka mechanizmu platformowego w układzie aktywnej redukcji drgań, XXI Ogólnopolska Konferencja Naukowo-Dydaktyczna Teorii Maszyn i Mechanizmów, Wydawnictwo ATH w Bielsku Białej ISBN 978-83-60714-57-7: 357-366.

Section 2

Vibration Control Case Studies

Chapter 6

A Computational Approach to Vibration Control of Vehicle Engine-Body Systems

Hamid Reza Karimi

Additional information is available at the end of the chapter

1. Introduction

In recent years, the noise and vibration of cars have become increasingly important [20, 23, 29, 30, 35]. A major comfort aspect is the transmission of engine-induced vibrations through powertrain mounts into the chassis (see Figure 1). Engine and powertrain mounts are usually designed according to criteria that incorporate a trade-off between the isolation of the engine from the chassis and the restriction of engine movements. The engine mount is an efficient passive means to isolate the car chassis structure from the engine vibration. However, the passive means for isolation is efficient only in the high frequency range. However the vibration disturbance generated by the engine occurs mainly in the low frequency range [8, 19, 23, 30]. These vibrations are result of the fuel explosion in the cylinder and the rotation of the different parts of the engine (see Figure 2). In order to attenuate the low frequency disturbances of the engine vibration while keeping the space and price constant, active vibration means are necessary.

A variety of control techniques, such as Proportional-Integral-Derivative (PID) or Lead-Lag compensation, Linear Quadratic Gaussian (LQG), H_2, H_∞, μ-synthesis and feedforward control have been used in active vibration systems [1, 3, 4, 10, 11, 15, 24, 26, 31, 32, 34, 35]. The main characteristic of feedforward control is that information about the disturbance source is available and is usually realised with the Filtered-X Least-Mean-Squares (Fx-LMS) algorithms. However, the disturbance source is assumed to be unknown in feedback control, then different strategies of feedback control for vibration attenuation of unknown disturbance exist ranging from classical methods to a more advanced methods. Recently, the performance result obtained by H_∞ feedback controller with the result obtained by feedforward controller using Fx-LMS algorithms for vehicle engine-body vibration system was compared in [30, 35].

On the other hand, wavelet theory is a relatively new and an emerging area in mathematical research [2]. It has been applied in a wide range of engineering disciplines such as signal processing, pattern recognition and computational graphics. Recently, some of the attempts are made in solving surface integral equations, improving the finite difference time domain method, solving linear differential equations and nonlinear partial differential equations and modelling nonlinear semiconductor devices [5, 6, 7, 13, 16, 17, 18, 21, 27].

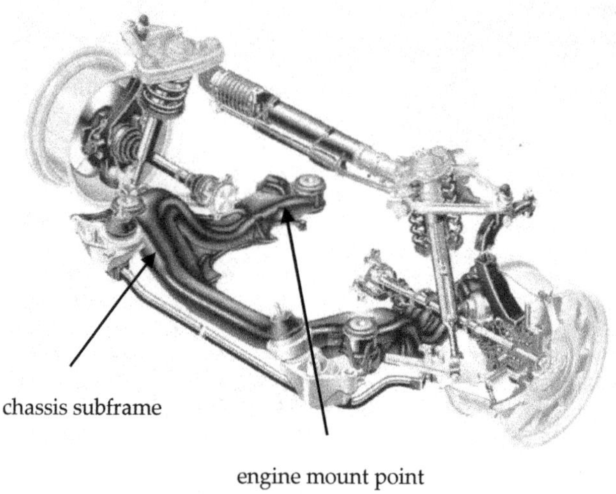

Figure 1. Front axis of AUDI A 8 from [22, 30] (Werkbild Audi AG).

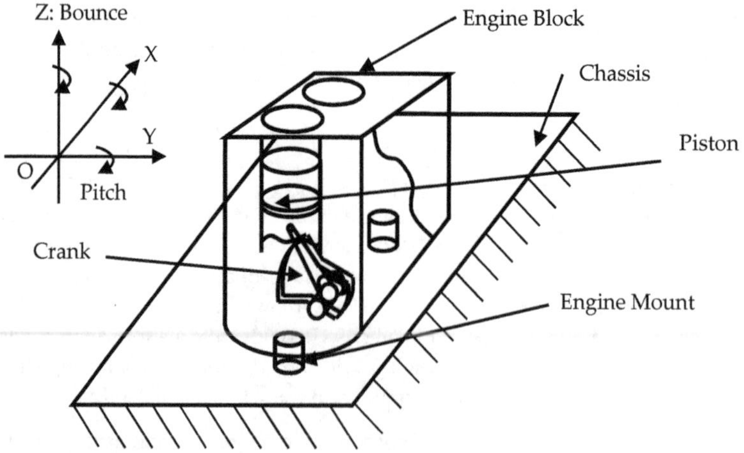

Figure 2. Chassis excited by the engine vibration.

Orthogonal functions like Haar wavelets (HWs) [13, 16], Walsh functions [7], block pulse functions [27], Laguerre polynomials [14], Legendre polynomials [5], Chebyshev functions [12] and Fourier series [28], often used to represent an arbitrary time functions, have received considerable attention in dealing with various problems of dynamic systems. The main characteristic of this technique is that it reduces these problems to those of solving a system of algebraic equations for the solution of problems described by differential equations, such as analysis of linear time-invariant, time-varying systems, model reduction, optimal control and system identification. Thus, the solution, identification and optimisation procedure are either greatly reduced or much simplified accordingly. The available sets of orthogonal functions can be divided into three classes such as piecewise constant basis functions (PCBFs) like HWs, Walsh functions and block pulse functions; orthogonal polynomials like Laguerre, Legendre and Chebyshev as well as sine-cosine functions in Fourier series [21].

In the present paper, we, for the first time, introduce a computational solution to the finite-time robust optimal control problem of the vehicle engine-body vibration system based on HWs. To this aim, mathematical model of the engine-body vibration structure is presented such the actuators and sensors used to investigate the robust optimal control are selected to be collocated. Moreover, the properties of HWs, Haar wavelet integral operational matrix and Haar wavelet product operational matrix are given and are utilized to provide a systematic computational framework to find the approximated robust optimal trajectory and finite-time H_∞ control of the vehicle engine-body vibration system with respect to a H_∞ performance by solving only the linear algebraic equations instead of solving the differential equations. One of the main advantages is solving linear algebraic equations instead of solving nonlinear differential Riccati equation to optimize the control problem of the vehicle engine-body vibration system. We demonstrate the applicability of the technique by the simulation results.

The rest of this paper is organized as fallows. Section 2 introduces properties of the HWs. Mathematical model of the engine-body vibration structure is stated in Section 3. Algebraic solution of the engine-body system is given in Section 4 and Haar wavelet-based optimal trajectories and robust optimal control are presented in Sections 5 and 6, respectively. Simulation results of the robust optimal control of the vehicle engine-body vibration system are shown in Section 7 and finally the conclusion is discussed.

The notations used throughout the paper are fairly standard. The matrices I_r, 0_r and $0_{r\times s}$ are the identity matrix with dimension $r\times r$ and the zero matrices with dimensions $r\times r$ and $r\times s$, respectively. The symbol $\otimes$ and $tr(A)$ denote Kronecker product and trace of the matrix A, respectively. Also, operator $vec(X)$ denotes the vector obtained by putting matrix X into one column. Finally, given a signal $x(t)$, $\|x(t)\|_2$ denotes the L_2 norm of $x(t)$; i.e.,

$$\|x(t)\|_2^2 = \int_0^\infty x(t)^T x(t)\, dt .$$

2. Properties of Haar Wavelets

Properties of HWs, which will be used in the next sections, are introduced in this section.

2.1. Haar Wavelets (HWs)

The oldest and most basic of the wavelet systems is named Haar wavelet that is a group of square waves with magnitude of . ± 1 . in the interval $[0,1)$ [6]. In other words, the HWs are defined on the interval $[0,1)$ as

$$\psi_0(t)=1, \quad t\in[0,1), \qquad \psi_1(t)=\begin{cases}1, & \text{for } t\in[0,\frac{1}{2}),\\ -1, & \text{for } t\in[\frac{1}{2},1),\end{cases} \tag{1}$$

and $\psi_i(t)=\psi_1(2^j t-k)$ for $i\geq 1$ and we write $i=2^j+k$ for $j\geq 0$ and $0\leq k<2^j$. We can easily see that the $\psi_0(t)$ and $\psi_1(t)$ are compactly supported, they give a local description, at different scales j, of the considered function.

2.2. Function approximation

The finite series representation of any square integrable function $y(t)$ in terms of an orthogonal basis in the interval $[0,1)$, namely $\hat{y}(t)$, is given by

$$\hat{y}(t)=\sum_{i=0}^{m-1} a_i\psi_i(t) := a^T\Psi_m(t) \tag{2}$$

where $a:=\left[a_0\ a_1 \cdots a_{m-1}\right]^T$ and $\Psi_m(t):=\left[\psi_0(t)\,\psi_1(t)\cdots\psi_{m-1}(t)\right]^T$ for $m=2^j$ and the Haar coefficients a_i are determined to minimize the mean integral square error $\varepsilon=\int_0^1 (y(t)-a^T\Psi_m(t))^2\,dt$ and are given by

$$a_i=2^j\int_0^1 y(t)\,\psi_i(t)\,dt \tag{3}$$

Remark 1. The approximation error, $\Xi_y(m):=y(t)-\hat{y}(t)$, is depending on the resolution m and is approaching zero by increasing parameter of the resolution.

The matrix H_m can be defined as

$$H_m=\left[\Psi_m(t_0),\Psi_m(t_1),\cdots,\Psi_m(t_{m-1})\right] \tag{4}$$

where $i/_m \le t_i < {}^{i+1}/_m$ and using (2), we get

$$\left[\hat{y}(t_0)\,\hat{y}(t_1)\ldots\hat{y}(t_{m-1})\right] = a^T H_m . \tag{5}$$

The integration of the vector $\Psi_m(t)$ can be approximated by

$$\int_0^t \Psi_m(t)\,dt = P_m \Psi_m(t) \tag{6}$$

where the matrix

$$P_m = <\int_0^t \Psi_m(\tau)\,d\tau, \Psi_m(t)> = \int_0^1\int_0^t \Psi_m(r)\,dr\,\Psi_m^T(t)\,dt$$

represents the integral operator matrix for PCBFs on the interval $\left[0,1\right)$ at the resolution m. For HWs, the square matrix P_m satisfies the following recursive formula [13]:

$$P_m = \frac{1}{2m}\begin{bmatrix} 2m P_{\frac{m}{2}} & -H_{\frac{m}{2}} \\ H_{\frac{m}{2}}^{-1} & 0_{\frac{m}{2}} \end{bmatrix} \tag{7}$$

with $P_1 = \frac{1}{2}$ and $H_m^{-1} = \frac{1}{m} H_m^T\, diag\,(r)$ where the matrix H_m defined in (4) and also the vector r is represented by

$$r{:=}(1,1,2,2,4,4,4,4,\cdots,\underbrace{(\frac{m}{2}),(\frac{m}{2}),\cdots,(\frac{m}{2})}_{(\frac{m}{2})\ elements})^T$$

for $m > 2$. For example, at resolution scale $j = 3$, the matrices H_8 and P_8 are represented as

$$H_8 = \begin{bmatrix} \psi_0(t_0) & \psi_0(t_1) & & \psi_0(t_7) \\ \psi_1(t_0) & \psi_1(t_1) & & \psi_1(t_7) \\ \psi_2(t_0) & \psi_2(t_1) & & \psi_2(t_7) \\ \psi_3(t_0) & \psi_3(t_1) & \cdots & \psi_3(t_7) \\ \psi_4(t_0) & \psi_4(t_1) & & \psi_4(t_7) \\ \psi_5(t_0) & \psi_5(t_1) & & \psi_5(t_7) \\ \psi_6(t_0) & \psi_6(t_1) & & \psi_6(t_7) \\ \psi_7(t_0) & \psi_7(t_1) & & \psi_7(t_7) \end{bmatrix} = \begin{bmatrix} 1 & 1 & 1 & 1 & 1 & 1 & 1 & 1 \\ 1 & 1 & 1 & 1 & -1 & -1 & -1 & -1 \\ 1 & 1 & -1 & -1 & 0 & 0 & 0 & 0 \\ 0 & 0 & 0 & 0 & 1 & 1 & -1 & -1 \\ 1 & -1 & 0 & 0 & 0 & 0 & 0 & 0 \\ 0 & 0 & 1 & -1 & 0 & 0 & 0 & 0 \\ 0 & 0 & 0 & 0 & 1 & -1 & 0 & 0 \\ 0 & 0 & 0 & 0 & 0 & 0 & 1 & -1 \end{bmatrix},$$

and

$$P_8 = \frac{1}{16}\left[\begin{array}{c:c}\begin{array}{c:c}\begin{array}{c:c} 8 & -4H_1 \\ \hdashline 4H_1^{-1} & 0 \end{array} & -2H_2 \\ \hdashline 4H_2^{-1} & 0 \end{array} & -H_4 \\ \hdashline H_4^{-1} & 0 \end{array}\right] = \frac{1}{16}\begin{bmatrix} 16P_4 & -H_4 \\ H_4^{-1} & 0 \end{bmatrix}$$

$$= \frac{1}{64}\begin{bmatrix} 32 & -16 & -8 & -8 & -4 & -4 & -4 & -4 \\ 16 & 0 & -8 & 8 & -4 & -4 & 4 & 4 \\ 4 & 4 & 0 & 0 & -4 & 4 & 0 & 0 \\ 4 & -4 & 0 & 0 & 0 & 0 & -4 & 4 \\ 1 & 1 & 2 & 0 & 0 & 0 & 0 & 0 \\ 1 & 1 & -2 & 0 & 0 & 0 & 0 & 0 \\ 1 & -1 & 0 & 2 & 0 & 0 & 0 & 0 \\ 1 & -1 & 0 & -2 & 0 & 0 & 0 & 0 \end{bmatrix},$$

for further information see [13, 25].

2.3. The product operational matrix

In the study of time-varying state-delayed systems, it is usually necessary to evaluate the product of two Haar function vectors [13]. Let us define

$$R_m(t) := \Psi_m(t)\Psi_m^T(t) \tag{8}$$

where $R_m(t)$ satisfies the following recursive formula

$$R_m(t) = \frac{1}{2m}\begin{bmatrix} R_{\frac{m}{2}}(t) & H_{\frac{m}{2}}\, diag(\Psi_b(t)) \\ (H_{\frac{m}{2}}\, diag(\Psi_b(t)))^T & diag(H_{\frac{m}{2}}^{-1}\Psi_a(t)) \end{bmatrix} \tag{9}$$

with $R_1(t) = \psi_0(t)\psi_0^T(t)$ and

$$\begin{cases} \Psi_a(t) := \left[\psi_0(t), \psi_1(t), \cdots, \psi_{\frac{m}{2}-1}(t)\right]^T = \Psi_{\frac{m}{2}}(t) \\ \Psi_b(t) := \left[\psi_{\frac{m}{2}}(t), \psi_{\frac{m}{2}+1}(t), \cdots, \psi_{m-1}(t)\right]^T. \end{cases} \tag{10}$$

Moreover, the following relation is important for solving optimal control problem of time-varying state-delayed system:

$$R_m(t)a_m = \tilde{a}_m \Psi_m(t) \tag{11}$$

where $\tilde{a}_1 = a_0$ and

$$\tilde{a}_m = \begin{bmatrix} \tilde{a}_{\frac{m}{2}} & H_{\frac{m}{2}} diag(a_b) \\ diag(a_b) H_{\frac{m}{2}}^{-1} & diag(a_a^T H_{\frac{m}{2}}) \end{bmatrix} \tag{12}$$

with

$$\begin{cases} a_a := \left[a_0, a_1, \cdots, a_{\frac{m}{2}-1} \right]^T = a_{\frac{m}{2}}(t) \\ a_b := \left[a_{\frac{m}{2}}(t), a_{\frac{m}{2}+1}(t), \cdots, a_{m-1}(t) \right]^T. \end{cases} \tag{13}$$

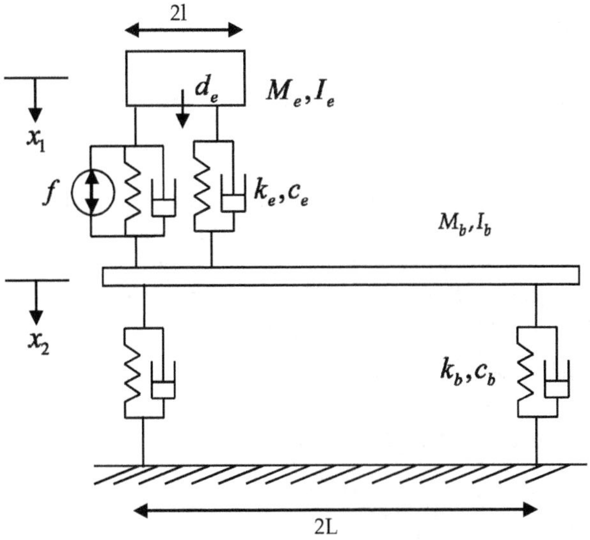

Figure 3. The sketch of engine-body vibration system

3. Mathematical model description

A schematic of the vehicle engine-body vibration structure is shown in Figure 3. The actuator and sensor used to this control framework are selected to be collocated, since this arrangement is ideal to ensure the stability of the closed loop system for a slightly damped structure [26]. In our study, only the bounce and pitch vibrations in the engine and body are considered [35]. The engine with mass M_e and inertia moment I_e is mounted in the body by the engine mounts k_e and c_e. The front mount is the active mount, the output force of which can be controlled by an electric signal. The active mount consists of a main chamber where an oscillating mass (inertia mass) is moving up and down. The inertia mass is driven by an electro-magnetic force generated by a magnetic coil which is controlled by the input current.

The vehicle body with mass M_b and inertia moment I_b is supported by front and rear tires, each of which is modeled as a system consisting of a spring k_b and a damping device c_b. Therefore, a four degree-of-freedom vibration suspension model shown in Figure 3 can be described by the following equations

$$\begin{cases} M_e\ddot{x}_1 + 2c_e\dot{x}_1 + 2k_e x_1 - 2c_e\dot{x}_2 - 2k_e x_2 - 2(L-l)c_e\dot{x}_4 - 2(L-l)k_e x_4 = f(t) + d_e(t) \\ M_b\ddot{x}_2 + 2(c_e + c_b)\dot{x}_2 + 2(k_e + k_b)x_2 - 2c_e\dot{x}_1 - 2k_e x_1 + 2(L-l)c_e\dot{x}_4 + 2(L-l)k_e x_4 = -f(t) \\ I_e\ddot{x}_3 + 2l^2c_e\dot{x}_3 + 2l^2k_e x_3 - 2l^2c_e\dot{x}_4 - 2l^2k_e x_4 = l\,f(t) \\ I_b\ddot{x}_4 + ((L^2 + (L-2l)^2)c_e + 2L^2c_b)\dot{x}_4 + ((L^2 + (L-2l)^2)k_e + 2L^2k_b)x_4 - 2l^2c_e\dot{x}_3 \\ \quad - 2l^2k_e x_3 - 2lc_e\dot{x}_1 - 2lk_e x_1 + 2(L-l)c_e\dot{x}_2 + 2(L-l)k_e x_2 = -L\,f(t) \end{cases} \tag{14}$$

where the states $x_1(t), x_2(t), x_3(t)$ and $x_4(t)$ are the bounces and pitches of the engine and body, respectively, where displacement of the chassis $(x_2(t))$ is usually taken as an output. Input force, $f(t)$, is used as the active force to compensate the vibration transmitted to vehicle body. Moreover, engine disturbance $d_e(t)$ can be the excitation, generated by the motion up/down of the different parts inside the engine;

The system Eq. (14) can be represented in the following state-space form

$$\begin{cases} M\ddot{x}(t) + C\dot{x}(t) + Kx(t) = B_f f(t) + B_d d_e(t), \qquad t \in \left[0, T_f\right] \\ z(t) = \begin{bmatrix} C_1 x(t) \\ C_2 \dot{x}(t) \\ C_3 f(t) \end{bmatrix} \end{cases} \tag{15}$$

where $x(t) \in \Re^4$ is the state; $f(t) \in \Re$ is the control input; $d_e(t) \in \Re$ is the disturbance input which belongs to $L_2[0,\infty)$; and $z(t) \in \Re^3$ is the controlled output with $C_1 \in \Re^{1\times 4}$, $C_2 \in \Re^{1\times 4}$ and C_3 is a positive scalar. The state-space matrices are also defined as

$$M = \begin{bmatrix} M_e & 0 & 0 & 0 \\ 0 & M_b & 0 & 0 \\ 0 & 0 & I_e & 0 \\ 0 & 0 & 0 & I_b \end{bmatrix}, C = \begin{bmatrix} 2c_e & -2c_e & 0 & -2(L-l)c_e \\ -2c_e & 2(c_e + c_b) & 0 & 2(L-l)c_e \\ 0 & 0 & 2l^2c_e & -2l^2c_e \\ -2lc_e & 2(L-l)c_e & -2l^2c_e & 0 \end{bmatrix},$$

$$B_f = \begin{bmatrix} 1 \\ -1 \\ l \\ -L \end{bmatrix}, \quad B_d = \begin{bmatrix} 1 \\ 0 \\ 0 \\ 0 \end{bmatrix},$$

$$K = \begin{bmatrix} 2k_e & -2k_e & 0 & -2(L-l)k_e \\ -2k_e & 2(k_e+k_b) & 0 & 2(L-l)k_e \\ 0 & 0 & 2l^2k_e & -2l^2k_e \\ -2lk_e & 2(L-l)k_e & -2l^2k_e & (L^2+(L-2l)^2)k_e+2L^2k_b \end{bmatrix}.$$

Taking displacement of the chassis $(x_2(t))$ as an output then a comparison of the displacement response respect to the input force $f(t)$ and the external disturbance $d_e(t)$ in the frequency range up to 1 KHz is depicted in Figure 4a) and 4b). Three relevant modes occur around the frequencies 1, 5 and 9 Hz, respectively, which represent the dynamics of the main degrees of freedom (DOFs) of the system.

4. Algebraic solution of system equations

In this section, we study the problem of solving the second-order differential equations of the engine-body system (14) in terms of the input control and exogenous disturbance using HWs and develop appropriate algebraic equations.

Based on HWs definition on the interval time $\left[0,1\right]$, we need to rescale the finite time interval $\left[0,T_f\right]$ into $\left[0,1\right]$ by considering $t = \mathrm{T}_f\,\sigma$; normalizing the system Eq. (15) with the time scale would be as follows

$$M\ddot{x}(\sigma)+C\dot{x}(\sigma)+Kx(\sigma)=B_f\,f(\sigma)+B_d\,d_e(\sigma) \tag{16}$$

Now by integrating the system above in an interval $\left[0,\sigma\right]$, we obtain

$$\begin{aligned} M(x(\sigma)-x(0))+\mathrm{T}_f C\int_0^{\sigma} x(\tau)\,d\tau+\mathrm{T}_f^2 K\int_0^{\sigma}\int_0^{\xi} x(\tau)\,d\tau\,d\xi = \mathrm{T}_f^2 B_f\int_0^{\sigma}\int_0^{\xi} f(\tau)\,d\tau\,d\xi+\mathrm{T}_f^2 B_d\int_0^{\sigma}\int_0^{\xi} d_e(\tau)\,d\tau\,d\xi \\ +\int_0^{\sigma}(M\dot{x}(0)+\mathrm{T}_f C x(0))\,d\xi. \end{aligned} \tag{17}$$

By using the Haar wavelet expansion (2), we express the solution of Eq. (15), input force $f(\sigma)$ and engine disturbance $d_e(\sigma)$ in terms of HWs in the forms

$$x(\sigma)=X\,\Psi_m(\sigma), \tag{18}$$

$$f(\sigma)=F\,\Psi_m(\sigma), \tag{19}$$

$$d_e(\sigma) = D_e\, \Psi_m(\sigma), \tag{20}$$

where $X \in \Re^{4\times m}$, $F \in \Re^{1\times m}$ and $D_e \in \Re^{1\times m}$ denote the wavelet coefficients of $x(\sigma)$, $f(\sigma)$ and $d_e(\sigma)$, respectively. The initial conditions of $x(0)$ and $\dot{x}(0)$ are also represented by $x(0) = X_0\, \Psi_m(\sigma)$ and $\dot{x}(0) = \overline{X}_0\, \Psi_m(\sigma)$, where the matrices $\{X_0, \overline{X}_0\} \in \Re^{4\times m}$ are defined, respectively, as

$$X_0 := \left[x(0)\ \underbrace{0_{4\times1}\ \ \cdots\ \ 0_{4\times1}}_{(m-1)} \right] \tag{21}$$

$$\overline{X}_0 := \left[\dot{x}(0)\ \underbrace{0_{4\times1}\ \ \cdots\ \ 0_{4\times1}}_{(m-1)} \right] \tag{22}$$

Therefore, using the wavelet expansions (18)-(20), the relation (17) becomes

$$M(X - X_0) + \mathrm{T}_f C X P_m + \mathrm{T}_f^2 K X P_m^2 = \mathrm{T}_f^2 B_f F P_m^2 + \mathrm{T}_f^2 B_d D_e P_m^2 + (M \overline{X}_0 + \mathrm{T}_f C X_0)\, P_m \tag{23}$$

For calculating the matrix X, we apply the operator $vec(.)$ to Eq. (23) and according to the property of the Kronecker product, i.e. $vec(ABC) = (C^T \otimes A)\, vec(B)$, we have:

$$\begin{aligned}(I_m \otimes M)(vec(X) - vec(X_0)) + \mathrm{T}_f (P_m^T \otimes C) vec(X) + \mathrm{T}_f^2 (P_m^{2T} \otimes K) vec(X) \\ = \mathrm{T}_f^2 (P_m^{2T} \otimes B_f) vec(F) + \mathrm{T}_f^2 (P_m^{2T} \otimes B_d) vec(D_e) \\ + \mathrm{T}_f (P_m^T \otimes C) vec(X_0) + (P_m^T \otimes M) vec(\overline{X}_0).\end{aligned} \tag{24}$$

Solving Eq. (24) for $vec(X)$ leads to

$$vec(X) = \Delta_1 vec(F) + \Delta_2\, vec(D_e) + \Delta_3\, vec(X_0) + \Delta_4\, vec(\overline{X}_0) \tag{25}$$

where the matrices $\{\Delta_1, \Delta_2\} \in \Re^{4m\times m}$ and $\{\Delta_3, \Delta_4\} \in \Re^{4m\times 4m}$ are defined as

$$\begin{cases} \Delta_1 = \mathrm{T}_f^2 (\mathrm{T}_f (P_m^T \otimes C) + \mathrm{T}_f^2 (P_m^{2T} \otimes K) + I_m \otimes M)^{-1} (P_m^{2T} \otimes B_f) \\ \Delta_2 = \mathrm{T}_f^2 (\mathrm{T}_f (P_m^T \otimes C) + \mathrm{T}_f^2 (P_m^{2T} \otimes K) + I_m \otimes M)^{-1} (P_m^{2T} \otimes B_d) \\ \Delta_3 = (\mathrm{T}_f (P_m^T \otimes C) + \mathrm{T}_f^2 (P_m^{2T} \otimes K) + I_m \otimes M)^{-1} (I_m \otimes M + \mathrm{T}_f P_m^T \otimes C) \\ \Delta_4 = (\mathrm{T}_f (P_m^T \otimes C) + \mathrm{T}_f^2 (P_m^{2T} \otimes K) + I_m \otimes M)^{-1} (P_m^T \otimes M). \end{cases} \tag{26}$$

Consequently, using (25) and (26) and the properties of the Kronecker product, the solution of system (15) is

$$x(\sigma) = (\Psi_m^T(\sigma) \otimes I_4)\, vec(X) \tag{27}$$

and it is also clear that to find the approximated solution of the system, we have to calculate the inverse of the matrix $T_f(P_m^T \otimes C)+T_f^2(P_m^{2T} \otimes K)+I_m \otimes M$ with dimension $4m \times 4m$ only once.

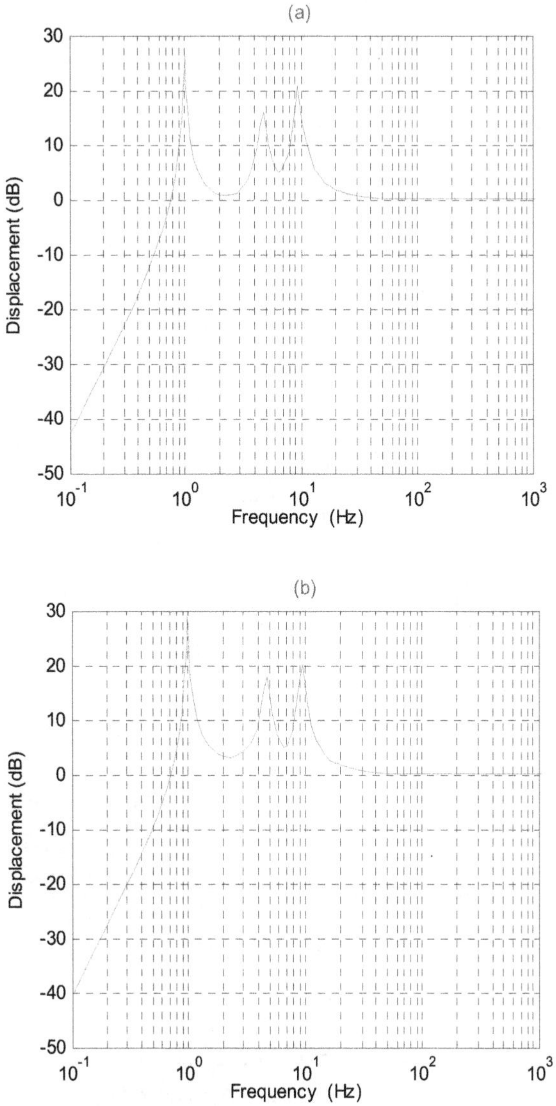

Figure 4. Displacement of the chassis respect to $f(t)$ (a) and $d_e(t)$ (b).

5. Optimal control design

The control objective is to find the optimal control $f(t)$ with respect to a quadratic cost functional approximately such acts as the active force to compensate the vibration transmitted to vehicle body. The quadratic cost functional weights the states and their derivatives with respect to time in the cost function as follows:

$$J=\frac{1}{2}x^T(\mathrm{T}_f)S_1\,x(\mathrm{T}_f)+\frac{1}{2}\dot{x}^T(\mathrm{T}_f)S_2\,\dot{x}(\mathrm{T}_f)+\frac{1}{2}\int_0^{\mathrm{T}_f}(x^T(t)Q_1\,x(t)+\dot{x}^T(t)Q_2\,\dot{x}(t)+R\,f(t)^2)\,dt \quad (28)$$

where $S_1:4\times 4$, $S_2:4\times 4$, $Q_1:4\times 4$ and $Q_2:4\times 4$ are positive-definite matrices and R is a positive scalar. We can rewrite the cost function (28) as follows:

$$J=\frac{1}{2}[x^T(1)\quad \mathrm{T}_f^{-1}\dot{x}^T(1)]\tilde{S}\begin{bmatrix} x(1) \\ \mathrm{T}_f^{-1}\dot{x}(1)\end{bmatrix}+\frac{\mathrm{T}_f}{2}\int_0^1([x^T(\sigma)\quad \mathrm{T}_f^{-1}\dot{x}^T(\sigma)]\tilde{Q}\begin{bmatrix} x(\sigma) \\ \mathrm{T}_f^{-1}\dot{x}(\sigma)\end{bmatrix}+R\,f(\sigma)^2)\,d\sigma\,. \quad (29)$$

where $\tilde{S}=diag(S_1,S_2)$ and $\tilde{Q}=diag(Q_1,Q_2)$ with the time scale $t=\mathrm{T}_f\,\sigma$.

From (15) and the relation $\dot{x}(\sigma)=\overline{X}\,\Psi_m(\sigma)$, where $\overline{X}:4\times m$ denotes the wavelet coefficients of $\dot{x}(\sigma)$ after its expansion in terms of HFs, we read

$$\begin{bmatrix} x(\sigma) \\ \mathrm{T}_f^{-1}\dot{x}(\sigma)\end{bmatrix}=\begin{bmatrix} \mathrm{X} \\ \mathrm{T}_f^{-1}\overline{\mathrm{X}}\end{bmatrix}\Psi_m(\sigma):=X_{aug}\,\Psi_m(\sigma) \quad (30)$$

where $X_{aug}=\begin{bmatrix} \mathrm{X} \\ \mathrm{T}_f^{-1}\overline{\mathrm{X}}\end{bmatrix}$ and

$$vec(X_{aug})=\begin{bmatrix} vec^T(\mathrm{X}) & \mathrm{T}_f^{-1}vec^T(\overline{\mathrm{X}})\end{bmatrix}^T \quad (31)$$

Remark 2. By substituting $\dot{x}(\sigma)=\overline{X}\,\Psi_m(\sigma)$ into $x(\sigma)-x(0)=\int_0^{\sigma}\dot{x}(t)\,dt$, we have:

$$\mathrm{X}\,\Psi_m(\sigma)-\mathrm{X}_0\,\Psi_m(\sigma)=\int_0^{\sigma}\overline{X}\,\Psi_m(\tau)\,d\tau\,, \quad (32)$$

and using (4), we read $X-X_0=\overline{X}\,P_m$. Then, by applying the operator of $vec(.)$ and according to the properties of Kronecker product in Appendix A1, we obtain

$$vec(\mathrm{X})-vec(\mathrm{X}_0)=(P_m^T\otimes I_n)\,vec(\overline{\mathrm{X}}) \quad (33)$$

By substituting the definition (31) in (33) and using the properties of the operator $tr(.)$ in Appendix A1, the cost function (28) is given by

$$J = \frac{1}{2}(vec^T(X_{aug})\Pi_{m1}\, vec(X_{aug}) + vec^T(F)\Pi_{m2}\, vec(F)) \tag{34}$$

where the matrices $\Pi_{m1} : 8m \times 8m$ and $\Pi_{m2} : m \times m$ are defined as

$$\Pi_{m1} = M_f^T \otimes \tilde{S} + \mathrm{T}_f (M^T \otimes \tilde{Q}) \text{ and } \Pi_{m2} = R\mathrm{T}_f M_m \text{, respectively,}$$

and the matrices $M_m : m \times m$ and $M_{mf} : m \times m$ are defined as

$$M_m := \int_0^1 \Psi_m(\sigma)\Psi_m^T(\sigma)\, d\sigma \text{ and } M_{mf} := \Psi_m(1)\Psi_m^T(1) \text{, respectively.}$$

It is clear that the cost function of $J(.)$ is a function of $\frac{i}{m} \le \sigma_i < \frac{i+1}{m}$, then for finding the optimal control law, which minimizes the cost functional $J(.)$, the following necessary condition should be satisfied

$$\frac{\partial J}{\partial vec(F)} = 0 \tag{35}$$

By considering $vec(X_{aug})$, which is a function of $vec(F)$, and using the properties of derivatives of inner product of Kronecker product in Appendix A2, we find

$$\frac{\partial J}{\partial vec(F)} = [\Delta_1^T \quad \mathrm{T}_f^{-1}\Delta_1^T(P_m^{-1} \otimes I_4)]\, \Pi_{m1}\, vec(X_{aug}) + \Pi_{m2}\, vec(F) \tag{36}$$

Then the wavelet coefficients of the optimal control law will be in vector form as

$$vec(F) = -\Pi_{m2}^{-1}[\Delta_1^T \quad \mathrm{T}_f^{-1}\Delta_1^T(P_m^{-1} \otimes I_4)]\, \Pi_{m1}\, vec(X_{aug}) \tag{37}$$

Consequently, the optimal vectors of $vec(X)$ and $vec(F)$ are found, respectively, in the following forms

$$\begin{aligned} vec(X) = {} & (I_{4m} + \Delta_1(\Pi_{m2}^{-1}[\Delta_1^T \quad \mathrm{T}_f^{-1}\Delta_1^T(P_m^{-1} \otimes I_4)]\, \Pi_{m1} \begin{bmatrix} I_{4m} \\ \mathrm{T}_f^{-1}(P_m^T \otimes I_4)^{-1} \end{bmatrix})^{-1} (\Delta_2\, vec(D_e) + (\Delta_1 \Pi_{m2}^{-1} \\ & \times [\Delta_1^T \quad \mathrm{T}_f^{-1}\Delta_1^T(P_m^{-1} \otimes I_4)]\, \Pi_{m1} \begin{bmatrix} 0_{4m} \\ \mathrm{T}_f^{-1}(P_m^T \otimes I_4)^{-1} \end{bmatrix} + \Delta_3)\, vec(X_0) + \Delta_4\, vec(\bar{X}_0)), \end{aligned} \tag{38}$$

and

$$vec(F) = -\Pi_{m2}^{-1}[\Delta_1^T \quad \mathrm{T}_f^{-1}\Delta_1^T(P_m^{-1}\otimes I_4)]\,\Pi_{m1}\{\begin{bmatrix} I_{4m} \\ \mathrm{T}_f^{-1}(P_m^T\otimes I_4)^{-1}\end{bmatrix}$$

$$\times(I_{4m} + \Delta_1\Pi_{m2}^{-1}[\Delta_1^T \quad \mathrm{T}_f^{-1}\Delta_1^T(P_m^{-1}\otimes I_4)]\,\Pi_{m1}\begin{bmatrix} I_{4m} \\ \mathrm{T}_f^{-1}(P_m^T\otimes I_4)^{-1}\end{bmatrix})^{-1}$$

$$\times(\Delta_2\, vec(D_e) + (\Delta_1\Pi_{m2}^{-1}[\Delta_1^T \quad \mathrm{T}_f^{-1}\Delta_1^T(P_m^{-1}\otimes I_4)]\,\Pi_{m1}\begin{bmatrix} 0_{4m} \\ \mathrm{T}_f^{-1}(P_m^T\otimes I_4)^{-1}\end{bmatrix} + \Delta_3)\, vec(X_0)$$

$$+\Delta_4\, vec(\bar{X}_0)) - \begin{bmatrix} 0_{4m} \\ \mathrm{T}_f^{-1}(P_m^T\otimes I_4)^{-1}\end{bmatrix} vec(X_0)\}. \tag{39}$$

Finally, the Haar function-based optimal trajectories and optimal control are obtained approximately from Eq. (27) and $f(t) = \Psi_m^T(t)\, vec(F)$.

6. Robust optimal control design

In this section, an optimal state feedback controller is to be determined computationally such that the following requirements are satisfied:

i. the closed-loop system is asymptotically stable;
ii. under zero initial condition, the closed-loop system satisfies $\|z(t)\|_2 < \gamma\|d_e(t)\|_2$ for any non-zero $d_e(t)\in[0,\infty)$ where $\gamma > 0$ is a prescribed scalar.

The control objective is to find the approximated robust optimal control $f(t)$ with H_∞ performance such $f(t)$ acts as the active force to compensate the vibration transmitted to vehicle body, i.e. guarantees desired L_2 gain performance. Next, we shall establish the H_∞ performance of the system (15) under zero initial condition. To this end, we introduce

$$J = \tfrac{1}{2}x^T(\mathrm{T}_f)S_1 x(\mathrm{T}_f) + \frac{1}{2}\dot{x}^T(\mathrm{T}_f)S_2\dot{x}(\mathrm{T}_f) + \tfrac{1}{2}\int_0^{T_f}(z^T(t)z(t) - \gamma^2 d_e^2(t))\,dt. \tag{40}$$

It is well known that a sufficient condition for achieving robust disturbance attenuation is that the inequality $J < 0$ for every $d_e(t)\in L_2[0,\infty)$ [33, 36]. Therefore, we will establish conditions under which

$$\underset{vec(F)}{Inf}\ \underset{vec(D_e)}{Sup}\ J(vec(F), vec(D_e)) \le 0 \tag{41}$$

From (15), the Eq. (40) can be represented as

$$
\begin{aligned}
J &= \tfrac{1}{2}\left(x^T(1) \quad \frac{1}{T_f}\dot{x}^T(1)\right)\tilde{S}\begin{pmatrix} x(1) \\ T_f^{-1}\dot{x}(1)\end{pmatrix} \\
&+\frac{T_f}{2}\int_0^1 \left(\left(x^T(\sigma) \quad T_f^{-1}\dot{x}^T(\sigma)\right)\tilde{C}\begin{pmatrix} x(\sigma) \\ T_f^{-1}\dot{x}(\sigma)\end{pmatrix} + C_3^2 f^2(\sigma) - \gamma^2 d_e^2(\sigma)\right) d\sigma
\end{aligned}
\tag{42}
$$

where $t = T_f\,\sigma$, $\tilde{S} = diag(S_1, S_2)$ and $\tilde{C} = diag(C_1^T C_1, C_2^T C_2)$.

Using the relation $\dot{x}(\sigma) = \bar{X}\,\Psi_m(\sigma)$, we read

$$
\begin{bmatrix} x(\sigma) \\ T_f^{-1}\dot{x}(\sigma)\end{bmatrix} = \begin{bmatrix} X \\ T_f^{-1}\bar{X}\end{bmatrix}\Psi_m(\sigma) := X_{aug}\,\Psi_m(\sigma) \tag{43}
$$

where $X_{aug} = \begin{bmatrix} X \\ T_f^{-1}\bar{X}\end{bmatrix}$ and

$$
vec(X_{aug}) = \begin{bmatrix} vec^T(X) & T_f^{-1} vec^T(\bar{X})\end{bmatrix}^T \tag{44}
$$

Moreover, according to Remark 2 in [18], the following relation is already satisfied between $vec(X)$ and $vec(\bar{X})$

$$
vec(X) - vec(X_0) = (P_m^T \otimes I_4)\, vec(\bar{X}) \tag{45}
$$

By using the definition (44) in Eq. (45), we have

$$
J = \tfrac{1}{2}(tr(M_{mf} X_{aug}^T \tilde{S} X_{aug})) + \frac{T_f}{2}(tr(M_m X_{aug}^T \tilde{C} X_{aug}) + tr(C_3^2 M_m F^T F) - \gamma^2 tr(M_m D_e^T D_e)) \tag{46}
$$

Using the property of the Kronecker product, i.e. $tr(ABC) = vec^T(A^T)(I_p \otimes B)\, vec(C)$, $(A\otimes C)(D\otimes B) = AD \otimes CB$ and $vec(ABC) = (C^T \otimes A)\, vec(B)$, we can write (42) as

$$
J = \frac{1}{2}(vec^T(X_{aug})\Pi_{m1} vec(X_{aug}) + C_3^2\, vec^T(F)\Pi_{m2}\, vec(F) - \gamma^2 vec^T(D_e)\Pi_{m2}\, vec(D_e)) \tag{47}
$$

where the matrices $\Pi_{m1} \in \Re^{8m \times 8m}$, $\Pi_{m2} \in \Re^{m \times m}$ are defined as $\Pi_{m1} = M_{mf} \otimes \tilde{S} + {}^{T_f}\!/_2 (M_m \otimes \tilde{C})$ and $\Pi_{m2} = {}^{T_f}\!/_2\, M_m$, respectively.

It is easy to show that the worst-case disturbance in Eq. (47) occurs when

$$
vec^*(D_e) = \gamma^{-2}\Pi_{m2}^{-1}\begin{bmatrix}\Delta_2^T & T_f^{-1}\Delta_2^T(P_m^{-1}\otimes I_4)\end{bmatrix}\Pi_{m1}\, vec(X_{aug}) := \gamma^{-2}\Pi_{md}\, vec(X_{aug}) \tag{48}
$$

By substituting Eq. (48) into Eq. (47) we obtain

$$\underset{vec(F)}{Inf}\ \underset{vec(D_e)}{Sup}\ J(vec(F),vec(D_e)) = \underset{vec(F)}{Inf}\ J(vec(F),vec^*(D_e)) \tag{49}$$

Minimizing the right-hand side of Eq. (49) results in the algebraic relation between wavelet coefficients of the robust optimal control and of the optimal state trajectories in the following closed form

$$\begin{aligned} vec(F) &= -C_3^{-2}\Pi_{m2}^{-1}\begin{bmatrix} \Delta_1^T & \mathrm{T}_f^{-1}\Delta_1^T(P_m^{-1}\otimes I_4)\end{bmatrix}(\Pi_{m1} - \gamma^{-2}\Pi_{md}^T\Pi_{m2}\Pi_{md})\, vec(X_{aug}) \\ &:= \Pi_{mf}\ vec(X_{aug}). \end{aligned} \tag{50}$$

As a result we have

$$\underset{vec(F)}{Inf}\ \underset{vec(D_e)}{Sup}\ J(vec(F),vec(D_e)) \le vec^T(X_{aug})(\Pi_{m1} + R\Pi_{mf}^T\Pi_{m2}\Pi_{mf} - \gamma^2\Pi_{md}^T\Pi_{m2}\Pi_{md})vec(X_{aug}) \tag{51}$$

Consequently, if there exists positive scalar γ to the matrix inequality

$$\Pi_{m1} + C_3^2\Pi_{mf}^T\Pi_{m2}\Pi_{mf} - \gamma^2\Pi_{md}^T\Pi_{m2}\Pi_{md} \le 0 \tag{52}$$

then inequality (41) is concluded.

From the relations above we obtain the robust optimal vectors of $vec(X)$ and $vec(F)$ after some matrix calculations, respectively, in the following forms

$$\begin{aligned} vec(X) = {}& (I_{4m} - (\Delta_1\Pi_{mf} + \gamma^{-2}\Delta_2\Pi_{md})\begin{bmatrix} I_{4m} \\ \mathrm{T}_f^{-1}(P_m^T\otimes I_4)^{-1}\end{bmatrix})^{-1}((\Delta_3 - (\Delta_1\Pi_{mf} + \gamma^{-2}\Delta_2\Pi_{md}) \\ & \times \begin{bmatrix} 0_{4m} \\ \mathrm{T}_f^{-1}(P_m^T\otimes I_4)^{-1}\end{bmatrix})\, vec(X_0) + \Delta_4\, vec(\bar{X}_0)), \end{aligned} \tag{53}$$

and

$$\begin{aligned} vec(F) = \Pi_{mf}\{ & (\begin{bmatrix} I_{4m} \\ \mathrm{T}_f^{-1}(P_m^T\otimes I_4)^{-1}\end{bmatrix}((I_{4m} - (\Delta_1\Pi_{mf} + \gamma^{-2}\Delta_2\Pi_{md})\begin{bmatrix} I_{4m} \\ \mathrm{T}_f^{-1}(P_m^T\otimes I_4)^{-1}\end{bmatrix})^{-1} \\ & \times(\Delta_3 - (\Delta_1\Pi_{mf} + \gamma^{-2}\Delta_2\Pi_{md})\begin{bmatrix} 0_{4m} \\ \mathrm{T}_f^{-1}(P_m^T\otimes I_4)^{-1}\end{bmatrix}) - \begin{bmatrix} 0_{4m} \\ \mathrm{T}_f^{-1}(P_m^T\otimes I_4)^{-1}\end{bmatrix})\, vec(X_0) \\ & + \begin{bmatrix} I_{4m} \\ \mathrm{T}_f^{-1}(P_m^T\otimes I_4)^{-1}\end{bmatrix}(I_{4m} - (\Delta_1\Pi_{mf} + \gamma^{-2}\Delta_2\Pi_{md})\begin{bmatrix} I_{4m} \\ \mathrm{T}_f^{-1}(P_m^T\otimes I_4)^{-1}\end{bmatrix})^{-1}\Delta_4\, vec(\bar{X}_0))\} \end{aligned} \tag{54}$$

Finally, the Haar wavelet-based robust optimal trajectories and robust optimal control are obtained approximately from Eq. (27) and $f(t) = \Psi_m^T(t)\, vec(F)$, respectively.

7. Numerical results

In this section the proposed computational methodology is applied to the vehicle engine-body vibration system (15) such the exogenous disturbance $d_e(t)$ is assumed to be a $Sin\,(.)$ function at the frequency of $10\,Hz$. The system parameters, used for the design and simulation are given in Tables 1 and 2 in the *Appendix B*. Table 3 in the *Appendix* gives the pole-zero locations of 8th –order model of the vehicle engine-body vibration system. It is clear that the vehicle engine-body vibration system is unstable and has the nonminimum phase property. The objective is to find the approximated robust optimal displacement of the chassis and robust optimal input force with H_∞ performance using HWs at the finite time interval $\left[0,1\right]$. Moreover, the matrices $\{S_1, S_1\} \in \Re^{4\times4}$ and the vectors C_1, C_2 and the scalar C_3 in the controlled output $z(t)$ in Eq. (15) are chosen as $S_1 = S_2 = 0_4$, $C_1 = [0, 1, 1, 2]$, $C_2 = [3, -1, 0, 1]$ and $C_3 = 1$.

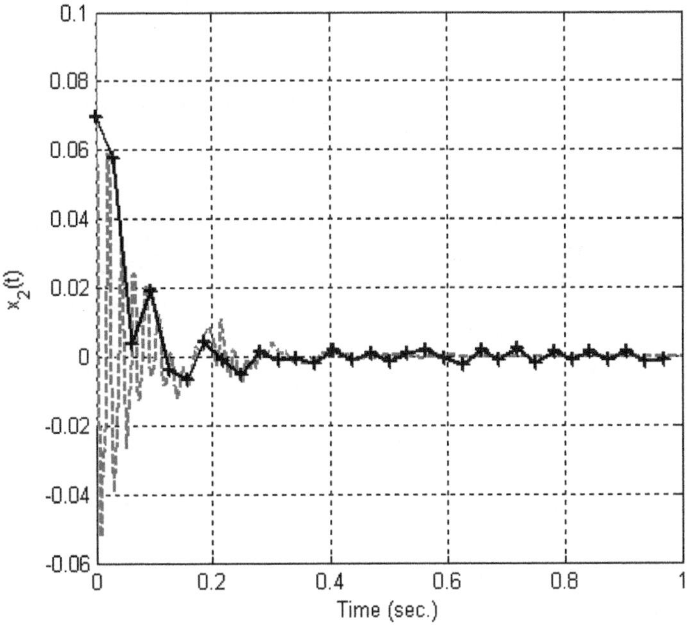

Figure 5. Comparison of displacement of the chassis found by HWs at resolution level $j = 5$ (solid) and by analytical solution (dashed).

To compare the approximate solutions $x_2(t)$ and $f(t)$, found by HWs, to the analytical solution found by Theorem 1 in the Appendix C, we choose the performance bound and the resolution level equal to 3.15 and 5, respectively, i.e. $\gamma = 3.15$ and $j = 5$. The time curves found are plotted in Figures 5 and 6. It is clear that the effect of the engine disturbance is attenuated onto the displacement of the chassis as the output as well. In other words, $f(t)$ compensates the vibration transmitted to the chassis. Compare the Haar wavelet based solutions to the continuous solutions using the differential Riccati equation, the approximate solutions (53) and (54) deliver both, robust control $f(t)$ and state trajectory $x(t)$ in one step by solving linear algebraic equations instead of solving nonlinear differential Riccati equation, while accuracy can easily be improved by increasing the resolution level j.

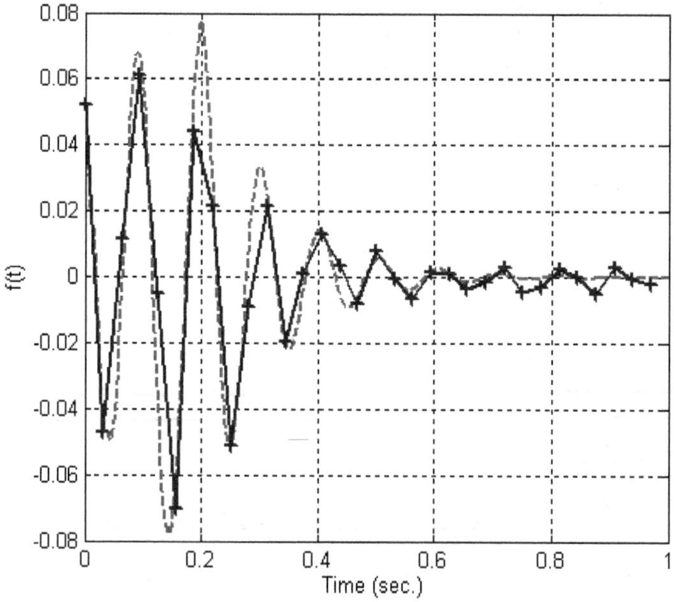

Figure 6. Comparison of input force found by HWs at resolution level $j = 5$ (solid) and by analytical solution (dashed).

8. Conclusion

This paper presented the modelling of engine-body vibration structure to control of bounce and pitch vibrations using HWs. To this aim, the Haar wavelet-based optimal control for vibration reduction of the engine-body system was developed computationally. The Haar wavelet properties were introduced and utilized to find the approximate solutions of optimal trajectories and robust optimal control by solving only algebraic equations instead of solving the Riccati differential equation. Numerical results were presented to illustrate the advantage of the approach.

9. Appendix

9.1. Appendix A

A1. *Some properties of Kronecker product*

Let $A: p \times q$, $B: q \times r$, $C: r \times s$ and $D: q \times t$ be fixed matrices, then we have:

$$
\begin{aligned}
&vec(ABC) = (C^T \otimes A)\, vec(B),\\
&tr(ABC) = vec^T(A^T)\,(I_p \otimes B)\, vec(C),\\
&tr(ABC) = vec^T(A^T)\,(I_p \otimes B)\, vec(C),\\
&(A \otimes C)(D \otimes B) = AD \otimes CB.
\end{aligned}
$$

A2. *Derivatives of inner products of Kronecker product*

Let $A: n \times n$ be fixed constants and $x: n \times 1$ be a vector of variables. Then, the following results can be established:

$$
\begin{aligned}
&\frac{\partial (Ax)}{\partial x} = vec(A),\\
&\frac{\partial (Ax)}{\partial x^T} = A,\\
&\frac{\partial (x^T A x)}{\partial x} = Ax + A^T x
\end{aligned}
$$

A3. *Chain rule for matrix derivatives using Kronecker product*

Let Z be a $p \times q$ matrix whose entries are a matrix function of the elements of $Y: s \times t$, where Y is a function of matrix $X: m \times n$. That is, $Z = H_1(Y)$, where $Y = H_2(X)$. The matrix of derivatives of Z with respect to X is given by

$$
\frac{\partial Z}{\partial X} = \left\{ \frac{\partial\, vec^T(Y)}{\partial X} \otimes I_p \right\} \left\{ I_n \otimes \frac{\partial Z}{\partial\, vec(Y)} \right\}.
$$

9.2. Appendix B

Parameters	Values
M_b	1000 $[kg]$
I_b	810 $[kg\ m^2]$
k_b	20000 $[N/m]$
c_b	300 $[N/m/s]$
L_b	2.5 $[m]$

Table 1. The vehicle body parameter.

Parameters	Values
M_e	250 [kg]
I_e	8.10 [$kg\ m^2$]
k_e	200000 [N/m]
c_e	200 [$N/m/s$]
L_e	0.5 [m]

Table 2. The engine parameters.

Poles	Zeros
-6.2313 $\pm\ i$ 111.62	-6.23 $\pm\ i$ 111.69
-1.09 $\pm\ i$ 58.17	-0.97 $\pm\ i$ 62.59
0.14 $\pm\ i$ 29.48	0.03 $\pm\ i$ 26.86
-0.29 $\pm\ i$ 6.19	-1.10

Table 3. Pole-zero locations of the 8th -order model.

9.3. Appendix C

Theorem 1 (State Feedback) [9]. Consider dynamical system

$$\begin{cases} \dot{x}(t) = A\,x(t) + B_1\,u(t) + B_2\,w(t) \\ z(t) = C\,x(t) + Du(t) \end{cases}$$

under assumption (A, B_1, C) is stabilizable. For a given $\gamma > 0$, the differential Riccati equation

$$\dot{X} = A^T\,X + X\,A + X(\gamma^{-2}B_2\,B_2^T - B_1\,B_1^T\,)X + C^T C$$

has a positive semi-definite solution $X(t)$ such that $A - (B_1\,B_1^T - \gamma^{-2}B_2\,B_2^T)X(t)$ is asymptotically stable. Then the control law $u(t) = -B_1^T X(t)x(t) := K(t)\ X(t)$ is stabilizing and satisfies $\left\| z(t) \right\|_2 < \gamma \left\| w(t) \right\|_2$.

Author details

Hamid Reza Karimi

Department of Engineering, Faculty of Engineering and Science, University of Agder, Grimstad, Norway

10. References

[1] Aglietti G., Stoustrup J., Rogers E., Langley R., and Gabriel S., 'LTR Control Methodologies for Micro vibrations' *Proc. IEEE CCA*, Sept. 1998.

[2] Burrus C.S., Gopinath R.A. and Guo H., '*Introduction to Wavelets and Wavelet Transforms.*' Prentice Hall, Upper Saddle River, New Jersey, 1998.

[3] Cao T., Chen L., He F., and Sammut K., 'Active Vibration Absorber Design via Sliding Mode Control' *Proc. ACC*, June 2000.

[4] Cavallo A., Maria G., and Setola R., 'A Sliding Manifold approach for Vibration Reduction of Flexible Systems' *Automatica*, vol. 35, pp. 1689-1696, 1999.

[5] Chang R.Y. and Wang M.L., 'Legendre Polynomials Approximation to Dynamical Linear State-Space Equations with Initial and Boundary Value Conditions' *Int. J. Control*, 40, 215-232, 1984.

[6] Chen C.F. and Hsiao C.H., 'Haar Wavelet Method for Solving Lumped and Distributed-Parameter Systems' *IEE Proc. Control Theory Appl.*, 144(1), 87-94, 1997.

[7] Chen C.F. and Hsiao C.H., 'A State-Space Approach to Walsh Series Solution of Linear Systems' *Int. J. System Sci.*, 1965, 6(9), 833-858, 1965.

[8] Elliott S.J. and Nelson P.A., 'Active Noise Control' *IEEE Signal Processing Magazine*, 12-35, October, 1993.

[9] Green M. and Limebeer D.J.N., 'Linear Robust Control.' *Prentice Hall*, 1996.

[10] Hino M., Iwai Z., Mizumoto I., and Kohzawa R., 'Active Vibration Control of a Multi-Degree-of-Freedom Structure by the Use of Robust Decentralized Simple Adaptive Control' *Proc. IEEE CCA*, Sept. 1996.

[11] Hong J., and Bernstein D. S., 'Bode Integral Constraints, Collocation and Spill Over in Active Noise and Vibration Control' *IEEE Trans. on Control Systems Technology*, 6(1), 1998.

[12] Horng I.R., and Chou J.H., 'Analysis, Parameter Estimation and Optimal Control of Time-Delay Systems via Chebyshev series' *Int. J. Control*, 41, 1221-1234, 1985.

[13] Hsiao C.H. and Wang W.J., 'State Analysis and Parameter Estimation of Bilinear Systems via Haar Wavelets' *IEEE Trans. Circuits and Systems I: Fundamental Theory and Applications*, 47(2), 246-250, 2000.

[14] Hwang C. and Shin Y.P., 'Laguerre Operational matrices for Fractional Calculus and Applications' *Int. J. Control*, 34, 557-584, 1981.

[15] Kamman J.W. and Naghshineh K., 'A Comparison of Open-Loop Feedforward and Closed-Loop Methods for Active Noise Control Using Volume Velocity Minimization' *Applied Acoustics*, 57, 29 - 37, 1999.

[16] Karimi H.R., Lohmann B., Jabehdar Maralani P. and Moshiri B. 'A Computational Method for Solving Optimal Control and Parameter Estimation of Linear Systems Using Haar Wavelets' *Int. J. of Computer Mathematics*, 81(9), 1121-1132, 2004.

[17] Karimi H.R., Jabehdar Maralani P., Moshiri B., Lohmann B., 'Numerically Efficient Approximations to the Optimal Control of Linear Singularly Perturbed Systems Based on Haar Wavelets' *Int. J. of Computer Mathematics*, 82(4), 495-507, April 2005.

[18] Karimi H.R., Moshiri B., Lohmann B., and Jabehdar Maralani P. 'Haar Wavelet-Based Approach for Optimal Control of Second-Order Linear Systems in Time Domain' *J. of Dynamical and Control Systems*, 11(2), 237-252, 2005.

[19] Karkosch H.J., Svaricek F., Shoureshi R. and Vance, J.L., 'Automotive Applications of Active Vibration Control' *Proc. ECC*, 2000.
[20] Krtolica R., and Hrovat D., 'Optimal Active Suspension Control Based on A Half-Car Model' *Proc. 29th CDC*, pp. 2238-2243, 1990.
[21] Marzban H.R., and Razzaghi M., 'Solution of Time-Varying Delay Systems by Hybrid Functions' *Mathematics and Computers in Simulation*, 64, 597-607, 2004.
[22] Matschinsky W., 'Radführungen der Straßenfahrzeuge, Kinematik, Elasto-kinematik und Konstruktion' Springer, 1998.
[23] McDonald A.M., Elliott S.J. and Stokes M.A., 'Active Noise and Vibration Control within the Automobile' *in Proc. Active Control of Sound and Vibration*, 147 - 157, Tokyo 1991.
[24] Nonami K., and Sivriogu S., 'Active Vibration Control Using LMI-Based Mixed H_2 / H_∞ State and Output Feedback Control with Nonlinearity' *Proc. CDC*, Dec. 1996.
[25] Ohkita M. and Kobayashi Y. 'An Application of Rationalized Haar Functions to Solution of Linear Differential Equations' *IEEE Trans. on Circuit and Systems*, 9, 853-862, 1986.
[26] Preumont A., '*Vibration Control of Active Structures: An Introduction*' Kluwer Academic Publishers, 1997.
[27] Rao G.P., '*Piecewise Constant Orthogonal Functions and Their Application to Systems and Control*' Springer-Verlag, Berlin, Heidelberg, 1983.
[28] Razzaghi M., Razzaghi M., 'Fourier Series Direct Method for Variational Problems' *Int. J. Control*, 48, 887-895, 1988.
[29] Riley B., and Bodie M., 'An adaptive strategy for Vehicle Vibration and Noise Cancellation' *Proc. CDC.*, 1996.
[30] Seba B., Nedeljkovic N., Paschedag J. and Lohmann B., 'Feedback Control and FX-LMS Feedforward Control for Car Engine Vibration Attenuation' *Applied Acoustics*, 66, 277-296, 2005.
[31] Shoureshi R., and Bell M., 'Hybrid Adaptive Robust Structural Vibration Control' *Proc. ACC*, June 1999.
[32] Sievers L. and Flotow A., 'Linear Control Design For Active Vibration Isolation of Narrow Band Disturbances' *Proc. 27th CDC.*, Texas, 1988.
[33] Wang L.Y. and Zhan W. 'Robust disturbance attenuation with stability for linear systems with norm-bounded Nonlinear uncertainties.' *IEEE Trans. on Automatic Control*, 41, 886-888, 1996.
[34] Weng M., Lu X. and Tumper D., 'Vibration Control of Flexible Beams Using Sensor Averaging and Actuator Averaging Methods' *IEEE Trans. Control Systems Technology*, 10(4), July 2002.
[35] Yang J., Suematsu Y., and Kang Z., 'Two-Degree-of-Freedom Controller to reduce the Vibration of Vehicle Engine-Body System' *IEEE Trans. Control Systems Technology*, 9(2), 295-304, March 2001.
[36] Zhou K. and Khargonekar P.P., 'Robust stabilization of linear systems with norm-bounded time-varying uncertainty.' *System Control Letters*, 10, 17-20, 1988.

Chapter 7

On Variable Structure Control Approaches to Semiactive Control of a Quarter Car System

Mauricio Zapateiro, Francesc Pozo and Ningsu Luo

Additional information is available at the end of the chapter

1. Introduction

Vehicle suspension systems are one of the most critical components of a vehicle and it have been a hot research topic due to their importance in vehicle performance. These systems are designed to provide comfort to the passengers to protect the chassis and the freigt [28]. However, ride comfort, road holding and suspension deflection are often conflicting and a compromise of the requirements must be considered. Among the proposed solutions, active suspension is an approach to improve ride comfort while keeping suspension stroke and tire deflection within an acceptable level [11, 21].

In semiactive suspension, the value of the damper coefficient can be controlled and can show reasonable performance as compared to that of an active suspension control. Besides, it does not require external energy. For instance, in the work by [18] a semiactive suspension control of a quarter-car model using a hybrid-fuzzy-logic-based controlled is developed and implemented. [23] formulated a force-tracking PI controller for an MR-damper controlled quarter-car system. The preliminary results showed that the proposed semiactive force tracking PI control scheme could provide effective control of the sprung mass resonance as well as the wheel-hop control. Furthermore, the proposed control yields lower magnitudes of mass acceleration in the ride zone. [25] designed a semi active suspension system using a magnetorhelogical damper. The control law was formulated following the sky-hook technique in which the direction of the relative velocity between the sprung and unsprung masses is compared to that of the velocity of the unsprung mass. Depending on this result, an on-off action is performed. [8] designed a semiactive static output H_∞ controller for a quarter car system equipped with a magnetorheological damper. In this case, the control law was formulated in order to regulate the vertical acceleration as a measure to keep passengers' comfort within acceptable limits. They also added a constraint in order to keep the transfer function form road disturbance to suspension deflection small enough to prevent excessive suspension bottoming.

Backstepping is a recursive design for systems with nonlinearities not constrained by linear bounds. The ease with which backstepping incorporated uncertainties and unknown parameters contributed to its instant popularity and rapid acceptance. Applications of this technique have been recently reported ranging from robotics to industry or aerospace [6, 7, 15, 22, 24]. Backstepping control has also been explored in some works about suspension systems. For example, [26] designed a semiactive backstepping control combined with neural network (NN) techniques for a system with MR damper. In that work, the controller was formulated for an experimental platform, whose MR damper was modeled by means of an artificial neural network. The control input was updated with a backstepping controller. On the other hand, [16] studied a hybrid control of active suspension systems for quarter-car models with two-degrees-of-freedom. This hybrid control was implemented by controlling the linear part with H_∞ techniques and the nonlinear part with an adaptive controller based on backstepping.

Some works on Quantitative Feedback Theory (QFT) applied to the control of suspension systems can be found in the literature. For instance, [1] analyzed H_∞ and QFT controllers designed for an active suspension system in order to account for the structured and unstructured uncertainties of the system. As a result, the vertical body acceleration in QFT-controlled is lower than that of the H_∞-controlled and its performance is superior. In the presence of a hydraulic actuator, the QFT-controlled system performance degrades but it is still comparable to that of the H_∞-control. [28] addressed a study leading to compare the performance of backstepping and QFT controllers in active and semiactive control of suspension systems. In this case, the nonlinearities were treated as uncertainties in the model so that the linear QFT could be applied to the control formulation. As a result, similar performances between both classes of controllers were achieved.

In this chapter, we will analyze three model-free variable structure controllers for a class of semiactive vehicle suspension systems equipped with MR dampers. The variable structure control (VSC) is a control scheme which is well suited for nonlinear dynamic systems [12]. VSC was firstly studied in the early 1950's for systems represented by single-input high-order differential equations. A rise of interest became in the 1970's because the robustness of VSC were step by step recognized. This control method can make the system completely insensitive to time-varying parameter uncertainties, multiple delayed state perturbations and external disturbances [17]. Nowadays, research and development continue to apply VSC control to a wide variety of engineering areas, such as aeronautics (guidance law of small bodies [29]), electric and electronic engineering (speed control of an induction motor drive [3]). By using this kind of controllers, it is possible to take the best out of several different systems by switching from one to the other. The first strategy that we propose in this work, σ_1, is based on the difference between the body angular velocity and the wheel angular velocity. The second strategy, σ_2, more complex, is based both on the difference between the body angular velocity and the wheel angular velocity, and on the difference between the body angular position and the wheel angular position. In this case, the resulting algorithm can be viewed as the clipped control in [9], but with some differences. Finally, the last strategy presented is based on a time variable depending on the absolute value of the difference between the body angular velocity and the wheel angular velocity, and on the difference between the body angular position and the wheel angular position. The study of the three variable structure controllers will be complemented with the comparison of a model-based controller which has been successfully applied by the authors in other works: backstepping. As it was mentioned earlier, backstepping is well suited to this kind of problems because it can account for robustness and nonlinearities. It has been used by the authors to analyze this particular problem [28] with interesting results.

The chapter is organized as follows. Section 2 presents the mathematical details of the system to be controlled. In Section 3, the three variable structure controllers are developed. In Section 4, the backstepping control formulation details are outlined. Section 5 shows the numerical results, and in Section 6, the conclusions are drawn.

2. Suspension system model

The suspension system can be modeled as a quarter car model, as shown in Figure 1. The system can be viewed as a composition of two subsystems: the tyre subsystem and the suspension subsystem. The tyre subsystem is represented by the wheel mass m_u while the suspension subsystem consists of a sprung mass, m_s, that resembles the vehicle mass. This way of seeing the system will be useful later on when designing the model-based semi active controller. The compressibility of the wheel pneumatic is k_t, while c_s and k_s are the damping and stiffness of the uncontrolled suspension system. The quarter car model equations are given by:

$$m_s\dot{x}_4 + c_s(x_4 - x_2) + k_s x_3 - f_{\mathrm{mr}} = 0 \tag{1}$$

$$m_u\dot{x}_2 - c_s(x_4 - x_2) - k_s x_3 + k_t x_1 + f_{\mathrm{mr}} = 0 \tag{2}$$

where:

- x_1 is the tyre deflection
- x_2 is the unsprung mass velocity
- x_3 is the suspension deflection
- x_4 is the sprung mass velocity.

Taking x_1, x_2, x_3 and x_4 as state variables allows us to formulate the following state-space representation:

- Tyre subsystem:

$$\begin{aligned} \dot{x}_1 &= x_2 - d \\ \dot{x}_2 &= -\frac{k_t}{m_u}x_1 + \rho u \end{aligned} \tag{3}$$

- Suspension subsystem:

$$\begin{aligned} \dot{x}_3 &= -x_2 + x_4 \\ \dot{x}_4 &= -u \end{aligned} \tag{4}$$

where $\rho = m_s/m_u$, d is the velocity of the disturbance input and u is the acceleration input due to the damping subsystem. The input u is given by:

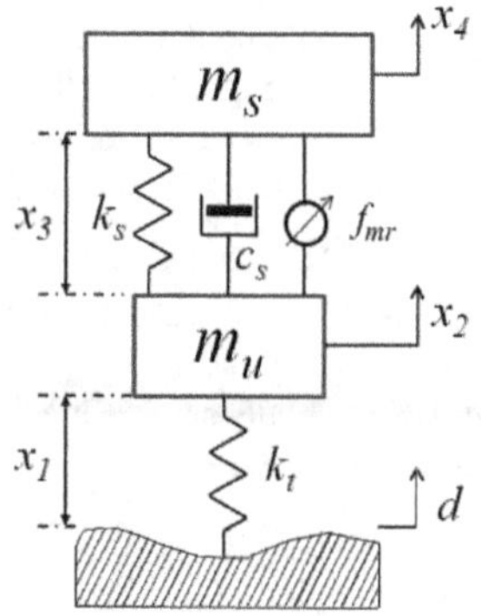

Figure 1. Quarter car suspension model

$$u = \frac{1}{m_s}(k_s x_3 + c_s(x_4 - x_2) - f_{\rm mr}) \tag{5}$$

where $f_{\rm mr}$ is the damping force generated by the semiactive device. In this study, we assume that the semiactive device is magnetorheological (MR) damper. It is modeled according to the following Bouc-Wen model [19]:

$$f_{\rm mr} = c_0(v)z_4 + k_0(v)z_3 + \alpha(v)\zeta \tag{6}$$

$$\dot{\zeta} = -\delta|z_4|\zeta|\zeta|^{n-1} - \beta z_4|\zeta|^n + \kappa z_4 \tag{7}$$

where ζ is an evolutionary variable that describes the hysteretic behavior of the damper, z_4 is the piston velocity, z_3 is the piston deflection and v is a voltage input that controls the current that generates the magnetic field; δ, β, κ and n are parameters that are chosen so to adjust the hysteretic dynamics of the damper; $c_0(v) = c_{0a} + c_{0b}v$ represents the voltage-dependent damping, $k_0(v) = k_{0a} + k_{0b}v$ represents the voltage-dependent stiffness and $\alpha(v) = \alpha_a + \alpha_b v$ is a voltage-dependent scaling factor.

3. Variable structure controller formulation

Feedback control radically alters the dynamics of a system: it affects its natural frequencies, its transient response as well as its stability. The MR damper of the quarter-car model considered in this study is voltage-controlled, so the voltage (v) is updated by a feedback control loop.

It is well known that the force generated by the MR damper cannot be commanded; only the voltage v applied to the current driver for the MR damper can be directly changed. One of the first control approaches involving an MR damper was proposed by [9] and called it clipped optimal control. In this approach, the command voltage takes one of two possible values: zero or the maximum. This is chosen according to the following algorithm:

$$v = V_{\rm max} H\{(f_{\rm d} - f_{\rm mr})f_{\rm mr}\} \tag{8}$$

$$= \frac{V_{\rm max}}{2}\left[\operatorname{sgn}\left[(f_{\rm d} - f_{\rm mr})f_{\rm mr}\right] + 1\right], \tag{9}$$

where $V_{\max}$ is the maximum voltage to the current driver associated with saturation of the magnetic field in the MR damper, $H(\cdot)$ is the Heaviside step function, f_{d} is the desired control force and f_{mr} is the measured force of the MR damper.

The sign part of equation (9) can be transformed in the following way:

$$\begin{aligned}
\operatorname{sgn}\left[(f_{\mathrm{d}}-f_{\mathrm{mr}})\,f_{\mathrm{mr}}\right] &= \begin{cases} 1, & (f_{\mathrm{d}}-f_{\mathrm{mr}})\,f_{\mathrm{mr}}>0 \\ -1, & (f_{\mathrm{d}}-f_{\mathrm{mr}})\,f_{\mathrm{mr}}<0 \end{cases} \\
&= \begin{cases} 1, & [(f_{\mathrm{d}}-f_{\mathrm{mr}})>0 \text{ and } f_{\mathrm{mr}}>0] \text{ or } [(f_{\mathrm{d}}-f_{\mathrm{mr}})<0 \text{ and } f_{\mathrm{mr}}<0] \\ -1, & [(f_{\mathrm{d}}-f_{\mathrm{mr}})>0 \text{ and } f_{\mathrm{mr}}<0] \text{ or } [(f_{\mathrm{d}}-f_{\mathrm{mr}})<0 \text{ and } f_{\mathrm{mr}}>0] \end{cases} \\
&= \begin{cases} 1, & [f_{\mathrm{d}}>f_{\mathrm{mr}} \text{ and } f_{\mathrm{mr}}>0] \text{ or } [f_{\mathrm{d}}<f_{\mathrm{mr}} \text{ and } f_{\mathrm{mr}}<0] \\ -1, & [f_{\mathrm{d}}>f_{\mathrm{mr}} \text{ and } f_{\mathrm{mr}}<0] \text{ or } [f_{\mathrm{d}}<f_{\mathrm{mr}} \text{ and } f_{\mathrm{mr}}>0] \end{cases} \\
&= \begin{cases} 1, & f_{\mathrm{d}}>f_{\mathrm{mr}} \text{ and } f_{\mathrm{mr}}>0 \\ 1, & f_{\mathrm{d}}<f_{\mathrm{mr}} \text{ and } f_{\mathrm{mr}}<0 \\ -1, & f_{\mathrm{d}}>f_{\mathrm{mr}} \text{ and } f_{\mathrm{mr}}<0 \\ -1, & f_{\mathrm{d}}<f_{\mathrm{mr}} \text{ and } f_{\mathrm{mr}}>0 \end{cases}
\end{aligned}$$

Finally, the full expression in equation (9) can be rewritten as a piecewise function in the following way:

$$\frac{V_{\max}}{2}\left[\operatorname{sgn}\left[(f_{\mathrm{d}}-f_{\mathrm{mr}})f_{\mathrm{mr}}\right]+1\right] = \begin{cases} V_{\max}, & f_{\mathrm{d}}>f_{\mathrm{mr}} \text{ and } f_{\mathrm{mr}}>0 \\ V_{\max}, & f_{\mathrm{d}}<f_{\mathrm{mr}} \text{ and } f_{\mathrm{mr}}<0 \\ 0, & f_{\mathrm{d}}>f_{\mathrm{mr}} \text{ and } f_{\mathrm{mr}}<0 \\ 0, & f_{\mathrm{d}}<f_{\mathrm{mr}} \text{ and } f_{\mathrm{mr}}>0 \end{cases}$$

This algorithm for selecting the command signal is graphically represented in Figure 2. More precisely, the shadowed area in Figure 2 is the area where $f_{\mathrm{d}} > f_{\mathrm{mr}}$ and $f_{\mathrm{mr}} > 0$, or $f_{\mathrm{d}} < f_{\mathrm{mr}}$ and $f_{\mathrm{mr}} < 0$. Note that in that particular work, they used the voltage as the control signal because that is the way that current driver can be controlled.

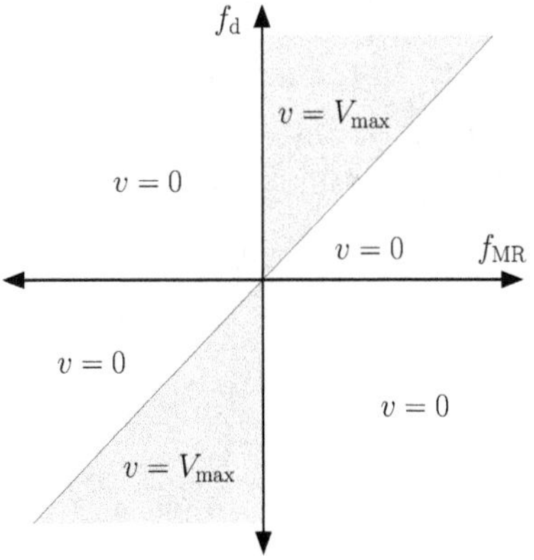

Figure 2. Graphical representation of the algorithm in equation (8) for selecting the command signal.

In this paper we consider the same idea of changing the voltage. This control signal is computed according to the following control strategies, computed as a function of the sprung mass velocity (x_4), the unsprung mass velocity (x_2), and the suspension deflection (x_3):

$$\sigma_1: \quad v(x_2, x_4) = \frac{V_{\max}}{2}\left[\operatorname{sgn}(x_4 - x_2) + 1\right] \tag{10}$$

$$\sigma_2: \quad v(x_2, x_3, x_4) = \frac{V_{\max}}{2}\left[\operatorname{sgn}\left(\operatorname{sgn}(x_4 - x_2) + x_3\right) + 1\right] \tag{11}$$

$$\sigma_3: \quad v(x_2, x_4) = \frac{V_{\max}}{2}\left[\operatorname{sgn}(r) + 1\right], \ \frac{dr}{dt} = -100r|x_4 - x_2| - 10(x_4 - x_2) \tag{12}$$

Variable structure controllers (VSC) are a very large class of robust controllers [10]. The distinctive feature of VSC is that the structure of the system is intentionally changed according to an assigned law. This can be obtained by switching on or cutting off feedback loops, scheduling gains and so forth. By using VSC, it is possible to take the best out of several different systems (more precisely structures), by switching from one to the other. The control law defines various regions in the phase space and the controller switches between a structure and another at the boundary between two different regions according to the control law.

The three strategies presented in this section can be viewed as variable structure controllers, since the value of the control signal is set to be zero or one, as can be seen in the following transformations:

$$\sigma_1: \quad v(x_2, x_4) = \frac{V_{\max}}{2}\left[\operatorname{sgn}(x_4 - x_2) + 1\right] \tag{13}$$

$$= \begin{cases} 0, & \text{if } \Delta\omega < 0, \\ V_{\max}, & \text{if } \Delta\omega \geq 0 \end{cases} \tag{14}$$

$$\sigma_2: \quad v(x_2, x_3, x_4) = \frac{V_{\max}}{2}\left[\operatorname{sgn}\left(\operatorname{sgn}(x_4 - x_2) + x_3\right) + 1\right] \tag{15}$$

$$= \begin{cases} 0, & \text{if } \Delta\omega < 0,\ x_3 < 1 \text{ (region IV)}, \\ 0, & \text{if } \Delta\omega > 0,\ x_3 < -1 \text{ (region I)}, \\ V_{\max}, & \text{if } \Delta\omega < 0,\ x_3 \geq 1 \text{ (region II)}, \\ V_{\max}, & \text{if } \Delta\omega \geq 0,\ x_3 \geq -1 \text{ (region III)} \end{cases} \tag{16}$$

$$\sigma_3: \quad v(x_2, x_4) = \frac{V_{\max}}{2}\left[\operatorname{sgn}(r) + 1\right] \tag{17}$$

$$= \begin{cases} 0, & \text{if } r < 0, \\ V_{\max}, & \text{if } r \geq 0 \end{cases}, \ \frac{dr}{dt} = -100r|x_4 - x_2| - 10(x_4 - x_2) \tag{18}$$

where $\Delta\omega = x_4 - x_2$. In Figure 3 we hace depicted the graphical representation of the strategy σ_2 for selecting the command signal.

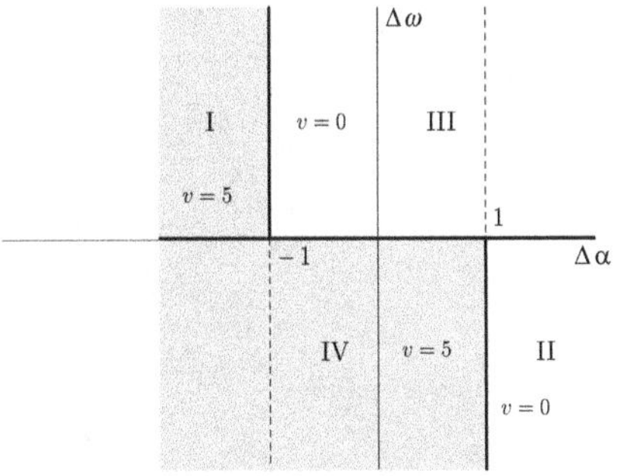

Figure 3. Graphical representation of the strategy σ_2 for selecting the command signal.

Semi-active control have two essential characteristics. The first is that the these devices offer the adaptability of active control devices without requiring the associated large power sources. The second is that the device cannot inject energy into the system; hence semi-active control devices do not have the potential to destabilize (in the bounded input–bounded output sense) the system [20]. As a consequence, the stability of the closed-loop system is guaranteed.

4. Backstepping controller formulation

In this section we present the formulation of a model-based controller. The objective, as explained in the Introduction, is to make a comparison between this model-based controller and the VSC controllers. We will appeal to the backstepping technique that has been developed in previous works for this kind of systems.The objective is to design an adaptive backstepping controller to regulate the suspension deflection with the aid of an MR damper thus providing safety and comfort while on the road. The adaptive backstepping controller will be designed in such a way that, for a given $\gamma > 0$, the state-dependent error variables e_1 and e_2 (to be defined later) accomplish the following H_∞ performance $J_\infty < 0$:

$$J_\infty = \int_0^\infty (\mathbf{e}^T \mathbf{R} \mathbf{e} - \gamma^2 \mathbf{w}^T \mathbf{w}) dt \tag{19}$$

where $\mathbf{e} = (e_1, e_2)^{\mathrm{T}}$ is a vector of controlled signals, $\mathbf{R} = \mathrm{diag}\{r_1, r_2\}$ is a positive definite matrix and $\mathbf{w}$ is an energy-bounded disturbance.

In order to formulate the backstepping controller, the state space model (3) - (4) must be first written in strict feedback form [14]. Therefore, the following coordinate transformation is performed [13]:

$$\begin{aligned} z_1 &= x_1 + \frac{\rho}{\rho+1} x_3 \\ z_2 &= \frac{1}{\rho+1} x_2 + \frac{\rho}{\rho+1} x_4 \\ z_3 &= x_3 \\ z_4 &= -x_2 + x_4 \end{aligned} \tag{20}$$

The system, represented in the new coordinates, is given by:

- Tyre subsystem:

$$\begin{aligned}\dot{z}_1 &= z_2 - d \\ \dot{z}_2 &= -k_t[m_u(\rho+1)]^{-1}z_1 + \rho k_t[m_u(\rho+1)^2]^{-1}z_3\end{aligned} \tag{21}$$

- Suspension subsystem:

$$\begin{aligned}\dot{z}_3 &= z_4 \\ \dot{z}_4 &= k_t m_u^{-1}z_1 - k_t\rho[m_u(\rho+1)]^{-1}z_3 - (\rho+1)u\end{aligned} \tag{22}$$

Substitution of the expression for u (5) into (22) yields:

$$\begin{aligned}\dot{z}_3 =& z_4 \\ \dot{z}_4 =& k_t m_u^{-1}z_1 - k_t\rho[m_u(\rho+1)]^{-1}z_3 - \\ & (\rho+1)m_s^{-1}[k_s x_3 + c_s(x_4 - x_2) - f_{\text{mr}}] \\ =& -[k_t m_s\rho(\rho+1)^{-1} + (\rho+1)k_s m_u](m_u m_s)^{-1}z_3 + \\ & k_t m_u^{-1}z_1 - (\rho+1)m_s^{-1}c_s z_4 + (\rho+1)m_s^{-1}f_{\text{mr}} \\ =& d_i - a_k z_3 - a_c z_4 + a_f f_{\text{mr}}\end{aligned} \tag{23}$$

where $a_k = [k_t m_s \rho(\rho+1)^{-1} + (\rho+1)k_s m_u](m_u m_s)^{-1}$, $a_c = (\rho+1)m_s^{-1}c_s$ and $a_f = (\rho+1)m_s^{-1}$; $d_i = k_t m_s^{-1} z_1$ reflects the fact that the disturbance enters to the suspension subsystem through the tyre subsystem.

Assume that a_k and a_c in (23) are uncertain constant parameters whose estimated values are $\hat{a}_k$ and $\hat{a}_c$, respectively. Thus, the errors between the estimates and the actual values are given by:

$$\tilde{a}_k = a_k - \hat{a}_k \tag{24}$$

$$\tilde{a}_c = a_c - \hat{a}_c \tag{25}$$

Let $a_d = k_t[m_u(\rho+1)]^{-1}$, $a_n = \rho k_t[m_u(\rho+1)^2]^{-1}$ and $a_m = k_t m_u^{-1}$. From (21) - (22), it can be shown that the transfer functions from $d(t)$ and $f_{\text{mr}}(t)$ to $z_1(t)$ are:

$$\frac{Z_1(s)}{D(s)} = \frac{-s(s^2 + a_c s + a_k)}{s^4 + a_c s^3 + (a_d + a_k)s^2 + a_d a_c s + a_d a_k - a_m a_n} \tag{26}$$

$$\frac{Z_1(s)}{F_{\text{mr}}(s)} = \frac{a_n a_f}{s^4 + a_c s^3 + (a_d + a_k)s^2 + a_d a_c s + a_d a_k - a_m a_n} \tag{27}$$

If the poles of the transfer functions (26) and (27) are in the left side of the s plane, then we can guarantee the bounded input - bounded output (BIBO) stability of $Z_1(s)$ for any bounded input $D(s)$ and $F_{\text{mr}}(s)$. Thus, the disturbance input $d_i(t)$ in (23) is also bounded. This boundedness condition will be necessary later in the controller stability condition.

Finally, since $d_i(t)$ is the only disturbance input to the suspension subsystem, the vector **w** of the H_∞ performance objective as given in (19) becomes:

$$J_\infty = \int_0^\infty (\mathbf{e}^T\mathbf{R}\mathbf{e} - \gamma^2 d_i^2)dt \tag{28}$$

In order to begin with the adaptive backstepping design, we firstly define the following error variable and its derivative:

$$e_1 = z_3 \tag{29}$$

$$\dot{e}_1 = \dot{z}_3 = z_4 \tag{30}$$

Now, the following Lyapunov function candidate is chosen:

$$V_1 = \frac{1}{2}e_1^2 \tag{31}$$

whose first-order derivative is:

$$\dot{V}_1 = e_1\dot{e}_1 = e_1 z_4 \tag{32}$$

Equation (30) can be stabilized with the following virtual control input:

$$z_{4d} = -r_1 e_1 \tag{33}$$

$$\dot{z}_{4d} = -r_1\dot{e}_1 = -r_1 z_4 \tag{34}$$

where $r_1 > 0$. Now define a second error variable and its derivative:

$$e_2 = z_4 - z_{4d} \tag{35}$$

$$\dot{e}_2 = \dot{z}_4 - \dot{z}_{4d} \tag{36}$$

Therefore,

$$\dot{V}_1 = e_2 z_4 = e_1(e_2 - r_1 e_1) = e_1 e_2 - r_1 e_1^2 \tag{37}$$

On the other hand, the derivatives of the errors of the uncertain parameter estimations are given by:

$$\dot{\tilde{a}}_k = -\dot{\hat{a}}_k \tag{38}$$

$$\dot{\tilde{a}}_c = -\dot{\hat{a}}_c \tag{39}$$

Now, an augmented Lyapunov function candidate is chosen:

$$V = V_1 + \frac{1}{2}e_2^2 + \frac{1}{2r_k}\tilde{a}_k^2 + \frac{1}{2r_c}\tilde{a}_c^2 \tag{40}$$

Thus, by using (35) - (39) and the fact that $a_k = \tilde{a}_k + \hat{a}_k$ and $a_c = \tilde{a}_c + \hat{a}_c$, the derivative of V yields:

$$\begin{aligned}
\dot{V} =& e_1\dot{e}_1 + e_2\dot{e}_2 + r_k^{-1}\tilde{a}_k\dot{\tilde{a}}_k + r_c^{-1}\tilde{a}_c\dot{\tilde{a}}_c \\
=& e_1e_2 - r_1e_1^2 + e_2d_i - a_kz_3e_2 - a_cz_4e_2 + a_ff_{\text{mr}}e_2 - r_1z_4e_2 - r_k^{-1}\tilde{a}_k\dot{\hat{a}}_k - r_c^{-1}\tilde{a}_c\dot{a}_c \\
=& e_1e_2 - r_1e_1^2 + e_2d_i + a_ff_{\text{mr}}e_2 - r_1z_4e_2 - r_k^{-1}\tilde{a}_k\dot{\hat{a}}_k - (\tilde{a}_k + \hat{a}_k)z_3e_2 - (\tilde{a}_c + \hat{a}_c)z_4e_2 - r_c^{-1}\tilde{a}_c\dot{\hat{a}}_c \\
=& e_1e_2 - r_1e_1^2 + e_2d_i - \tilde{a}_k(z_3e_3 + r_k^{-1}\dot{\hat{a}}_k) - \hat{a}_kz_3e_2 - \tilde{a}_c(z_4e_2 + r_c^{-1}\dot{\hat{a}}_c) - \hat{a}_cz_4e_2 + a_f + \\
& f_{\text{mr}}e_2 - r_1z_4e_2
\end{aligned} \tag{41}$$

Now consider the following adaptation laws:

$$z_3 e_1 + r_k^{-1}\dot{\hat{a}}_k = 0 \tag{42}$$

$$z_4 e_2 + r_c^{-1}\dot{\hat{a}}_c = 0 \tag{43}$$

Substitution of (42) and (43) into (41) yields:

$$\dot{V} = -r_1 e_1^2 + e_2 d_i + e_2(e_1 - \hat{a}_k z_3 - \hat{a}_c z_4 + a_f f_{\mathrm{mr}} - r_1 z_4) \tag{44}$$

By choosing the following control law:

$$f_{\mathrm{mr}} = -\frac{e_1 - \hat{a}_k z_3 - \hat{a}_c z_4 - r_1 z_4 + r_2 e_2 + e_2(2\gamma)^{-2}}{a_f} \tag{45}$$

with $\gamma > 0$ and $r_2 > 0$, we get:

$$\begin{aligned} \dot{V} &= -r_1 e_1^2 + e_2 d_i - r_2 e_2^2 - e_2^2(2\gamma)^{-2} \\ &= -r_1 e_1^2 + e_2 d_i - r_2 e_2^2 - e_2^2(2\gamma)^{-2} + \gamma^2 d_i^2 - \gamma^2 d_i^2 \\ &= -r_1 e_1^2 - r_2 e_2^2 + \gamma^2 d_i^2 - (\gamma d_i - e_2(2\gamma)^{-2})^2 \\ \dot{V} &\leq -r_1 e_1^2 - r_2 e_2^2 + \gamma^2 d_i^2 \end{aligned} \tag{46}$$

The objective of guaranteeing global boundedness of trajectories is equivalently expressed as rendering $\dot{V}$ negative outside a compact region. As stated earlier, the disturbance input d_i is bounded as long as the poles of the transfer functions (26) and (27) are in the left side of the s plane. When this is the case, the boundedness of the input disturbance d_i guarantees the existence of a small compact region $D \subset \mathbb{R}^2$ (depending on γ and d_i itself) such that $\dot{V}$ is negative outside this set. More precisely, when $r_1 e_1^2 + r_2 e_2^2 < \gamma^2 d_i^2$, $\dot{V}$ is positive and then the error variables are increasing values. Finally, when the expression $r_1 e_1^2 + r_2 e_2^2$ is greater than $\gamma^2 d_i^2$, $\dot{V}$ is then negative. This implies that all the closed-loop trajectories have to remain bounded, as we wanted to show. Now, under zero initial conditions, from 46 we can write:

$$\int_0^\infty \dot{V}\,dt \leq -\int_0^\infty r_1 e_1^2\,dt - \int_0^\infty r_2 e_2^2\,dt + \int_0^\infty \gamma^2 d_i^2\,dt \tag{47}$$

or, equivanlently,

$$V|_{t=\infty} - V|_{t=0} \leq -\int_0^\infty \mathbf{e}^T\mathbf{R}\mathbf{e}\,dt + \gamma^2\int_0^\infty d_i^2\,dt \tag{48}$$

Then, it can be shown that

$$J_\infty = \int_0^\infty (\mathbf{e}^T\mathbf{R}\mathbf{e} - \gamma^2 d_i^2)\,dt \leq -V|_{t=\infty} \leq 0 \tag{49}$$

Thus, the adaptive backstepping controller satisfies both the H_∞ performance and the asymptotic stability of the system.

The control force given by (45) can be used to drive an actively controlled damper. However, the fact that semiactive devices cannot inject energy into a system, makes necessary the modification of this control law in order to implement it with a semiactive damper; that is, semiactive dampers cannot apply force to the system, only absorb it. There are different ways to perform this [2, 27]. In this work, we will calculate the MR damper voltage making use of its mathematical model. Thus, the following control law is proposed:

$$v = \frac{-e_1 - \hat{a}_z z_3 + \hat{a}_c z_4 + r_1 z_4 - r_2 e_2 - e_2(2\gamma)^{-2} + a_f(c_{0a} z_4 + k_{0a} z_3 + \alpha_a \zeta)}{a_f(c_{0b} z_4 + k_{0b} z_3 + \alpha_b \zeta)} \tag{50}$$

provided that $a_f(c_{0b} z_4 + k_{0b} z_3 + \alpha_b \zeta) \neq 0$; otherwise, $v = 0$.

The same process followed to obtain the control law (45) can be used to demonstrate that the control law (4) does stabilize the system. Begin by replacing (6) into (44) in order to obtain:

$$\begin{aligned} \dot{V} = & - r_1 e_1^2 + e_2 d_i + e_2[e_1 - \hat{a}_k z_3 - \hat{a}_c z_4 + a_f(c_{0a} z_4 + k_{0a} z_3 + \alpha_a \zeta) + \\ & a_f(c_{0b} z_4 + k_{0b} z_3 + \alpha_b \zeta) v - r_1 z_4] \end{aligned} \tag{51}$$

Thus, by replacing the control law of (4) into (51) we also get $\dot{V} \leq -r_1 e_1^2 - r_2 e_2^2 + \gamma^2 d_i^2$ and, as previously stated, the stability of the system is guaranteed.

Finally, we can write the control law in terms of the state variables as follows:

$$\begin{aligned} v = & \frac{\left(-\hat{a}_c - r_1 + r_2 + (2\gamma)^{-2} + a_f c_{0a}\right) x_2 + \left(-1 - \hat{a}_z - r_1 r_2 + r_1 (2\gamma)^{-2} + a_f k_{0a}\right) x_3}{-a_f c_{0b} x_2 + a_f k_{0b} x_3 + a_f c_{0b} x_4 + a_f \alpha_b \zeta} + \\ & \frac{\left(\hat{z}_c + r_1 - r_2 - (2\gamma)^{-2} + a_f c_{0a}\right) x_4 + a_f \alpha_a \zeta}{-a_f c_{0b} x_2 + a_f k_{0b} x_3 + a_f c_{0b} x4 + a_f \alpha_b \zeta} \end{aligned} \tag{52}$$

5. Numerical simulations

In this section we will analyze the performance results obtained form simulations performed in Matlab/SImulink. The numerical values of the model that we used in this study. Thus: $\alpha_a = 332.7$ N/m, $\alpha_b = 1862.5$ N·V/m, $c_{0a} = 7544.1$ N·s/m, $c_{0b} = 7127.3$ N·s·V/m, $k_{0}a = 11375.7$ N/m, $k_{0b} = 14435.0$ N·V/m, $\delta = 4209.8$ m^{-2}, $\kappa = 10246$ and $n = 2$. This is a scaled version of the MR damper found in [5]. The parameter values of the suspension system are [13]: m_s=11739 kg, m_u=300 kg, k_s=252000 N/m, c_s=10000 N·s/m and k_t=300000 N/m. In order to facilitate the analysis, we will quantify the performance results by means of the indices shown in Table 1. Indices J_1 - J_3 show the ratio between the peak response of the controlled suspension system (displacement, velocity and acceleration) and that of the uncontrolled system. Indices J_4 - J_6 are the normalized ITSE (integral of the time squared error) signals that indicate how much the displacement, velocity and acceleration are attenuated compared to the uncontrolled case. Index J_7 is the relative maximum control effort with respect to the weight of the suspension system. Small indices indicate good control performance. Two scenarios are considered: an uneven road, simulated by random vibrations and the presence of a bump on the road.

We assume that the car has laser sensors that allow us to read the position of the sprung and unsprung masses. Since the velocities are needed for control implementation, these are obtained by first low-pass filtering the displacement readings and then applying a filter of the form $\frac{s}{(\lambda s+1)^q}$ that approximates the derivative of the signal. In this filter, λ is a sufficiently small constant that can be obtained from the ratio between the two-norm of the second derivative of the signal and the noise amplitude; q is the order of the filter which should be at lest equal to 2. Choosing parameters this way, allows for minimizing the error between the real and the estimated signal derivatives [4].

Index	Definition
$J_1 = \frac{max\|x_3(t)\|_{cont}}{max\|x_3(t)\|_{unc}}$	Norm. peak suspension deflection.
$J_2 = \frac{max\|x_4(t)\|_{cont}}{max\|x_4(t)\|_{unc}}$	Norm. peak sprung mass velocity.
$J_3 = \frac{max\|\dot{x}_4(t)\|_{cont}}{max\|\dot{x}_4(t)\|_{unc}}$	Norm. peak sprung mass acceleration.
$J_4 = \frac{\int_0^T tx_{3cont}^2(t)\,dt}{\int_0^T tx_{3unc}^2(t)\,dt}$	Norm. suspension deflection ITSE.
$J_5 = \frac{\int_0^T tx_{4cont}^2(t)\,dt}{\int_0^T tx_{4unc}^2(t)\,dt}$	Norm. sprung mass velocity ITSE.
$J_6 = \frac{\int_0^T t\dot{x}_{4cont}^2(t)\,dt}{\int_0^T t\dot{x}_{4unc}^2(t)\,dt}$	Norm. sprung mass acceleration ITSE.
$J_7 = \frac{max\|f_{mr}(t)\|}{w_s}$	Maximum control effort.

Table 1. Performance indices.

In the first scenario, the unevenness of the road was simulated by random vibration, as shown in Figure 4. This figure also compares the performance of the three σ controllers. What we can see for this figure, is that the three VSC controllers perform in a similar way and satisfactorily control the deflection of the tyre subsystem. In Figure 5, we see the performance of the same controllers at regulating the suspension deflection. Once again, the three controllers accomplish the objective in a similar way. This visual observations can be confirmed by analyzing the performance indices of Table 2. In Figures 6 and 7, we can see a comparison of the σ_3 controller and the backstepping controller. A notable superiority of the VSC controller is observed over the backstepping controller. It can be due to the fact that this kind of controllers are more sensitive to the fast-changing dynamics of a signal, in this case, the velocity, which can make it react faster. The performance indices of Table 2 also show that it is harder for the backstepping controller to keep the peak acceleration, velocity and displacement under acceptable limits, despite its control effort is much higher than that of the VSC controllers.

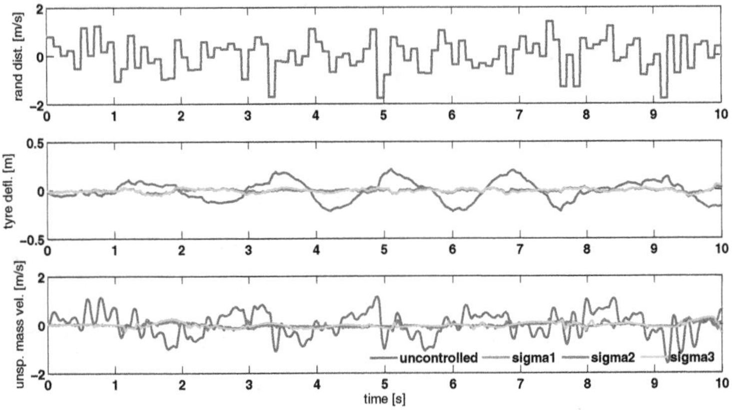

Figure 4. Uneven road disturbance and tyre subsystem response.

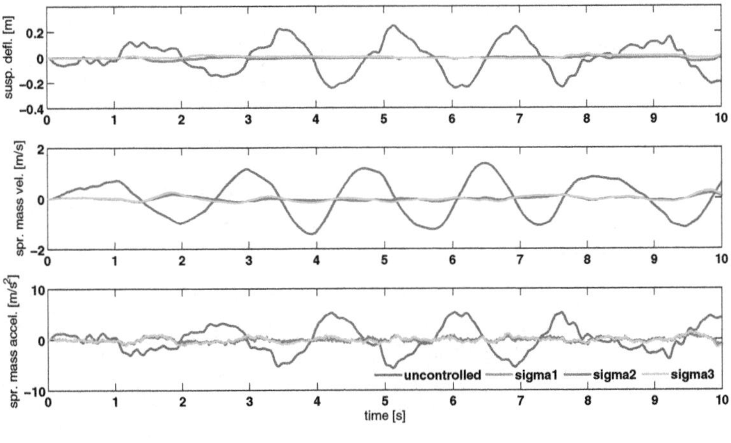

Figure 5. Uneven road disturbance and tyre subsystem response.

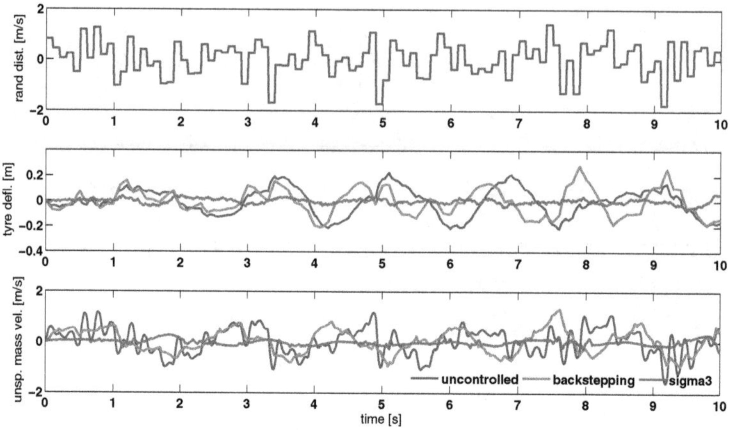

Figure 6. Uneven road disturbance and tyre subsystem response.

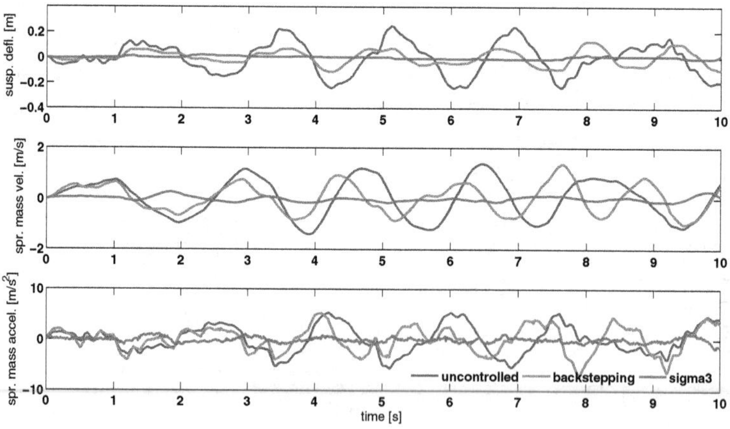

Figure 7. Uneven road disturbance and car subsystem response.

Index	σ_1	σ_2	σ_3	Backstepping
J_1	0.1288	0.1280	0.0993	0.5144
J_2	0.1485	0.1488	0.2157	0.9800
J_3	0.2205	0.2803	0.2803	1.2586
J_4	0.0059	0.0058	0.0053	0.2317
J_5	0.0090	0.0089	0.0181	0.5852
J_6	0.0189	0.0187	0.0310	0.9615
J_7	0.1268	0.1279	0.1471	0.4538

Table 2. Performance indices of the random unevenness disturbance case.

In the second scenario a bump on the road is simulated as seen in Figure 8. In this case, the VSC controllers have a similar performance and it happened in the previous scenario. The performance indices of Table 3 confirm this fact. In comparison, the σ_3 controller seems to perform slightly better, specially at reusing the peak response of the suspicion and tyre deflections as can be seen in Figure 10 and 11 where a comparison between then σ_3 and backstepping controllers is illustrated. These results are in the line than those of the first scenario. It is worth noting the fact that the VSC controllers perform better with less control effort.

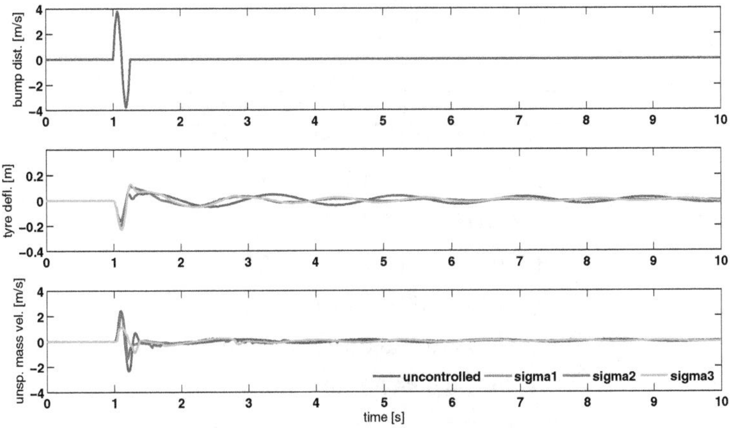

Figure 8. Bump on the road disturbance and tyre subsystem response.

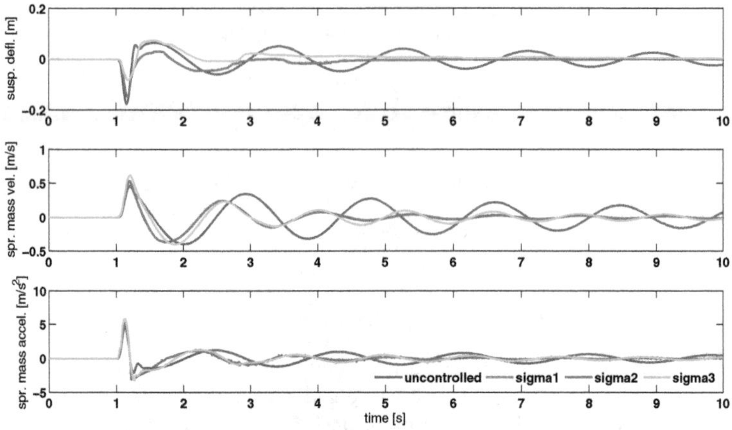

Figure 9. Bump on the road disturbance and tyre subsystem response.

Index	σ_1	σ_2	σ_3	Backstepping
J_1	0.8317	0.8325	0.4584	0.4271
J_2	1.1507	1.1505	1.3430	1.3892
J_3	1.1157	1.2623	1.2623	1.3007
J_4	0.1605	0.1625	0.2241	0.0703
J_5	0.1827	0.1797	0.2681	0.4702
J_6	0.4168	0.4113	0.5884	1.0308
J_7	0.3613	0.3614	0.4100	0.4431

Table 3. Performance indices of the road bump disturbance case.

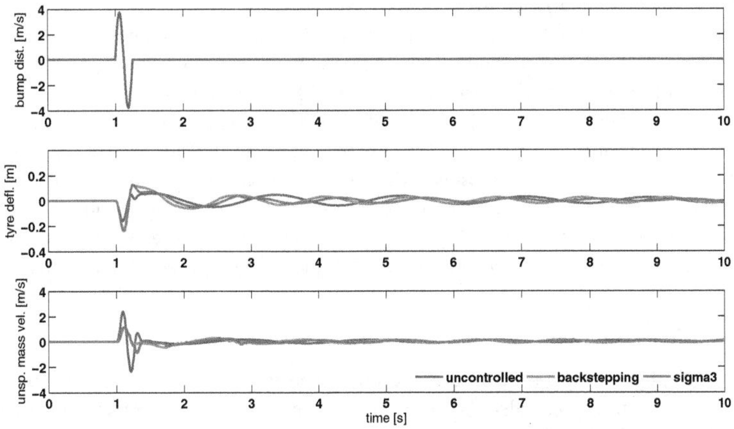

Figure 10. Bump on the road disturbance and tyre subsystem response.

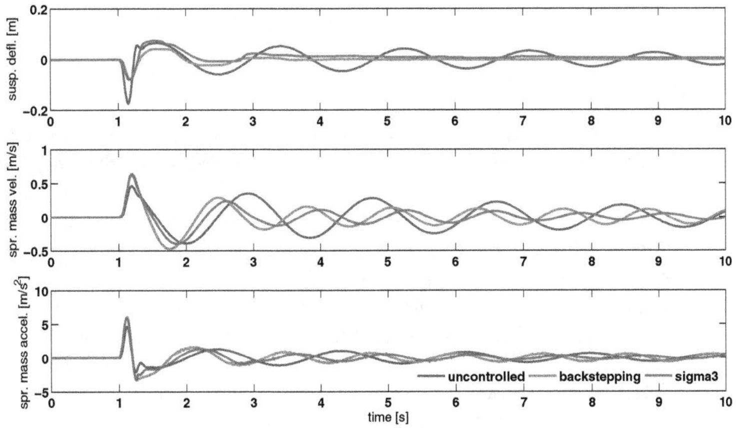

Figure 11. Bump on the road disturbance and car subsystem response.

6. Conclusions

In this chapter we presented the problem of the vibration control in vehicles. One model-based and three variable structure controllers were analyzed and compared in order to study their performance during typical road disturbances. The performance of the controller were also analyzed for the particular situation in which the suspension system is made up of a magnetorheological damper, which is well-known to be a nonlinear device. All of the controllers performed satisfactorily at regulating the suspension deflection while keeping the acceleration, velocity and displacement variables within acceptable limits. One important result obtained in this work was that despite the simplicity of these controllers, they performed significantly better than the model-based controller. It is to be noted that further studies -theoretical and experimental- should be performed in order to get a better insight of the performance of such controllers and the possibilities of being used in real systems.

Acknowledgements

M. Zapateiro has been supported by the 'Juan de la Cierva' fellowship from the Government of Spain. This study has also been partially supported by Secretaría de Estado de Investigación, Desarrollo e Innovación (formerly, Ministry of Science and Innovation) through DPI2011-27567-C02-02, DPI2011-28033-C03-01 and DPI 2011-27567-C02-01.

Author details

Mauricio Zapateiro and Francesc Pozo
Department of Applied Mathematics III, Universitat Politècnica de Catalunya, Barcelona, Spain

Ningsu Luo
Institute of Informatics and Applications, University of Girona, Girona, Spain

7. References

[1] A.M. Amani, A.K. Sedigh, M.J. Yazdampanah. A QFT approach to robust control of automobiles active suspension, *2004 5th Asian Control Conference*, Melbourne, Australia, 2004.

[2] A. Bahar, F. Pozo, L. Acho, J. Rodellar, A. Barbat. Hierarchical semi-active control of base-isolated structures using a new inverse model of magnetorheological dampers, *Computers and Structures*, Vol. 88, pp. 483-496, 2010.

[3] O. Barambones and P. Alkorta, A robust vector control for induction motor drives with an adaptive sliding-mode control law, *Journal of The Franklin Institute*, vol. 348, no. 2, pp. 300–314, 2011.

[4] M. Braci, S. Diop. On numerical differentiation algorithms for nonlinear estimation, *Proc. of the 42nd IEEE Conference on Decision and Control*, Hawaii, USA, 2003.

[5] J. Carrion, B.F. Spencer, Jr. Model-based strategies for real-time hybrid testing, *Technical Report, University of Illinois at Urbana-Champaign*, Available at: http://hdl.handle.net/2142/3629, 2007.

[6] C.Y. Chen, M.H.M. Chen, C.H. Lee. Backstepping controller design for a manipulator with compliance, *ICIC Express Letters*, vol. 4, no. 5(A), pp. 1723-1728, 2010.
[7] S.-L. Chen, C.-C. Weng. Robust control of a voltage-controlled three-pole active magnetic bearing system, *IEEE/ASME Transactions on Mechatronics*, Vol. 15, no. 3, pp. 381-388, 2010.
[8] H. Du, K.Y. Sze, J. Lam. Semi-active H_∞ control of vehicle suspension with magneto-rheological dampers, *Journal of Sound and Vibration*, vol. 283, pp. 981-996.
[9] S.J. Dyke, B.F. Spencer, M.K. Sain, J.D. Carlson. Modeling and control of magnetorheological dampers for seismic response reduction, *Smart Materials and Structures*, vol. 5, no. 5, pp. 565–575, 1996.
[10] W. Gao and J.C. Hung. Variable structure control of nonlinear systems: a new approach, *IEEE Transactions on Industrial Electronics*, vol. 40, no. 1, pp. 45–55, 1993.
[11] H. Gao, W. Sun, P. Shi. Robust sampled-data H_∞ control for vehicle active suspension systems, *IEEE Transactions on Control Systems Technology*, vol. 18, no. 1, pp. 238–245, 2010.
[12] V.Y. Glizer, V. Turetsky, L. Fridman, and J. Shinar. History-dependent modified sliding mode interception strategies with maximal capture zone, *Journal of The Franklin Institute*, vol. 349, no. 2, pp. 638–657, 2012.
[13] N. Karlsson, A. Teel, D. Hrovat. A backstepping approach to control of active suspensions, *Proceedings of the 40th IEEE Conference on Decision and Control*, Orlando, Florida, USA, 2001.
[14] M. Krstic, I. Kanellakopoulos, O. Kokotovic. *Nonlinear and Adaptive Control Design*, John Wiley and Sons, Inc., 1995.
[15] C. Liu, S, Tong, Y. Li, Adaptive fuzzy backstepping output feedback control for nonlinear systems with unknown sign of high-frequency gain, *ICIC Express Letters*, Vol. 4,no. 5(A), pp. 1698-1694, 2010.
[16] T.T. Nguyen, T.H. Bui, T.P. Tran, S.B. Kim. A hybrid control of active suspension system using H_∞ and nonlinear adaptive controls, *IEEE International Symposium on Industrial Electronics*, Pusan, Korea, 2001.
[17] M.C. Pai. Design of adaptive sliding mode controller for robust tracking and model following, *Journal of The Franklin Institute*, vol. 347, no. 10, pp. 1837–1849, 2010.
[18] M.M. Rashid, N. A. Rahim, M.A. Hussain, M.A. Rahman. Analysis and Experimental Study of Magnetorheological-Based Damper for Semiactive Suspension System Using Fuzzy Hybrids, *IEEE Transactions on Industry Applications*, vol. 47, no. 2, pp. 1051–1059, 2011.
[19] B.F. Spencer, Jr., S.J. Dyke, M. Sain, J.D. Carlson, Phenomenological model of a magnetorheological damper, *ASCE Journal of Engineering Mechanics*, Vol. 123, pp. 230-238, 1997.
[20] T.T Soong and B.F. Spencer Jr. Supplemental energy dissipation: state-of-the-art and state-of-the-practice, *Engineering Structures*, vol. 24, no. 3, pp. 243–259, 2002.
[21] W. Sun, H. Gao, O. Kaynak. Finite Frequency H_∞ Control for Vehicle Active Suspension Systems, *IEEE Transactions on Control Systems Technology*, vol. 19, no. 2, pp. 416–422, 2011.
[22] S. Tong, Y. Li, T. Wang. Adaptive fuzzy backstepping fault-tolerant control for uncertain nonlinear systems based on dynamic surface, *International Journal of Innovative Computing, Information and Control*, Vol. 5, no. 10(A), pp. 3249-3261, 2009.

[23] E.R. Wang, X.Q. Ma, S. Rakheja, C.Y. Su. Semi-active control of vehicle vibration with MR-dampers, *Proc. of the 42nd IEEE Conference on Decision and Control*, Maui, Hawaii USA, 2003.

[24] T. Wang, S. Tong, Y. Li. Robust adaptive fuzzy control for nonlinear system with dynamic uncertainties based on backstepping, *International Journal of Innovative Computing, Information and Control*, Vol. 5, no. 9, pp. 2675-2688, 2009.

[25] G.Z. Yao, F.F: Yap, G. Chen, W.H. Li, S.H. Yeo. MR damper and its application for semi-active control of vehicle suspension system, *Mechatronics*, vol. 12(7), pp. 963-973.

[26] M. Zapateiro, N. Luo, H.R. Karimi, J. Vehí. Vibration control of a class of semiactive suspension system using neural network and backstepping techniques, *Mechanical Systems and Signal Processing*, Vol. 23, 1946-1953, 2009.

[27] M. Zapateiro, H.R. Karimi, N. Luo, B.F. Spencer, Jr. Real-time hybrid testing of semiactive control strategies for vibration reduction in a structure with MR damper, *Structural Control and Health Monitoring*, Vol. 17(4), pp.427-451, 2010.

[28] M. Zapateiro, F. Pozo, H.R. Karimi, N. Luo, Semiactive control methodologies for suspension control with magnetorheological dampers, *IEEE/ASME Transactions on Mechatronics*, vol. 17(2), pp. 370-380, 2012.

[29] Z. Zexu, W. Weidong, L. Litao, H. Xiangyu, C. Hutao, L. Shuang, and C. Pingyuan. Robust sliding mode guidance and control for soft landing on small bodies, *Journal of The Franklin Institute*, vol. 349, no. 2, pp. 493–509, 2012.

Chapter 8

Multi-Objective Control Design with Pole Placement Constraints for Wind Turbine System

Tore Bakka and Hamid Reza Karimi

Additional information is available at the end of the chapter

1. Introduction

The demand for energy world wide is increasing every day. And in these "green times" renewable energy is a hot topic all over the world. Wind energy is currently one of the most popular energy sectors. The growth in the wind power industry has been tremendous over the last decade, its been increasing every year and it is nowadays one of the most promising sources for renewable energy. Since the early 1990s wind power has enjoyed a renewed interest, particularly in the European Union where the annual growth rate is about 20%. It is also a growing interest in offshore wind turbines, either bottom fixed or floating. Offshore wind is higher and less turbulent than the conditions we find onshore. In order to sustain this growth in interest and industry, wind turbine performance must continue to be improved. The wind turbines are getting bigger and bigger which in turn leads to larger torques and loads on critical parts of the structure. This calls for a multi-objective control approach, which means we want to achieve several control objectives at the same time. E.g. maximize the power output while mitigating any unwanted oscillations in critical parts of the wind turbine structure. One of the major reasons the wind turbine is a challenging task to control is due to the nonlinearity in the relationship between turning wind into power. The power extracted from the wind is proportional to the cube of the wind speed.

A wind turbines power production capability is often presented in relation to wind speed, as shown in Fig 1. From the figure we see that the power capability vs. wind speed is divided into four regions of operation. Region I is the start up phase. As the wind accelerates beyond the cut-in speed, we enter region II. A common control strategy in this region is to keep the pitch angle constant while controlling the generator torque. At the point where the wind speed is higher than the rated wind speed of the turbine (rated speed), we enter region III. In this region the torque is kept constant and the controlling parameter is the pitching angle. This is the region we are concerned with in this paper, i.e. the above rated wind speed conditions. The last region is the shutting-down phase (cut-out). The expression for power produced by

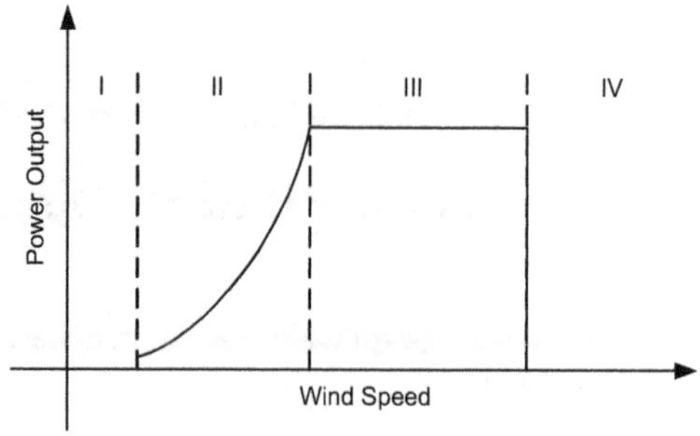

Figure 1. Range of operation for a typical wind turbine

the wind is given by ([1]):

$$P_a = \frac{1}{2}\rho\pi R^2 v^3 C_p(\lambda,\beta) \quad (1)$$

The dimensionless tip-speed ratio (TSR) λ is defined as:

$$\lambda = \frac{v_b}{v} \quad (2)$$

where v_b is the tip speed of the blade and v is the wind speed. From (1) we can find the aerodynamic torque and the thrust force acting on the tower:

$$T_a = \frac{1}{2}\rho\pi R^3 v^2 C_p(\lambda,\beta) \quad (3)$$

$$F_t = \frac{1}{2}\rho\pi R^2 v^2 C_T(\lambda,\beta) \quad (4)$$

where P_a is the aerodynamic power, ρ is the air density, R is the blade radius. C_p gives the relationship between how much power is available in the wind and how much can be converted to electrical power. Not all the available power can be converted, this is due to the fact that the wind can not be completely drained of energy. If it could, the wind speed at the rotor front would reduce to zero and the rotation of the rotor would stop. It can be proven that the theoretical upper limit of C_p is $16/27 \approx 0.59$, which is known as the Betz limit. A typical modern wind turbine has a maximum power coefficient of about 0.5. C_t is the thrust force coefficient, both these coefficients are dependent on the TSR λ and the pitch angle β.

1.1. Modeling and control of wind turbines

1.1.1. Modeling

Whenever we are dealing with control of a wind turbine generating system (WTGS), the turbine becomes a critical part of the discussion. There are many ways to model it, this can be done by simple one mass models ([2]-[3]) or multiple mass models ([4]). Several advanced wind turbine simulation softwares have emerged during the last decade. HAWC2 ([5]), Cp-Lambda ([6]) and FAST ([7]) are a few examples. They are developed at RISØ in Denmark, POLI-Wind in Italy and NREL in the US, respectively. In these codes the turbine and structure are considered as complex flexible mechanisms, and uses the finite-element-method (FEM) multibody approach. An aero-servo-elastic model is introduced, which consists of aerodynamic forces from the wind, the servo dynamics from the different actuators and the elasticity in the different joints and the structure. Both FAST and HAWC2 can simulate offshore and onshore cases while Cp-Lambda is limited to the onshore case.

1.1.2. Control

Recently, linear controllers have been extensively used for power regulation through the control of blade pitch angle in wind turbine systems e.g. [8]-[15]. However, the performance of these linear controllers are limited by the highly nonlinear characteristics of wind turbines. Advanced control is one research area where such improvement can be achieved.

1.2. Outline of chapter

The paper starts with describing the wind turbine model in section 2. Section 3 deals with control design of the wind turbine system. The system is formulated on a generalized form and the LMI constraints for $\mathcal{H}_2$, $\mathcal{H}_\infty$ and pole placement are developed. Simulation results are presented and evaluated in section 4. Section 5 summarizes the paper and gives the concluding remarks.

2. Wind turbine model

The floating wind turbine model used for the simulation is a 5MW turbine with three blades, which is an upscaled version of Statoils 2.3MW Hywind turbine situated off the Norwegian west coast. The turbine is specified by the National Renewable Energy Laboratory (NREL), more information about the turbine specifications can be found in ([16]). Table 1 shows some of the basic facts of the turbine. The simulation scenario is for above rated wind speed

Rated power	5 $[MW]$
Rated wind speed	11.6 $[m/s]$
Rated rotor speed	12.1 $[RPM]$
Rotor radius	63 $[m]$
Hub hight	90 $[m]$

Table 1. Basic facts of NREL's OC3 turbine

conditions, this means the turbine is operating in region III (see Fig. 1). In this region the major objectives are to maintain the turbines stability, calculate the collective pitch angle in order to prevent large oscillations in the drive train and in the tower and try to keep the rotor and generator at their rated speeds. If we can achieve this, then we keep the power output smooth. The model is obtained from the wind turbine simulation software FAST (Fatigue, Aerodynamics, Structures, and Turbulence). More information about the software can be found in the user's manual ([7]). FAST has two different forms of operation or analysis modes (Fig. 2). The first analysis mode is time-marching of the nonlinear equations of motion - which is, simulation. During simulation, wind turbine aerodynamic and structural response to wind-inflow conditions are determined in time. Outputs of simulation include time-series data on the aerodynamic loads as well as loads and deflections of the structural parts of the wind turbine. The second form of analysis provided in FAST is linearization. FAST has the capability of extracting linearized representations of the complete nonlinear aeroelastic wind turbine model. Three degrees of freedom are chosen, and they are; rotor, generator and tower

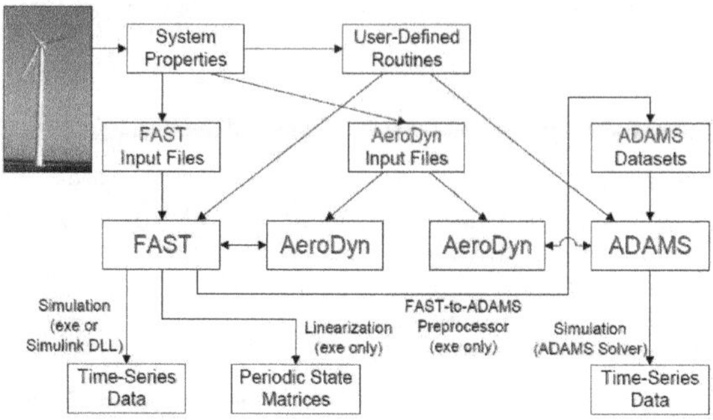

Figure 2. Modes of operation ([7])

dynamics. If desirable, the model can easily be expanded to include more degrees of freedom. The form of the linear model obtained from the software is stated in (5).

$$\begin{aligned} \dot{x} &= \bar{A}x + \bar{B}u \\ y &= \bar{C}x + \bar{D}u \end{aligned} \quad (5)$$

The bar over the matrices in the state space system tells us that these are the average matrices for the wind turbine. FAST calculates the matrices for a state space system at each desired azimuth angle and gives out the average values. The state space system is strictly proper, i.e. $D = 0$. This means that the system will approach zero as the frequency approaches infinity. These average matrices will later be used for the controller design.

The nonlinear model is linearized at a wind speed of 18 $[m/s]$ and at the rated rotors rotational speed 12.1 $[rpm]$. The system has one input and three outputs. Input is pitch angle and the outputs are rotor speed, generator speed and tower fore-aft displacement. The $\bar{A}$ matrix has dimensions 6×6 with the state vector $x = [x_1\ x_2\ x_3\ x_4\ x_5\ x_6\ ]^T$, where x_1 is fore-aft displacement, x_2 is generator position, x_3 is rotor position, x_4 is fore-aft speed, x_5 is generator speed and x_6 is rotor speed.

3. Control design

The control purpose of the $\mathcal{H}_\infty$ is to minimize the disturbance effect on the system output. The $\mathcal{H}_2$ purpose is to try to make the system more robust against random disturbances (LQG aspects). With the use of mixed $\mathcal{H}_2/\mathcal{H}_\infty$ control we can benefit from both control synthesis, minimizing possible disturbances effects while rejecting stochastic noise. The control design is solved with the use of linear matrix inequalities (LMIs). These robust control designs mostly deal with frequency domain aspects of the closed loop system, but it is well known that the location of the closed loop poles play a large role in the transient behavior of the controlled system. In this way we can prevent fast poles and end up with a system which responds in a more realistic way. In the following sections we will use boldface letters to emphasize the LMI optimization variables.

3.1. System representation

Fig. 3 shows the output feedback control scheme. where $P(s)$ is the generalized plant and

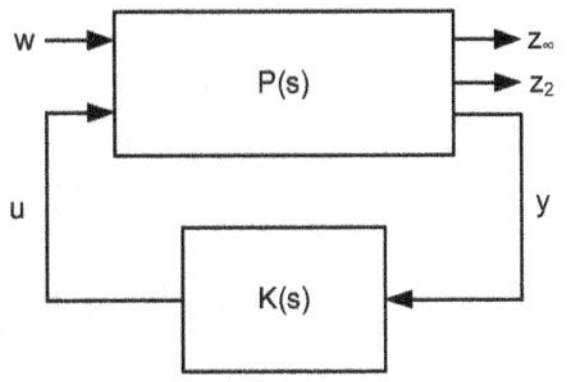

Figure 3. Output feedback block diagram

$K(s)$ is the controller. The two blocks represent the equations in (6)-(7). $P(s)$ includes the wind turbine model and the signals of interest

$$\begin{aligned} \dot{x} &= Ax + B_1 w + B_2 u \\ z_2 &= C_{21} x + D_1 w + D_2 u \\ z_\infty &= C_{1i} x + D_{1i} w + D_{2i} u \\ y &= C_2 x + D_{21} w + D_{22} u \end{aligned} \tag{6}$$

where A, B_2, C_2 and D_{22} represent the matrices from the standard state space system (5). The other matrices are considered with appropriate dimensions. u is the control input, w is the disturbance signal and y is the measured output. The signals z_2 and z_∞ are, respectively, controlled outputs for $\mathcal{H}_2$ and $\mathcal{H}_\infty$ performance measures. For system (6), the dynamic output feedback $u(s) = K(s)y(s)$, is on the following form:

$$K(s)\begin{cases} \dot{\zeta} = A_k \zeta + B_k y \\ u = C_k \zeta + D_k y. \end{cases} \tag{7}$$

The closed loop system is given in (8) with the states $x_{cl} = [x\ \zeta]^T$.

$$\begin{aligned} \dot{x}_{cl} &= A_{cl} x + B_{cl} w \\ z_2 &= C_{cl2} x + D_{cl2} w \\ z_\infty &= C_{cl\infty} x + D_{cl\infty} w \end{aligned} \tag{8}$$

where

$$\left(\begin{array}{c|c} A_{cl} & B_{cl} \\ \hline C_{cl2} & D_{cl2} \\ C_{cl\infty} & D_{cl\infty} \end{array}\right)$$

$$= \left(\begin{array}{cc|c} A + B_2D_kC_2 & B_2C_k & B_1 + B_2D_kD_{21} \\ B_kC_2 & A_k & B_kD_{21} \\ \hline C_{21} + D_2D_kC_2 & D_2C_k & D_1 + D_2D_kD_{21} \\ C_{2i} + D_{2i}D_kC_2 & D_{2i}C_k & D_{1i} + D_{2i}D_kD_{21} \end{array}\right). \tag{9}$$

The closed loop transfer functions from w to z_2 and from w to z_∞ are given in (10) and (11) respectively.

$$T_2 = C_{cl2}\left(sI - A_{cl}\right)^{-1} B_{cl} + D_{cl2} \tag{10}$$

$$T_\infty = C_{cl\infty}\left(sI - A_{cl}\right)^{-1} B_{cl} + D_{cl\infty} \tag{11}$$

3.2. $\mathcal{H}_2$ Control

The closed loop $\mathcal{H}_2$ norm can be computed as $||T_2||_2^2 = Trace(C_{cl2}\mathbf{S}_0C_{cl2}^T)$, where $\mathbf{S}_0$ solves the equation

$$A_{cl}\mathbf{S}_0 + \mathbf{S}_0A_{cl}^T + B_{cl}B_{cl}^T = 0. \tag{12}$$

Since $\mathbf{S}_0 < \mathbf{S}$ for any $\mathbf{S}$ satisfying

$$A_{cl}\mathbf{S} + \mathbf{S}A_{cl}^T + B_{cl}B_{cl}^T < 0 \tag{13}$$

it is verified that $||T_2||_2^2 < v$ if and only if there exists $\mathbf{S} > 0$ satisfying (13) and $Trace(C_{cl2}\mathbf{S}C_{cl2}^T) < v$. With an auxiliary parameter $\mathbf{Q}$, we obtain the following analysis result; A_{cl} is stable and $||T_2||_2^2 < v$ if and only if there exist symmetric $\mathbf{X} = \mathbf{S}^{-1}$ and $\mathbf{Q}$ such that

$$\begin{pmatrix} A_{cl}^T\mathbf{X} + \mathbf{X}A_{cl} & \mathbf{X}B_{cl} \\ * & -I \end{pmatrix} < 0$$

$$\begin{pmatrix} \mathbf{X} & C_{cl2}^T \\ * & \mathbf{Q} \end{pmatrix} > 0$$

$$trace(Q) < v. \tag{14}$$

3.3. $\mathcal{H}_\infty$ Control

The closed loop $\mathcal{H}_\infty$ norm is defined in (15).

$$||T_{cl}||_\infty < \gamma \tag{15}$$

In order to formulate the $\mathcal{H}_\infty$ norm in terms of a matrix inequality, we need to do some manipulations to its original expression. It is known that the ∞-norm of a closed loop system is the same as taking the 2-norm of the signals of interest z divided by the 2-norm of the systems exogenous input w. In this way we will end up with the famous Bounded Real Lemma (BRL). First, lets define Lyapunovs stability criteria. A linear state space system (8) is asymptotically stable if all real parts of the eigenvalues of the A_{cl}-matrix are negative. The Lyapunov criteria involves searching for the matrix $\mathbf{X}$, if it exists then the system is stable. A quadratic Lyapunov function is defined in (16) and its derivative in (17).

$$V(x) = x^T\mathbf{X}x \tag{16}$$

$$\dot{V}(x) = \dot{x}^T\mathbf{X}x + x^T\mathbf{X}\dot{x} \tag{17}$$

The first crucial steps in obtaining the BRL are shown next. Firstly we remove the roots by squaring both sides of the inequality sign. Secondly we collect everything in one integral expression and lastly we do the trick where we add and subtract the Lyapunov function to the inequality. We can do this because we know that $\int_0^\infty \dot{V}(x) - V(\infty) + V(0) = 0$ is true. By doing this we only end up with the integral expression.

$$\frac{||z||_2}{||w||_2} < \gamma \Rightarrow \left[\int_0^\infty z^T z\,\mathrm{d}t\right]^{\frac{1}{2}} < \gamma \left[\int_0^\infty w^T w\,\mathrm{d}t\right]^{\frac{1}{2}}$$

$$\Rightarrow \int_0^\infty \left(\frac{1}{\gamma} z^T z - \gamma w^T w + \dot{V}(x)\right) \mathrm{d}t + \underbrace{V(0)}_{Always\ zero} \underbrace{-V(\infty)}_{Always\ negative} < 0 \tag{18}$$

Now we insert both the expression for the signal of interest z from (8) and the derivative of the Lyapunov function $\dot{V}(x)$ (17) into (18). We do some algebraic and matrix manipulations and end up with the following matrix inequality (BRL). This is not yet an LMI because of the nonlinear terms which occur when the feedback loop is closed.

$$\begin{pmatrix} A_{cl}^T \mathbf{X} + \mathbf{X} A_{cl} & \mathbf{X} B_{cl} & C_{cl\infty}^T \\ * & -\gamma I & D_{cl\infty}^T \\ * & * & -\gamma I \end{pmatrix} < 0$$

$$\mathbf{X} > 0 \tag{19}$$

3.4. Change of variables

For the multi-objective case we want to create a feedback controller $u = K(s)y$ through minimization of the trade off criterion in (20)

$$J = \alpha ||T_\infty||_\infty^2 + \beta ||T_2||_2^2 \tag{20}$$

where α and β are the weights. As mentioned, the $\mathcal{H}_2$ and $\mathcal{H}_\infty$ constraints (14) and (19) are not yet LMIs. In order to transform these nonlinear terms into proper LMIs we linearize them with the use of change of variables. This is not as straight forward as for the state feedback case, where $\mathbf{X} = \mathcal{P}^{-1}$ and $\mathbf{F} = F\mathcal{P}^{-1}$ turn all constraints into LMIs. More details about the change of variables can be found in ([17])

The new matrices $\mathcal{P}$ and $\mathcal{P}^{-1}$ are partitioned as follows,

$$\mathcal{P} = \begin{bmatrix} \mathbf{X} & N \\ N^T & \# \end{bmatrix}, \quad \mathcal{P}^{-1} = \begin{bmatrix} \mathbf{Y} & M \\ M^T & \# \end{bmatrix} \tag{21}$$

where $\mathbf{X}$ and $\mathbf{Y}$ are symmetric matrices of dimension $n \times n$. N and M will be calculated on the basis of $\mathbf{X}$ and $\mathbf{Y}$ at the end of this section. The matrices noted as # are not necessary to be known. In addition, we define the following two matrices,

$$\Pi_1 = \begin{bmatrix} \mathbf{Y} & I \\ M^T & 0 \end{bmatrix}, \qquad \Pi_2 = \begin{bmatrix} I & \mathbf{X} \\ 0 & N^T \end{bmatrix} \tag{22}$$

as can be inferred from the identity $\mathcal{P}\mathcal{P}^{-1} = I$ satisfying

$$\mathcal{P}\Pi_1 = \Pi_2. \tag{23}$$

Now we are ready to convert the nonlinear matrix inequalities into LMIs. This is done by performing congruence transformation with $diag(\Pi_1, I, I)$ and $diag(\Pi_1, I)$ on the $\mathcal{H}_\infty$ and $\mathcal{H}_2$ constraints respectively. We now obtain the following LMIs

$$\begin{pmatrix} sym\left(A\mathbf{X} + B_2\hat{\mathbf{C}}\right) & \hat{\mathbf{A}}^T + A + B_2\hat{\mathbf{D}}C_2 & B_1 + B_2\hat{\mathbf{D}}D_{21} & \mathbf{X}C_{1i}^T + \hat{\mathbf{C}}^T D_{21}^T \\ * & sym\left(\mathbf{Y}A + \hat{\mathbf{B}}C_2\right) & \mathbf{Y}B_1 + \hat{\mathbf{B}}D_{21} & C_{1i}^T + C_2^T\hat{\mathbf{D}}^T + D_{2i}^T \\ * & * & -\gamma I & D_{1i}^T + D_{21}^T\hat{\mathbf{D}}D_{2i}^T \\ * & * & * & -\gamma I \end{pmatrix} < 0$$

$$\begin{pmatrix} sym\left(A\mathbf{X} + B_2\hat{\mathbf{C}}\right) & \hat{\mathbf{A}}^T + A + B_2\hat{\mathbf{D}}C_2 & B_1 + B_2\hat{\mathbf{D}}D_{21} \\ * & sym\left(\mathbf{Y}A + \hat{\mathbf{B}}C_2\right) & \mathbf{Y}B_1 + \hat{\mathbf{B}}D_{21} \\ * & * & -I \end{pmatrix} < 0$$

$$\begin{pmatrix} \mathbf{X} & I & \left(C_{12}\mathbf{X} + D_2\hat{\mathbf{C}}\right)^T \\ * & \mathbf{Y} & \left(C_{12} + D_2\hat{\mathbf{D}}C_2\right)^T \\ * & * & \mathbf{Q} \end{pmatrix} > 0$$

$$trace(\mathbf{Q}) < \nu_0$$

$$\gamma < \gamma_0 \tag{24}$$

where $sym(A)$ is defined as $A + A^T$. Now following identities can be obtained

$$\Pi_1^T \mathcal{P}\mathcal{A}\Pi_1 = \Pi_2^T \mathcal{A}\Pi_1 = \begin{bmatrix} A\mathbf{X} + B_2\hat{\mathbf{C}} & A + B_2\hat{\mathbf{D}}C_2 \\ \hat{\mathbf{A}} & \mathbf{Y}A + \hat{\mathbf{B}}C_2 \end{bmatrix} \tag{25}$$

$$\Pi_1^T \mathcal{P}\mathcal{B} = \Pi_2^T \mathcal{B} = \begin{bmatrix} B_1 + B_2\hat{\mathbf{D}}D_{21} \\ \mathbf{Y}B_1 \end{bmatrix} \tag{26}$$

$$\mathcal{C}\Pi_1 = \left[C_{1i}\mathbf{X} + D_{21}\hat{\mathbf{C}} \;\; C_{1i} + D_{21}\hat{\mathbf{D}}\right] \tag{27}$$

$$\Pi_1^T \mathcal{P}\Pi_1 = \Pi_1^T \Pi_2 = \begin{pmatrix} \mathbf{X} & I \\ I & \mathbf{Y} \end{pmatrix}. \tag{28}$$

In order for (28) to be true the following relationship must hold

$$MN^T = I - \mathbf{XY}. \tag{29}$$

This relationship can be solved by utilizing the singular value decomposition (SVD). We know that s_d is a diagonal matrix and we define a new matrix $\bar{s}_d$, which is the square root of all the entries in s_d. In this way we can find the matrices M and N as shown below.

$$svd(I - \mathbf{XY}) = us_d v^T \tag{30}$$

$$\bar{s}_d = diag(sqrt(s_d)) \tag{31}$$

$$M = u\bar{s}_d \tag{32}$$

$$N^T = \bar{s}_d v^T \tag{33}$$

3.5. LMI region

An LMI region is any convex subset D of the complex plane that can be characterized as an LMI in z and $\bar{z}$ ([18])

$$D = \{z \in C : L + \bar{M}z + \bar{M}^T\bar{z} < 0\} \tag{34}$$

for fixed real matrices $\bar{M}$ and $L = L^T$, where $\bar{z}$ is a complex number. This class of regions encompasses half planes, strips, conic sectors, disks, ellipses and any intersection of the above. From ([18]), we find that all eigenvalues of the matrix A is in the LMI region $\{z \in C : [l_{ij} + \bar{m}_{ij}z + \bar{m}_{ji}\bar{z}]_{i,j} < 0\}$ if and only if there exists a symmetric matrix $\mathbf{X}_D$ such that

$$\left[l_{ij}\mathbf{X}_D + \bar{m}_{ij}A^T\mathbf{X}_D + \bar{m}_{ji}\mathbf{X}_D A\right]_{i,j} < 0, \qquad \mathbf{X}_D > 0. \tag{35}$$

Also, here we need to include the change of variables. This is done in (36) where $\otimes$ denotes the Kronecker product.

$$\left(L \otimes \begin{pmatrix} \mathbf{X} & I \\ I & \mathbf{Y} \end{pmatrix} + \bar{M} \otimes \begin{pmatrix} A\mathbf{X} + B\hat{\mathbf{C}} & A + B\hat{\mathbf{D}}C \\ \hat{\mathbf{A}} & \mathbf{Y}A + \hat{\mathbf{B}}C \end{pmatrix}\right.$$

$$\left. + \bar{M}^T \otimes \begin{pmatrix} A\mathbf{X} + B\hat{\mathbf{C}} & A + B\hat{\mathbf{D}}C \\ \hat{\mathbf{A}} & \mathbf{Y}A + \hat{\mathbf{B}}C \end{pmatrix}^T\right) < 0 \tag{36}$$

As an example we define the desired region D as a disk (Fig. 4), with center located along the x-axis (distance q from the origin) and radius r. This determines the region

$$D = \begin{pmatrix} -r & q + z \\ q + \bar{z} & -r \end{pmatrix} \tag{37}$$

where the closed loop eigenvalues may be placed.

From this we can find that the matrices L and $\bar{M}$ have the following form

$$L = \begin{pmatrix} -r & q \\ q & -r \end{pmatrix}, \qquad \bar{M} = \begin{pmatrix} 0 & 1 \\ 0 & 0 \end{pmatrix}. \tag{38}$$

All the constraints in (24) and the constraint in (36) are subjected to the minimization of the

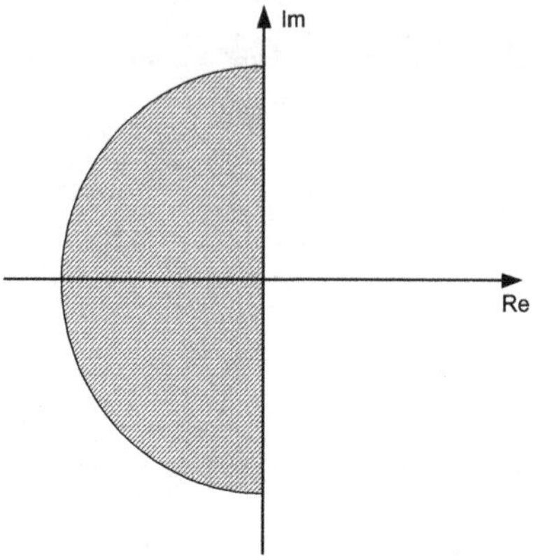

Figure 4. LMI region

objective function given in (20). They need to be solved in terms of $\left(\mathbf{X},\ \mathbf{Y},\ \mathbf{Q},\ \hat{\mathbf{A}},\ \hat{\mathbf{B}},\ \hat{\mathbf{C}},\ \hat{\mathbf{D}}\right)$. Finally, the controller matrices can be found by the following relationship

$$D_k = \hat{\mathbf{D}} \tag{39}$$

$$C_k = (\hat{\mathbf{C}} - D_k C_2 \mathbf{X}) \left(M^T\right)^{-1} \tag{40}$$

$$B_k = N^{-1} (\hat{\mathbf{B}} - \mathbf{Y} B_2 D_k) \tag{41}$$

$$A_k = N^{-1} \left(\hat{\mathbf{A}} - N B_k C_2 \mathbf{X} - \mathbf{Y} B_2 C_k M^T - \mathbf{Y}(A + B D_k C_2)\mathbf{X}\right) \left(M^T\right)^{-1}. \tag{42}$$

From the aforementioned expressions we are able to solve the mixed H_2/H_∞ control problem with pole placement constraints.

4. Simulation

All calculations and simulations are carried out in MatLab/Simulink ([19]) interfaced with YALMIP ([20]). The solver which is used for the LMI calculation is SeDuMi ([21]). FAST comes with a Simulink template which can be changed how ever the user may see fit. As described earlier the simulation scenario is for the above rated wind speed situation. We want to mitigate oscillations in the drive train and dampen tower movement while maintaining the rotors rated rotational speed. Satisfactory simulation results were found with $\alpha = 1$ and $\beta = 1$

as weights for the objective function and with the following performance measures:

$$z_2 = x_4 + x_5 - x_6 + u \tag{43}$$

$$z_\infty = x_1 + x_2 - x_3. \tag{44}$$

The norms obtained from the optimization are shown in (45) and the closed loop poles are shown in Fig. (5).

$$trace(\mathbf{Q}) \;=\; 9.995 \qquad \gamma \;=\; 27.7216 \tag{45}$$

In order to find the performance measures we argue that there are no oscillations in the drive

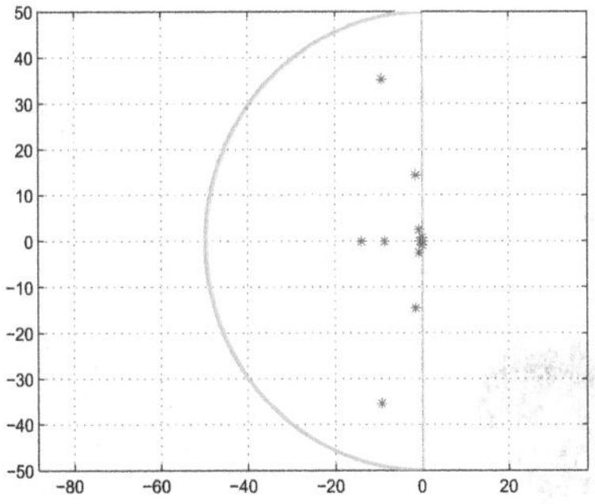

Figure 5. Closed loop poles

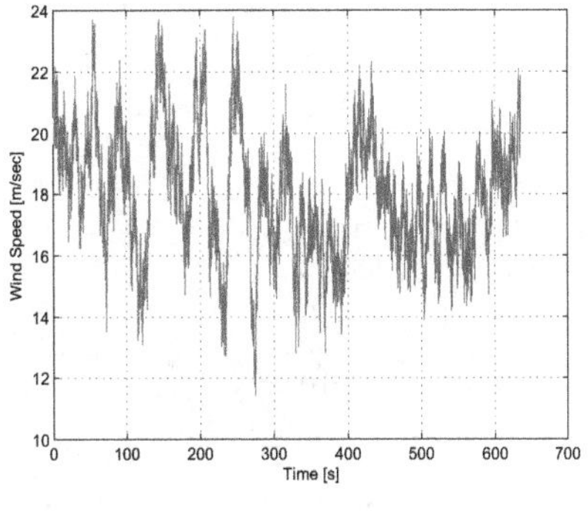

Figure 6. Wind profile

train if the position of the generator and the rotor are the same. In other words we must try to keep $x_2 - x_3$ close to zero. Similar procedure is done for the speed difference and the tower's speed and position. The wind profile is obtained from the software TurbSim ([22]), which is also developed at the NREL. This is a stochastic, full-field, turbulent-wind simulator, and the output can be used as input to FAST. The profile is a 50 year extreme wind condition, with an average speed of 18 $[m/s]$ and a turbulence intensity of 17 %. The waves have a significant wave hight of 6 $[m]$ and a peak wave period of 10 $[s]$.

As a consequence of having better performance the pitching activity has heavily increased, see Fig. 10. The pitching activity lies around 5 – 10 $[deg/s]$, which is within the limits of a modern wind turbine. The output torque is presented in Fig. 11, which also shows better performance.

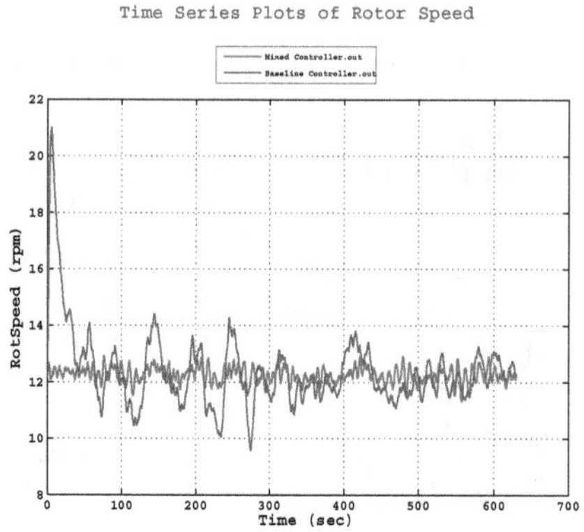

Figure 7. Rotor speed

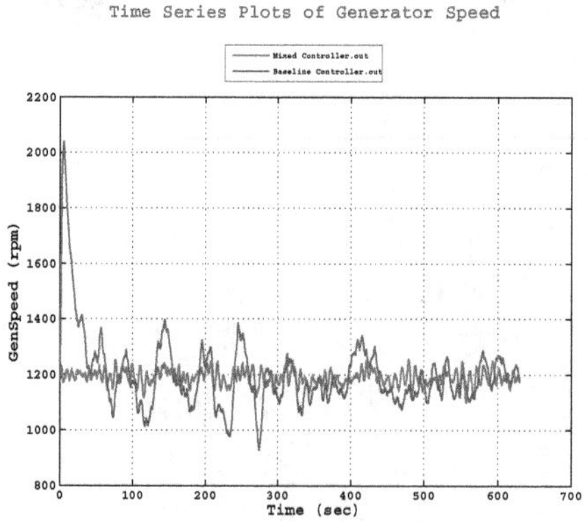

Figure 8. Generator speed

The controller is tested on the fully nonlinear system, where the only degrees of freedom which are left out are yaw and translational surge. The state space system of the controller is shown in the appendix. The plots included in this chapter are the same ones that were used as feedback to the controller. That is, rotor rotational speed (Fig. 7), generator rotational speed (Fig. 8) and tower fore-aft displacement (Fig. 9). The simulation results show a comparison between our simulations and simulations done with FAST's baseline controller. The baseline controller is a gain scheduled PI controller and is indicated on the plots as the blue line. The controller proposed in this chapter is indicated with the red line.

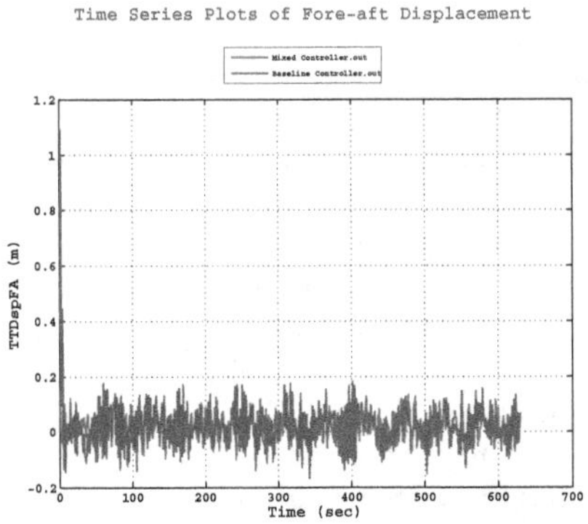

Figure 9. Tower fore-aft displacement

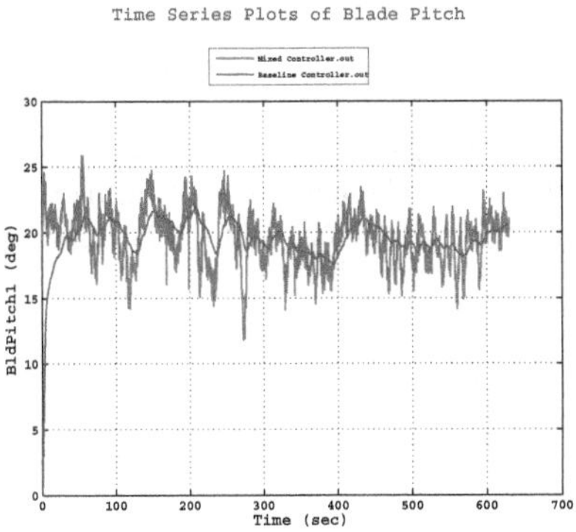

Figure 10. Blade pitch angle

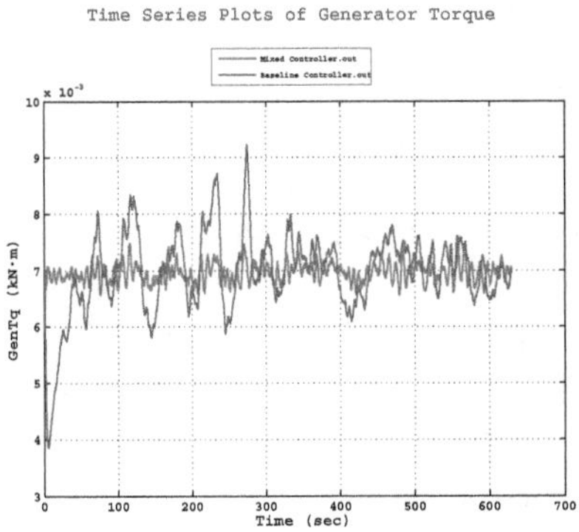

Figure 11. Generator output torque

5. Conclusion

In this chapter we have introduced a nonlinear model of an offshore floating wind turbine using the commercial wind turbine software FAST. By the use of its embedded routines a linear model is extracted. The linear model used in this chapter is built up with relatively few degrees of freedom. More degrees of freedom can easily be added to the model, depending on the control objectives. In our approach this relatively simple model serves its purpose in testing advanced control routines on an offshore floating wind turbine system. On the basis of this linear model a mixed $\mathcal{H}_2/\mathcal{H}_\infty$ control with pole placement constraint is designed, tested and compared with FAST's baseline controller. The proposed controller is tested on the nonlinear wind turbine model under the influence of a 50 year extreme wind condition, 6 $[m]$ significant wave hight and 10 $[s]$ peak wave period. The control objectives have been to mitigate unwanted oscillations in the drive train and tower, in addition to maintain the rotor and generator rotational speeds at their rated values. If any unwanted oscillations in the drive train are damped and rated speed is maintained, the torque output should be smooth. Our proposed control shows better performance than FAST's baseline controller.

Acknowledgment

This work has been (partially) funded by Norwegian Centre for Offshore Wind Energy (NORCOWE) under grant 193821/S60 from Research Council of Norway (RCN). NORCOWE is a consortium with partners from industry and science, hosted by Christian Michelsen Research.

Author details

Tore Bakka and Hamid Reza Karimi
University of Agder, Norway

Appendix

State space system of wind turbine model:

$$\dot{x} = \begin{bmatrix} 0 & 0 & 0 & 1 & 0 & 0 \\ 0 & 0 & 0 & 0 & 1 & 0 \\ 0 & 0 & 0 & 0 & 0 & 1 \\ -5.4263 & -1.7533\times10^{-5} & -1.8036\times10^{-5} & -0.29188 & -2.3801 & -2.3801 \\ 0 & 0 & 172.6 & 0 & 12.082\times10^{-5} & 1.237 \\ -1.0727\times10^{-3} & 6.6008\times10^{-7} & -195 & -0.027639 & -0.37886 & -1.7757 \end{bmatrix} x + \begin{bmatrix} 0 \\ 0 \\ 0 \\ -9.913 \\ 0 \\ -1.5151 \end{bmatrix} u$$

$$y = \begin{bmatrix} 0\ 0\ 0\ 0\ 9.549 & 9.549 \\ 0\ 0\ 0\ 0\ 926.3 & 0 \\ 1\ 0\ 0\ 0\ \ 0 & 0 \end{bmatrix} x + \begin{bmatrix} 0 \\ 0 \\ 0 \end{bmatrix} u \qquad (46)$$

State space system of the controller:

$$\dot{\zeta} = \begin{bmatrix} -26.6610 & -30.6408 & 0.8454 & -0.1632 & 0.6128 & -0.0186 \\ 101.0655 & 81.5111 & 0.8504 & 0.9437 & -2.5035 & 0.0951 \\ -196.7809 & -173.8849 & -1.1085 & -3.2463 & 6.4727 & -0.1600 \\ 42.9129 & 52.3436 & -0.1755 & -3.0062 & -0.2203 & 0.1970 \\ 1239.7 & 1063.4 & 11.6 & 20.4 & -39.6 & 1.1 \\ -47346.0 & -43037.0 & -91.0 & -306.0 & 1178.0 & -55.0 \end{bmatrix} \zeta + \begin{bmatrix} -0.5521 & -0.0185 & -12.2055 \\ -3.5502 & 0.0763 & 14.6491 \\ 12.5597 & -0.1696 & 0.2177 \\ 6.6856 & 0.0668 & -25.7465 \\ -95.4987 & 1.0100 & 6.3680 \\ 1988.8 & -44.3 & -2523.0 \end{bmatrix} u$$

$$u = \begin{bmatrix} 0.2036 & 1.3748 & -0.1657 & 0.0032 & -0.0026 & -0.0008 \end{bmatrix} \zeta + \begin{bmatrix} 0.2132 & -0.0001 & 1.3252 \end{bmatrix} u \quad (47)$$

6. References

[1] D. M. Eggleston and F. S. Stoddard. *Wind Turbine Engineering Design.* Van Nostrand Reinhold Co., New York, 1987.

[2] J. Tamura, T. Yamajaki, M. Ueno, Y. Matsumura, and S. Kimoto. Transient stability simulation of power system including wind generator by PSCAD/EMTDC. In *IEEE Porto Power Tech Conference*, September 2001.

[3] E. S. Abdin and W. Xu. Control design and dynamic performance analysis of a wind turbine induction generator unit. *IEEE Transactions on Energy Conversion*, Vol. 15(No. 1):91–96, March 2000.

[4] R. Takahashi T. Murata J. Tamura Y. Tomaki A. Sakahara S.M. Muyeen, Md. Hasan Ali and E. Sasano. Comparative study on transient stability analysis of wind turbine generator system using different drive train models. *IET Renewable Power Generation*, Vol. 1(No. 2):131 – 141, June 2007.

[5] T. J. Larsen. How 2 HAWC2, the user's manual. In *Risø-R-1597(ver. 3-9)(EN)*, September 2009.

[6] C.L. Bottasso and A. Croce. Cp-Lambda user manual. In *RisÃÿ-R-1597(ver. 3-9)(EN).* Dipartimento di Ingnegneria Aerospaziale, Politecnico di Milano, September 2009.

[7] J. Jonkman and M. L. Buhl. FAST User's Guide. In *Technical Report NREL/EL-500-38230*, September 2009.

[8] R. Eide and H. R. Karimi. *Control Design Methodologies for Vibration Mitigation on Wind Turbine Systems.* InTech, Vibration Analysis and Control - New Trends and Developments, Dr. Francisco Beltran-Carbajal (Ed.), pp. 217-242, ISBN: 978-953-307-433-7, 2011.

[9] H.R. Karimi. Wavelet-based optimal control of a wind turbine system: A computational approach. *J. Advanced Mechanical Design, Systems and Manufacturing,* Vol. 5(No. 3):171–186, 2011.

[10] B. Skaare, T. D. Hanson, and F. G. Nielsen. Importance of control strategies on fatigue life of floating wind turbines. In *Proceedings of 26th International Conference on Offshore Mechanics and Arctic Engineering*, pages 493–500, June 2007.

[11] J. Jonkman. Influence of control on the pitch damping of a floating wind turbine. In *ASME Wind Energy Symposium*, 2008.

[12] H. Namik and K. Stol. Individual blade pitch control of floating offshore wind turbines. *Wind Energy*, Vol. 13(No. 1):74 – 85, 2009.

[13] S. Christiansen, T. Knudsen, and T. Bak. Optimal control of a ballast- stabilized floating wind turbine. In *IEEE Multi-Conference on Systems and Control*, pages 1214 – 1219, Colorado, Denver, September 2011.

[14] E. Wayman, P. Sclavounos, S. Butterfield, J. Jonkman, and W. Musial. Coupled dynamic modeling of floating wind turbine systems. In *Offshore Technology Conference*, Houston, Texas, May 2006.

[15] T. D. Hanson, B. Skaare, R. Yttervik, F. G. Nielsen, and O. HavmÃÿller. Comparison of measured and simulated responses at the first full scale floating wind turbine hywind. In *EWEA*, Brussels, February 2011.

[16] J. Jonkman, S. Butterfield, W. Musial, and G. Scott. Definition of a 5-MW reference wind turbine for offshore system development. In *Technical Report NREL/TP-500-38060*. National Renewable Energy Laboratory, February 2009.

[17] C. Scherer, P. Gahinet, and M. Chilali. Multiobjective outputfeedback control via LMI optimization. *IEEE Transactions on Automatic Control*, Vol. 42(No. 7):896 – 911, July 1997.

[18] M. Chilali and P. Gahinet. h_∞ design with pole placement constraints: An LMI approach. *IEEE Transactions on Automatic Control*, Vol. 41(No. 3):358 – 367, March 1996.

[19] MATLAB. *version 7.10.0 (R2010a)*. The MathWorks Inc., Natick, Massachusetts, 2010.

[20] Johan Löfberg. Yalmip : a toolbox for modeling and optimization in MATLAB. In *2004 IEEE International Symposium on Computer Aided Control Systems Design*, pages 284 – 289, Taipei, Taiwan, February 2005.

[21] J. F. Strum. Using SeDuMi 1.02, a MatLab toolbox for optimization over symmetric cones. In *Optimization Methods and Software*, pages 625 – 653, 1999.

[22] B. J. Jonkman. TurmSim user'guide: Version 1.50. In *Technical Report NREL/EL-500-46198*. National Renewable Energy Laboratory, September 2009.

Chapter 9

Automatic Balancing of Rotor-Bearing Systems

Andrés Blanco-Ortega, Gerardo Silva-Navarro,
Jorge Colín-Ocampo, Marco Oliver-Salazar, Gerardo Vela-Valdés

Additional information is available at the end of the chapter

1. Introduction

Rotating machinery is commonly used in many mechanical systems, including electrical motors, machine tools, compressors, turbo machinery and aircraft gas turbine engines. Typically, these systems are affected by exogenous or endogenous vibrations produced by unbalance, misalignment, resonances, bowed shafts, material imperfections and cracks. Vibration can result from a number of conditions, acting alone or in combination. The vibration problems may be caused by auxiliary equipment, not just the primary equipment. Control of machinery vibration is essential in the industry today to increase running speeds and the requirement for rotating machinery to operate within specified levels of vibration.

Vibration caused by mass imbalance is a common problem in rotating machinery. Rotor imbalance occurs when the principal inertia axis of the rotor does not coincide with its geometrical axis and leads to synchronous vibrations and significant undesirable forces transmitted to the mechanical elements and supports. A heavy spot in a rotating component will cause vibration when the unbalanced weight rotates around the rotor axis, creating a centrifugal force. Imbalance could be caused by manufacturing defects (machining errors, casting flaws, etc.) or maintenance issues (deformed or dirty fan blades, missing balance weights, etc.). As rotor speed changes, the effects of imbalance may become higher. Imbalance can severely reduce bearing life-time as well as cause undue machine vibration. Shaft misalignment is a condition in which the shafts of the driving and driven machines are not on the same centre-line generating reaction forces and moments in the couplings. Flexible couplings are used to reduce the misalignment effects and transmit rotary power without torsional slip.

Many methods have been developed to reduce the unbalance-induced vibration by using different devices such as active balancing devices, electromagnetic bearings, active squeeze film dampers, lateral force actuators, pressurized bearings and movable bearings (see, e.g., Blanco et al., 2003, 2007, 2008, 2010a, 2010b; Chong-Won, 2006; Dyer et al., 2002; El-Shafei,

2002; Green et al., 2008; Guozhi et al. 2000; Hredzak et al., 2006; Sheu et al., 1997; Zhou y Shi, 2001, 2002). These active balancing control schemes require information of the eccentricity of the involved rotating machinery. On the other hand, there exists a vast literature on identification and estimation methods, which are essentially asymptotic, recursive or complex, which generally suffer from poor speed performance (see, e.g., Ljung, 1987; Soderstrom, 1989; and Sagara and Zhao, 1989, 1990).

Passive, semi-active and active control schemes have been proposed in order to cancel or attenuate the vibration amplitudes in rotating machinery. In passive control the rotating machinery is modified off-line, e.g. the rotor is stopped to adjust some of its parameters such as mass, stiffness or damping. Balancing consist of placing correction masses onto the rotating shaft (inertial disk) so that centrifugal forces due to these masses cancel out those caused by the residual imbalance mass.

Active vibration control (AVC) changes the dynamical properties of the system by using actuators or active devices during instantaneous operating conditions measured by the appropriate sensors. The main advantage of active control (compared to passive control) is the versatility in adapting to different load conditions, perturbations and configurations of the rotating machinery and hence, extending the system's life while greatly reducing operating costs.

Semiactive vibration control devices are increasingly being investigated and implemented. These devices change the system properties such as damping and stiffness while the rotor is operating. This control scheme is based on the analysis of the open loop response. Semi-active control devices have received a great deal of attention in recent years because they offer the adaptability of active control without requiring the associated large power sources.

This chapter deals with the active cancellation problem of mechanical vibrations in rotor-bearing systems. The use of an active disk is proposed for actively balancing a rotor by placing a balancing mass at a suitable position. Two nonlinear controllers with integral compensation are proposed to place the balancing mass at a specific position. Algebraic identification is used for on-line eccentricity estimation as the implementation of this active disk is based on knowledge of the eccentricity. An important property of this algebraic identification is that the eccentricity identification is not asymptotic but algebraic, in contrast to most of the traditional identification methods, which generally suffer of poor speed performance. In addition, a velocity control is designed to drive the rotor velocity to a desired operating point during the first critical speed.

The proposed results are strongly based on the algebraic parameter identification approach for linear systems reported in (Flies and Sira, 2003), which requires a priori knowledge of the mathematical model of the system. This approach has been used for parameter and signal estimation in nonlinear and linear vibrating mechanical systems, where numerical simulations and experimental results show that the algebraic identification provides high robustness against parameter uncertainty, frequency variations, small measurement errors and noise (Beltran et al., 2005, 2006, 2010).

2. Active balancing and vibration control of rotating machinery

Many methods for passive balancing have been proposed, such as single plane, two planes or multi-plane balancing. These off-line balancing methods are very common in industrial applications. In these methods, the rotor is modeled as a rigid shaft that without elastic deformation during operation. Rotors operating under 5000 rpm can be considered rigid rotors. For flexible rotors the modal balancing and influence coefficient methods were developed for off-line balancing. Figure 1 shows an inertial disk to be balanced by adding a mass in opposite direction to compensate the residual unbalance.

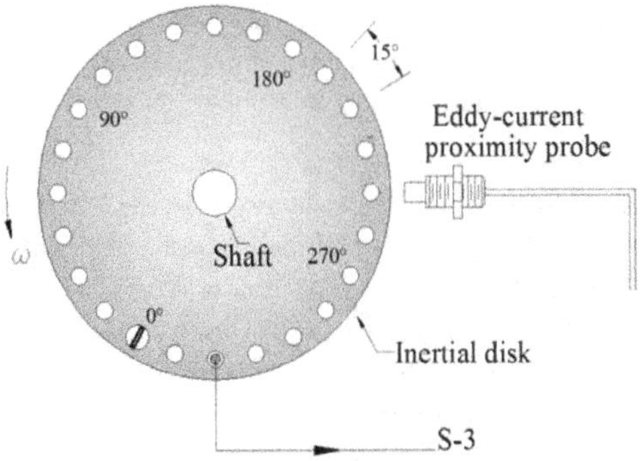

Figure 1. Inertial disk and eddy current probe displacement sensor.

Thearle (Thearle, 1932) developed a machine for dynamically balancing rotating elements or high speed rotors (figure 2), where an out-of-balance mass of a rotating element or body can quickly and easily be located, providing the exact amount and location of the balancing mass that should be placed or removed to reduce the vibration. The balancing machine contains a balancing head with a clutch which is first opened to release a set of balls to naturally take place in the balancing positions. Subsequently, the clutch is closed producing a clamping of the balls in the adjusted positions, while the body is being rotated above its critical speed and then released. Other automatic balancing devices have been proposed; essentially using one of the four balancing methods; two angular arms, two sliding arms, one angular and sliding arm, or, one spirally sliding arm (Chong-Won, 2006; Zhou y Shi, 2001).

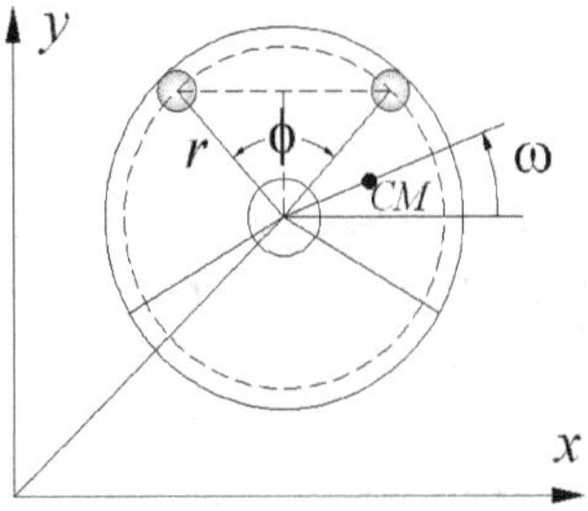

Figure 2. Diagram of the automatic balancer using two masses.

The use of piezoelectric actuators as active vibration dampers in rotating machines has been considered in the past. Palazzolo, et al. (Palazzolo et al., 1993) first used the piezoelectric pusher for active vibration control in rotating machinery as it is shown in Figure 3.a. The pusher is soft mounted to the machine case to improve the electromechanical stability and connected to the squirrel cage-ball bearing supports of a rotating shaft, to actively control the unbalance, transient and subsynchronous responses of the test rotor, using velocity feedback. The piezoelectric actuators are modeled as dampers and springs. Recently, Carmignani et al. (Carmignani et al., 2001) developed an adaptive hydrodynamic bearing made of a mobile housing mounted on piezoelectric actuators to attenuate the vibration amplitudes in constant speed below the first critical speed. The actuators, arranged at 90° on a perpendicular plane to the shaft axis, exert two sinusoidal forces with a tuned phase angle to produce a balancing or, alternatively, a dampering effect. The authors presented experimental and numerical results.

Active Magnetic Bearings (AMBs) are the mostly used devices but their use in the industrial field is still limited due to a low stiffness and the need of additional conventional bearings for fault emergency. An AMB system is a collection of electromagnets used to suspend an object and stabilization of the system is performed by feedback control, see Figure 3.b. In recent decades, AMBs has been widely used as a non-contact, lubrication-free, support in many machines and devices. Many researchers (Lee, 2001; Sheu-Yang, 1997) have proposed a variety of AMBs that are compact and simple-structured. The AMB system, which is open-loop unstable and highly coupled due to nonlinearities inherited in the system such as the gyroscopic effect and imbalance, requires a dynamic controller to stabilize the system.

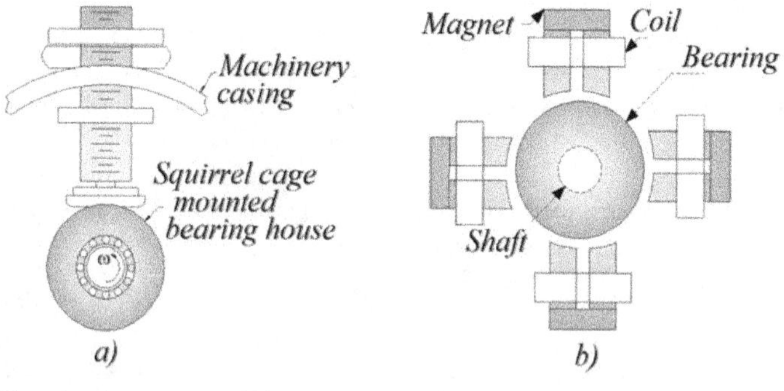

Figure 3. a) Piezoelectric actuator and b) active magnetic bearing.

Another device for AVC in rotating machinery is the one based on fluid film bearings. The dynamics of a rotor system supported by fluid film bearings is inherently a nonlinear problem and these fluid film bearings have been used in combination with other devices, such as piezoelectric actuators, magneto or electro-rheological fluids, etc. (see Figure 4).

Guozhi *et al.* (2000) proposed the use of a fluid bearing with rheological fluids to reduce the vibrations around the first critical speed. Magnetorheological (MR) or electrorheological

(ER) fluids are materials that respond to an applied magnetic or electric field with a dramatic change in rheological behavior. To attenuate the vibration amplitudes around the first critical speed an on/off control is proposed to control the large amplitude around the first critical speed.

Hathout and El-Shafei (Hathout and El-Shafei, 1997) proposed a hybrid squeeze film damper (HSFD), (see Figure 4.b), to attenuate the vibrations in rotating machinery for both sudden unbalance and transient run-up through critical speeds. El-Shafei (El-Shafei, 2000) have implemented different control algorithms (PID-type controllers, LQR, gain scheduling, adaptive and bang-bang controllers) for active control of rotor vibrations for HSFD-supported rotors. Controlling the fluid pressure in the chamber, the bearing properties of stiffness and damping can be changed.

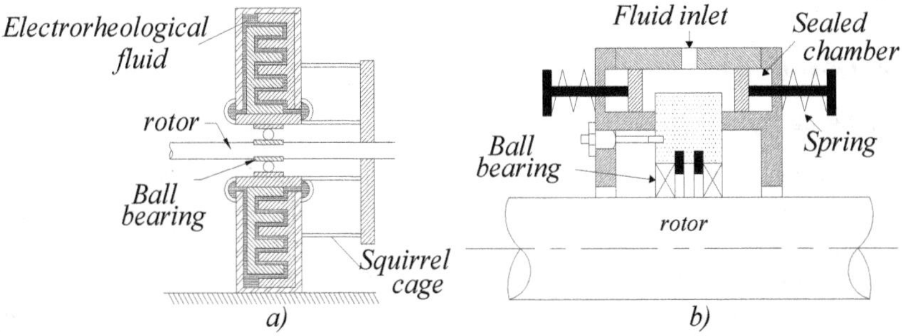

Figure 4. Fluid film bearings: a) using rheological fluids and b) using a pressure chamber.

Sun y Kroedkiewski (Sun and Krodkiewski, 1997, 1998) proposed a new type of active oil bearing, see Figure 5.a. The active bearing is supplied with a flexible sleeve whose deformation can be changed during rotor operation. The flexible sleeve is also a part of a hydraulic damper whose parameters can be controlled during operation as well. The oil film and the pressure chamber are separated by the flexible sealing. The equilibrium position of the flexible sleeve and the bearing journal is determined by load and pressure, which can be controlled during operation. Parameters of this damper can also be varied during operation to eliminate the self exciting vibration and increase the stability of the equilibrium position of the rotor-oil bearing system.

Recently, Dyer et al., (Dyer et al., 2002) developed an electromagnetically actuated unbalance compensator. The compensator consists of two rings as shown in Figure 5.b. These two rings are not balanced and can be viewed as two heavy spots. These two rings are held in place by permanent magnetic forces. When the balancer is activated, an electric current passes through the coil and the rings can be moved individually with respect to the spindle by the electromagnetic force. The combination of these two heavy spots is equivalent to a single heavy spot whose magnitude and position can change to attenuate the vibration amplitudes.

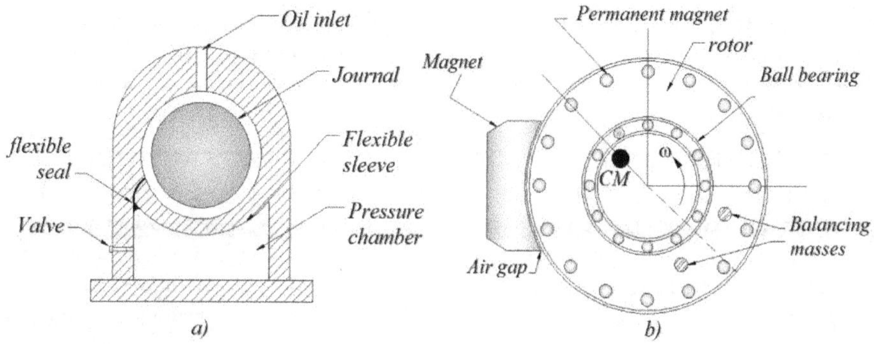

Figure 5. a) Fluid film bearing with flexible sleeve and b) electromagnetically actuated unbalance compensator.

3. Rotor-bearing system

3.1. Mathematical model

The rotor-bearing system consists of a planar and rigid disk of mass M mounted on a flexible shaft of negligible mass and stiffness k at the mid-span between two symmetric bearing supports (see Fig. 6 when a=b). Due to rotor imbalance the mass center is not located at the geometric center of the disk S but at the point G (center of mass of the unbalanced disk), the distance between these points is known as disk eccentricity or static unbalance u (see Vance, 1988; Dimarogonas, 1996).

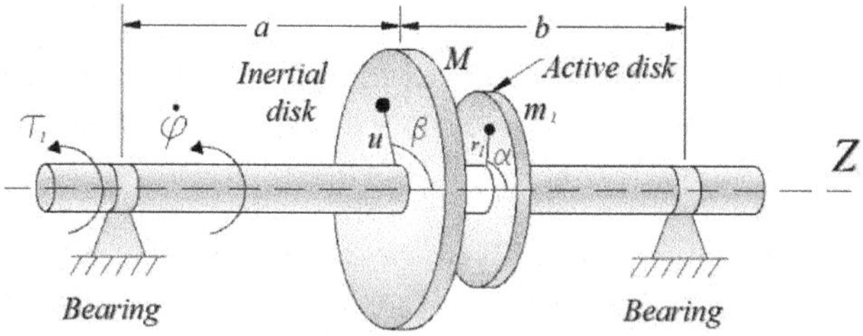

Figure 6. Rotor bearing system with active disk.

In the analysis, the rotor-bearing system has an active disk (Blanco et al., 2008) mounted on the shaft and near the main disk (see Fig. 6). The active disk is designed in order to move a mass m_1 in all angular and radial positions inside of the disk given by α and r_1, respectively. In fact, these movements can be obtained with some mechanical elements such as helical gears and a ball screw (see Fig. 7.a). The mass m_1 and the radial distance r_1 are designed in order to compensate the residual unbalance of the rotor bearing system by means of the correct angular position α of the balancing mass. The angular position of the unbalance is denote by β, see Fig. 7.b.

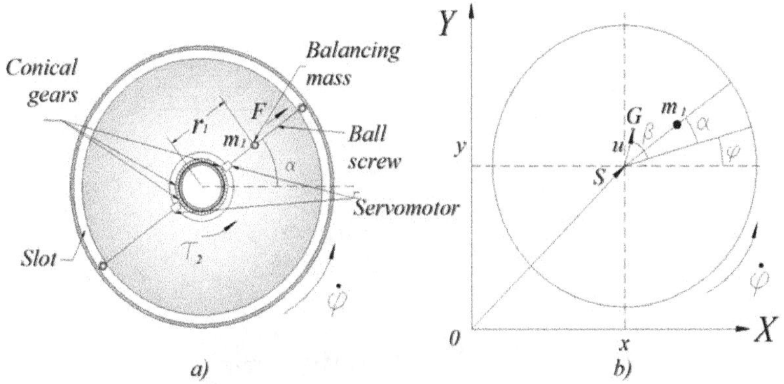

Figure 7. Schematic diagram and main elements of the active disk.

The mathematical model of the five degree-of-freedom rotor-bearing system with active disk was obtained using Euler-Lagrange equations, which is given by

$$(M + m_1)\ddot{x} + c\dot{x} + kx = p_x(t)$$

$$(M + m_1)\ddot{y} + c\dot{y} + ky = p_y(t)$$

$$J_e\ddot{\varphi} + c_\varphi\dot{\varphi} = \tau_1 + p_\varphi(t) \tag{1}$$

$$m_1 r_1^2\ddot{\alpha} + 2m_1 r_1\dot{r}_1\dot{\alpha} + m_1 g r_1 \cos\alpha = \tau_2$$

$$m_1\ddot{r}_1 - m_1 r_1\dot{\alpha}^2 + m_1 g \sin\alpha = F$$

with

$$p_x(t) = Mu[\ddot{\varphi}\sin(\varphi + \beta) + \dot{\varphi}^2\cos(\varphi + \beta)] + m_1 r_1[\ddot{\varphi}\sin(\varphi + \alpha) + \dot{\varphi}^2\cos(\varphi + \alpha)]$$

$$p_y(t) = Mu[\dot{\varphi}^2\sin(\varphi + \beta) - \ddot{\varphi}\cos(\varphi + \beta)] + m_1 r_1[\dot{\varphi}^2\sin(\varphi + \alpha) - \ddot{\varphi}\cos(\varphi + \alpha)]$$

$$p_\varphi(t) = Mu[\ddot{x}\sin(\varphi + \beta) - \ddot{y}\cos(\varphi + \beta)] + m_1 r_1[\ddot{x}\sin(\varphi + \alpha) - \ddot{y}\cos(\varphi + \alpha)]$$

$$J_e = J + Mu^2 + mr^2$$

Here c is the equivalent viscous damping provided by the isotropic bearings, J and c_φ are the inertia polar moment and the viscous damping of the rotor, respectively. $\tau_1(t)$ is the applied torque (control input) for rotor speed regulation, x and y are the orthogonal coordinates that describe the disk position, r_1 and α denote the radial and angular positions of the balancing mass, which is controlled by means of the control force $F(t)$ and the $\tau_2(t)$ control torque (servomechanism). The angular position of the rotor is denote by φ.

Defining the state variables as $z_1 = x, z_2 = \dot{x}$, $z_3 = y, z_4 = \dot{y}, z_5 = \varphi, z_6 = \dot{\varphi}, z_7 = r_1, z_8 = \dot{r}_1, z_9 = \alpha$ and $z_{10} = \dot{\alpha}$, the following state space description is obtained.

$$\dot{z}_1 = z_2$$

$$\dot{z}_2 = \frac{1}{\Delta}\left(\frac{1}{M_e}(b^2 + J_e M_e)f_1 + \frac{ab}{M_e}f_2 + a(\tau_1 - c_\varphi z_6)\right)$$

$$\dot{z}_3 = z_4$$

$$\dot{z}_4 = \frac{1}{\Delta}\left(\frac{ab}{M_e}f_1 + \frac{1}{M_e}(J_e M_e - a^2)f_2 + b(\tau_1 - c_\varphi z_6)\right)$$

$$\dot{z}_5 = z_6$$

$$\dot{z}_6 = \frac{1}{\Delta}\left(-af_1 - bf_2 - M_e(\tau_1 - c_\varphi z_6)\right) \quad (2)$$

$$\dot{z}_7 = z_8$$

$$\dot{z}_8 = \frac{1}{m_1}(F - gm_1 \sin z_9 + m_1 z_7 z_{10}^2)$$

$$\dot{z}_9 = z_{10}$$

$$\dot{z}_{10} = \frac{1}{m_1 z_7^2}(\tau_2 - gm_1 z_7 \cos z_9 - 2m_1 z_7 z_8 z_{10})$$

$$y = z_1^2 + z_3^2$$

with

$f_1 = c_\varphi z_2 + kz_1 - Mz_6^2 u_y - m_1 r_y z_6^2$, $f_2 = cz_4 + kz_3 - Mz_6^2 u_x - m_1 r_x z_6^2$, $a = -Mu_x - m_1 r_x$, $b = Mu_y + m_1 r_y$, $J_e = J + Mu^2 + m_1 r_1^2$, $M_e = M + m_1$ and $\Delta = a^2 + b^2 - J_e M_e$.

The rotor-bearing system with active disk is then described by the five degree-of-freedom, highly nonlinear and coupled model (2). The proposed control objective consists of reducing as much as possible the rotor vibration amplitude, denoted in non-dimensional units by

$$R = \frac{\sqrt{z_1^2 + z_3^2}}{u} \quad (3)$$

for run-up, coast-down or steady state operation of the rotor system, even in presence of small exogenous or endogenous perturbations.

In the following table the rotor system parameters used throughout the chapter are presented.

$M = 1.2kg$	$m_1 = 0.003kg$	$a = b = 0.3m$
$\beta = \frac{\pi}{6}rad$	$\alpha = 0rad$	$r_{disk} = 0.04m$
$u = 100\mu m$	$c_\varphi = 1.5 \times 10^{-3}\frac{Ns}{m}$	$D = 0.01m$

Table 1. System parameters

3.2. Active vibration control

3.2.1 Active disk control

Here it is proposed to use an active disk for actively balancing of the rotor (see Fig. 8). It can be seen that if the mass m_1 is located at the position $\left(\bar{r} = \frac{Mu}{m_1}, \bar{\alpha} = \beta + \pi\right)$ the unbalance can be cancelled because the centrifugal force due to this mass cancel out those caused by the residual imbalance mass. The balancing mass is placed at an angle of 180° to the unbalanced mass to restore the centre of rotation.

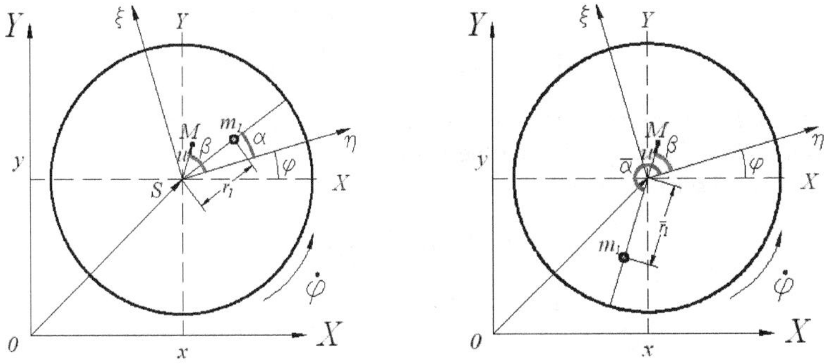

Figure 8. Main components of the active disk.

In order to design the position controllers for the balancing mass m_1, consider its associated dynamics:

$$\dot{z}_7 = z_8$$

$$\dot{z}_8 = \frac{1}{m_1}(F - gm_1 \sin z_9 + m_1 z_7 z_{10}^2)$$

$$\dot{z}_9 = z_{10}$$

$$\dot{z}_{10} = \frac{1}{m_1 z_7^2}(\tau_2 - gm_1 z_7 \cos z_9 - 2m_1 z_7 z_8 z_{10})$$

$$y_2 = z_7$$

$$y_3 = z_9$$

From these equations, the exact linearization method (Sira et al., 2005) is applied and the following nonlinear controllers can be obtained with integral compensation to take the balancing mass to the equilibrium position $\left(y_2 = \bar{r} = \frac{Mu}{m_1}, y_3 = \bar{\alpha} = \beta + \pi\right)$:

$$F = m_1 v_2 + gm_1 \operatorname{sen} z_9 - m_1 z_7 z_{10}^2 \tag{4}$$

$$\tau_2 = m_1 z_7^2 v_3 + gm_1 z_7 \cos z_9 + 2m_1 z_7 z_8 z_{10} \tag{5}$$

with

$$v_2 = \ddot{y}_2^*(t) - \gamma_{22}[\dot{y}_2 - \dot{y}_2^*(t)] - \gamma_{21}[y_2 - y_2^*(t)] - \gamma_{20} \int_0^t [y_2 - y_2^*(\sigma)]d\sigma$$

$$v_3 = \ddot{y}_3^*(t) - \gamma_{32}[\dot{y}_3 - \dot{y}_3^*(t)] - \gamma_{31}[y_3 - y_3^*(t)] - \gamma_{30} \int_0^t [y_3 - y_3^*(\sigma)]d\sigma$$

where $\dot{y}_2^*(t)$ and $\dot{y}_3^*(t)$ are desired trajectories for the outputs y_2 and y_3. Thus, the tracking errors $e_2 = y_2 - y_2^*(t)$ and $e_3 = y_3 - y_3^*(t)$, obey the following set of linear, decoupled, homogeneous differential equations:

$$e_2^{(3)} + \gamma_{22}\ddot{e}_2 + \gamma_{21}\dot{e}_2 + \gamma_{20}e_2 = 0$$

$$e_3^{(3)} + \gamma_{32}\ddot{e}_3 + \gamma_{31}\dot{e}_3 + \gamma_{30}e_3 = 0$$

which can be handle to have the point: $(e_2, e_3) = (0,0)$, as an exponentially asymptotically stable equilibrium point by selecting the design parameters $\{\gamma_{20}, \gamma_{21}, \gamma_{22}, \gamma_{30}, \gamma_{31}, \gamma_{32}\}$ such that the characteristic polynomials

$$p_2(s) = s^3 + \gamma_{22}s^2 + \gamma_{21}s + \gamma_{20}$$

$$p_3(s) = s^3 + \gamma_{32}s^2 + \gamma_{31}s + \gamma_{30}$$

are Hurwitz polynomials.

It is evident, however, that the controllers (4) and (5) require information of the disk eccentricity (u, β). In what follows the algebraic identification method to estimate the disk eccentricity (u, β) is applied.

3.2.2 *Angular velocity controller*

In order to control the speed of the rotor, consider its associated dynamics, under the assumption that the effect of the unbalance was cancelled out by the active disk and that the disk eccentricity (u, β) is perfectly known:

$$[J + (Mu^2 + m_1 r_1^2)] \ \dot{z}_6 + c_\varphi z_6 = \tau_1$$

$$y_1 = z_6 \tag{6}$$

From this equation, the following PI controller to asymptotically track a desired reference trajectory $\dot{y}_1^*(t)$ can be obtained:

$$\tau_1 = [J + (Mu^2 + m_1 r_1^2)]v_1 + c_\varphi z_6$$

$$v_1 = \dot{y}_1^*(t) - \gamma_{11}[y_1 - y_1^*(t)] - \gamma_{10} \int_0^t [y_1 - y_1^*(\sigma)]d\sigma. \tag{7}$$

The use of this controller yields the following closed-loop dynamics for the trajectory tracking error $e_1 = y_1 - y_1^*(t)$ as follows

$$\ddot{e}_1 + \gamma_{11}\dot{e}_1 + \gamma_{10}e_1 = 0 \tag{8}$$

Therefore, by selecting the design parameters $\{\gamma_{10}, \gamma_{11}\}$ such that the associated characteristic polynomial for (8) be Hurwitz, it is guaranteed that the error dynamics be globally asymptotically stable.

4. On-line algebraic identification of eccentricity

Consider the first two equations in (1), where measurements of the position coordinates of the disk (z_1, z_3) are available to be used in the on-line eccentricity identification scheme.

$$(M+m_1)\dot{z}_2 + cz_2 + kz_1 = Mu[\dot{z}_6 \operatorname{sen}(z_5+\beta) + \dot{z}_6^2 \cos(z_5+\beta)] +$$
$$+m_1 r_1[\dot{z}_6 \operatorname{sen}(z_5+\alpha) + \dot{z}_6^2 \cos(z_5+\alpha)]$$
$$(M+m_1)\dot{z}_4 + cz_4 + kz_3 = Mu[\dot{z}_6^2 \operatorname{sen}(z_5+\beta) - \dot{z}_6 \cos(z_5+\beta)] + \tag{9}$$
$$+m_1 r_1[\dot{z}_6^2 \operatorname{sen}(z_5+\alpha) - \dot{z}_6 \cos(z_5+\alpha)]$$

Multiplying (9) by the quantity t^2 and integrating the result twice with respect to time t, the following is obtained:

$$\int^{(2)}\left[(M+m_1)t^2\frac{dz_2}{dt} + ct^2 z_2 + kt^2 z_1\right] =$$
$$\int^{(2)} Mut^2\frac{d}{dt}[z_6 \operatorname{sen}(z_5+\beta)] + \int^{(2)} m_1 r_1 t^2 \frac{d}{dt}[z_6 \operatorname{sen}(z_5+\alpha)]$$
$$\int^{(2)}\left[(M+m_1)t^2\frac{dz_4}{dt} + ct^2 z_4 + kt^2 z_3\right] \tag{10}$$
$$= \int^{(2)} Mut^2\frac{d}{dt}[z_6 \cos(z_5+\beta)] + \int^{(2)} m_1 r_1 t^2 \frac{d}{dt}[z_6 \cos(z_5+\alpha)]$$

where $\int^{(n)} \varphi(t)$ are iterated integrals of the form $\int_0^t \int_0^{\sigma_1} \cdots \int_0^{\sigma_{n-1}} \varphi(\sigma_n) d\sigma_n \cdots d\sigma_1$, with $\int \varphi(t) = \int_0^t \varphi(\sigma) d\sigma$ and n a positive integer.

Using integration by parts, the following is obtained:

$$(M+m_1)\left[t^2 z_1 - 4\int t z_1 + 2\int^{(2)} z_1\right] + c_\varphi\left[\int t^2 z_1 - 2\int^{(2)} t z_1\right] + k\int^{(2)} t^2 z_1 =$$
$$= Mu\left[t^2 z_6 \sin(z_5+\beta) - 2\int^{(2)} t z_6 \sin(z_5+\beta)\right] +$$
$$+m_1 r_1\left[\int t^2 z_6 \sin(z_5+\alpha) - 2\int^{(2)} t z_6 \sin(z_5+\alpha)\right] \tag{11}$$
$$(M+m_1)\left[t^2 z_3 - 4\int t z_3 + 2\int^{(2)} z_3\right] + c_\varphi\left[\int t^2 z_3 - 2\int^{(2)} t z_3\right] + k\int^{(2)} t^2 z_3 =$$
$$= -Mu\left[t^2 z_6 \cos(z_5+\beta) - 2\int^{(2)} t z_6 \cos(z_5+\beta)\right] +$$
$$+m_1 r_1\left[\int t^2 z_6 \cos(z_5+\alpha) - 2\int^{(2)} t z_6 \cos(z_5+\alpha)\right]$$

The above integral-type equations (11), after some algebraic manipulations, lead to the following linear equations system:

$$A(t)\theta = b(t) \tag{12}$$

where $\theta = [u_\eta = u\cos\beta\, , u_\xi = u\sin\beta]^T$ denotes the eccentricity parameter vector to be identified and $A(t)$, $b(t)$ are 2x2 and 2x1 matrices, respectively, which are described by

$$A(t) = \begin{bmatrix} a_{11}(t) & a_{12}(t) \\ -a_{12}(t) & a_{11}(t) \end{bmatrix}, b(t) = \begin{bmatrix} b_1(t) \\ b_2(t) \end{bmatrix}$$

whose components are time functions specified as

$$a_{11} = M\left[\int t^2 z_6 \operatorname{sen} z_5 - 2\int^{(2)} t z_6 \operatorname{sen} z_5\right]$$

$$a_{12} = M\left[\int t^2 z_6 \cos z_5 - 2\int^{(2)} t z_6 \cos z_5\right]$$

$$b_1 = (M + m_1)t^2 z_1 + \int (ct^2 z_1 - 4(M + m_1)tz_1)$$

$$+\int (m_1 z_6 z_7 t^2 \operatorname{sen}(z_5 + \alpha)) + \int^{(2)} (2(M + m_1)z_1 - 2ctz_1 + kt^2 z_1) + \int^{(2)} 2m_1 z_6 z_7 t \operatorname{sen}(z_5 + \alpha)$$

$$b_2 = (M + m_1)t^2 z_3 + \int (ct^2 z_3 - 4(M + m_1)tz_3)$$

$$+\int (m_1 z_6 z_7 t^2 \cos(z_5 + \alpha)) - \int^{(2)} 2m_1 z_6 z_7 t \cos(z_5 + \alpha)$$

From equation (12) it can be concluded that the parameter vector θ is *algebraically identifiable* if, and only if, the trajectory of the dynamical system is *persistent* in the sense established by Fliess and Sira-Ramírez (Fliess and Sira-Ramírez, 2003), that is, the trajectories or dynamic behavior of the system satisfy the condition

$$\det A(t) \neq 0$$

In general, this condition holds at least in a small time interval $(t_0, t_0 + \delta_0]$, where δ_0 is a positive and sufficiently small value. The parameter identification is quickly performed and it is almost exact with respect to the real parameters. It is also evident the presence of singularities in the algebraic identifier, i.e., when the determinant *den*=det*(A)* is zero. The first singularity, however, occurs after the identification has been finished.

In (Beltran, 2010) is described the application of an on-line algebraic identification methodology for parameter and signal estimation in vibrating systems. The algebraic identification is employed to estimate the frequency and amplitude of exogenous vibrations affecting the mechanical system using only position measurements. Some simulations and experimental results are presented using the on-line algebraic identification scheme for an electromechanical platform (ECP™ rectilinear plant) with a single degree-of-freedom mass-spring-damper system.

By solving equations (12) the following algebraic identifier for the unknown eccentricity parameters is obtained:

$$\left.\begin{aligned} u_{\eta e} &= \frac{b_1 a_{11} - b_2 a_{12}}{a_{11}^2 + a_{12}^2} \\ u_{\xi e} &= \frac{b_1 a_{12} + b_2 a_{11}}{a_{11}^2 + a_{12}^2} \\ u_e &= \sqrt{u_{\eta e}^2 + u_{\xi e}^2} \\ \beta_e &= \cos\left(\frac{u_{\eta e}}{u_e}\right) \end{aligned}\right\} \quad \forall t \in (t_0, t_0 + \delta_0] \tag{13}$$

5. Simulation results

In Fig. 9 it is depicted the identification process of the eccentricity. A good and fast estimation ($t \ll 0.1s$) can be observed. Fig. 10 shows the dynamic behavior of the adaptive-like control scheme (7), which starts using the nominal value u= $0\mu m$. A desired reference trajectory was considered for regulating the evolution of the output variable y_1 towards the desired equilibrium $\bar{y} = \bar{z}_6 = 300 rad/s$, which is given by a Bezier type polynomial in time.

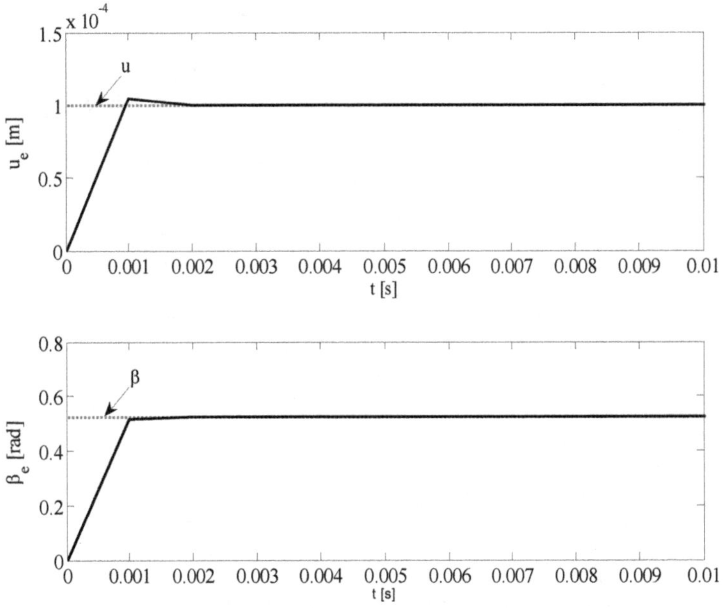

Figure 9. Eccentricity (u) and angular position (β) identification.

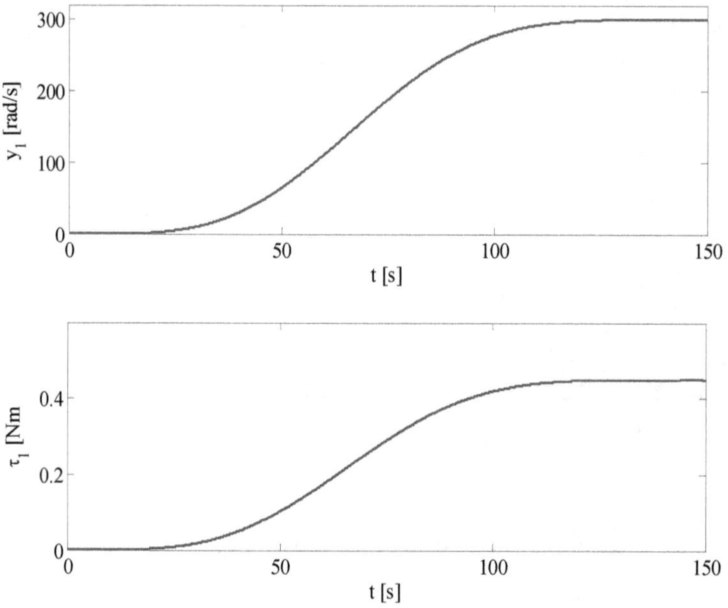

Figure 10. Rotor speed and control torque.

Fig. 11 shows the dynamic behavior of the active disk controllers when the balancing mass is driven to the equilibrium position $\left(\bar{r} = \frac{M}{m_1} u_e, \bar{\alpha} = \beta_e + \pi\right)$. In this position the active disk cancels the unbalance, as it is shown in the Fig. 12. The controllers are implemented when the eccentricity has been estimated.

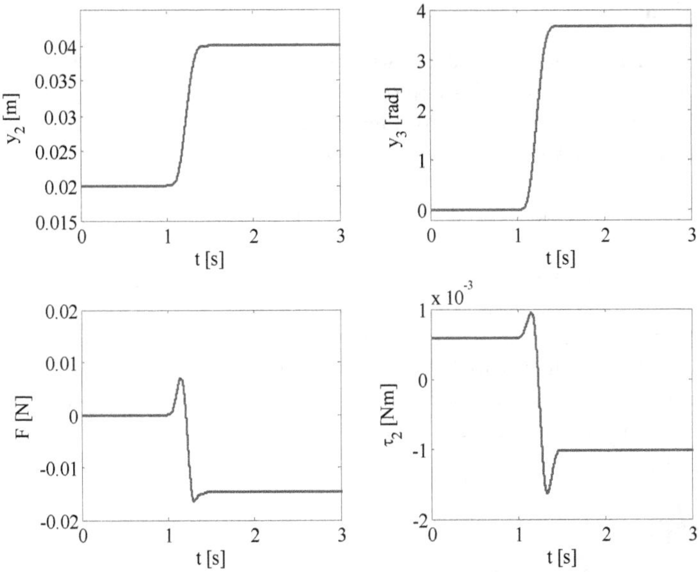

Figure 11. Dynamic response of the balancing mass: radial position (y_2=z_7), angular position (y_3=z_9), control force (F) and moment force (τ_2).

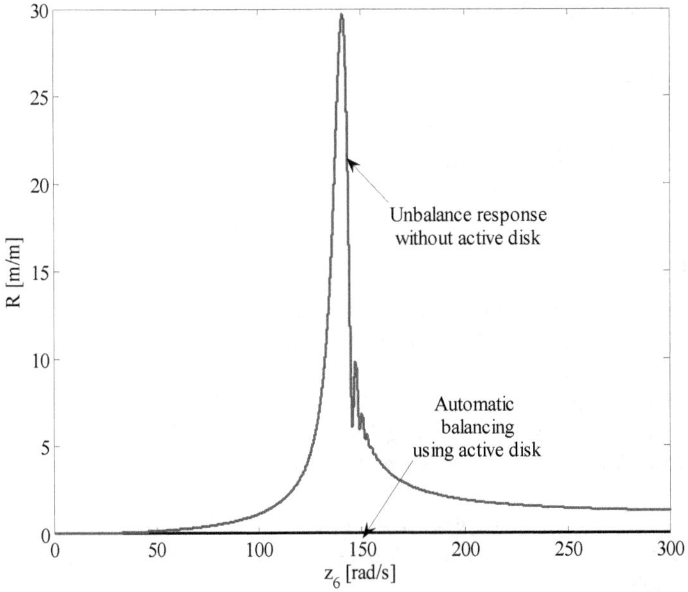

Figure 12. Unbalance response with automatic balancing and without active disk.

6. Conclusions

The active vibration control of rotor-bearing systems using active disks for actively balancing a rotor is addressed. This approach consists of locating a balancing mass at a suitable position. Since this active control scheme requires information of the eccentricity, a novel algebraic identification approach is proposed for the on-line estimation of the eccentricity parameters. This approach is quite promising, in the sense that from a theoretical point of view, the algebraic identification is practically instantaneous and robust with respect to parameter uncertainty, frequency variations, small measurement errors and noise. Thus the algebraic identification is combined with two control schemes to place the balancing mass in the correct position to cancel the unbalance of the rotor. A velocity control is designed to take the rotor velocity to a desired operating point over the first critical speed in order to show the vibration cancellation. The controllers were developed in the context of an off-line prespecified reference trajectory tracking problem. Numerical simulations were included to illustrate the proposed high dynamic performance of the active vibration control scheme proposed.

Author details

Andrés Blanco-Ortega, Jorge Colín-Ocampo, Marco Oliver-Salazar and Gerardo Vela-Valdés
Centro Nacional de Investigación y Desarrollo Tecnológico, CENIDET, México

Gerardo Silva-Navarro
Centro de Investigación y de Estudios Avanzados del IPN, CINVESTAV, México

Acknowledgement

Research reported here was supported by grants of the Dirección General de Educación Superior Tecnológica, DGEST through PROMEP under Project "Vibration control of rotating machinery".

7. References

Beltrán F.; Silva, G.; Sira, H. and Quezada, J. (2005). Active vibration control using on-line algebraic identification of harmonic vibrations, *Proceedings of American Control Conference*, pp. 4820 - 4825, ISBN 0-7803-9099-7, Portland, Oregon, USA, June 8-10, 2005.

Beltrán, F.; Sira, H. and Silva, G. (2006). Adaptive-like Active Vibration Suppression for a Nonlinear Mechanical System Using On-Line Algebraic Identification, *Proceedings. of the Thirteenth International Congress on Sound and Vibration*, Vienna, Austria, July 2-6, 2006.

Beltrán F.; Silva, G.; Sira, H. and Blanco, A. (2010). Active Vibration Control Using On-line Algebraic Identification and Sliding Modes. Computación y Sistemas. Vol. 13, No. 3, pp 313-330. ISSN 1405-5546.

Blanco, A.; Silva, G. and Gómez, J. C. (2003). Dynamic Stiffness Control and Acceleration Scheduling for the Active Balancing Control of a Jeffcott-Like Rotor System, *Proceedings of The tenth International Congress on Sound and Vibration*, pp. 227-234, Stockholm, Sweden, July 7-10, 2003.

Blanco, A.; Beltran, F. and Silva, G. (2007). On Line Algebraic Identification of Eccentricity in Active Vibration Control of Rotor Bearing Systems, *4th International Conference on Electrical and Electronics Engineering*, pp. 253 - 256, ISBN 978-1-4244-1166-5, México, Sept. 5-7, 2007.

Blanco, A.; Beltrán, F. and Silva, G. (2008). Active Disk for Automatic Balancing of Rotor-Bearing Systems. *American Control Conference, ACC 2008*. pp. 3023 - 3028, ISBN 978-1-4244-2079-7, Seattle, WA, USA, June 11-13, 2008.

Blanco, A.; Beltrán, F.; Silva, G. and Oliver, M. A. (2010). Active vibration control of a rotor-bearing system based on dynamic stiffness, *Revista de la Facultad de Ingeniería. Universidad de Antioquia*. No. 55, pp. 125-133. ISSN. 0120-6230.

Blanco, A.; Beltrán, F.; Silva, G. and Méndez, H. (2010). Control de Vibraciones en Sistemas Rotatorios, Revista Iberoamericana de Automática e Informática Industrial. Vol. 7, No. 4, (Octubre 2010.) pp. 36-43, ISSN 1697-7912.

Carmignani, C.; Forte, P. and Rustighi, E. (2001). Active Control of Rotors by Means of Piezoelectric Actuators, *Proceedings of Design Engineering Technical Conference and Computers and Information in Engineering Conference*, p.757-764, ISBN 0791835413, Vol. 6, Pittsburgh, Pennsylvania, USA, September 9-12, 2001.

Chong-Won, L. (2006). Mechatronics in Rotating Machinery. 7th IFToMM-Conference on Rotor Dynamics, pp. 25-28, Vienna, Austria, September, 2006.

Dimarogonas, A. (1996). *Vibration for Engineers*. Prentice Hall, ISBN 978-0134562292, 1996.

Dyer,, S. W.; Ni, J.; Shi, J. and Shin, K. (2002). Robust Optimal Influence-Coefficient Control of Multiple-Plane Active Rotor Balancing Systems, *Journal of Dynamic Systems, Measuremente, and Control*, pp. 41-46. ISSN 0022-0434, Vol. 124.

El-Shafei, A. (2000). Active Control Algorithms for the Control of Rotor Vibrations Using HSFDS. Proc. of ASME TURBOEXPO 2000, Munich Germany.

Fliess, M. and Sira-Ramírez, H. (2003) An algebraic framework for linear identification, *ESAIM: Control, Optimization and Calculus of Variations*, pp. 151-168, Vol. 9, 2003.

Green K, Champneys A.R., Friswell M.I. y Muñoz (2008) A.M.Investigation of a multi-ball, automatic dynamic balancing mechanism for eccentric rotors. *Royal Society Publishing*, pp. 705-728, Vol. 366, No. 1866.

Guozhi, Y., Fah, Y.F., Guang, C., Guang, M., Tong, F. and yang, Q., Electro-Rheological Multi-layer Squeeze Film Damper and its Application to Vibration Control of Rotor System, *Journal of Vibration and Acoustics*, pp. 7-11, Vol. 122, No. 1, 2000.

Hathout, J. P and El-Shafei, A. (1997). PI Control of HSFDs for Active Control of Rotor-Bearing Systems, *Journal of Engineering for Gas Turbines and Power*, pp. 658-667, ISSN 0742-4795, Vol. 119, No. 3.

Hredzak, B. and Guo, G. (2006). Adjustable Balancer With Electromagnetic Release of Balancing Members. *IEEE Transactions on Magnetics*, pp. 1591-1596, Vol. 42, No. 5.

Krodkiewski, J. M. and Sun L. (1998). Modelling of Multi-Bearing Rotor Systems Incorporating an Active Journal Bearing, 215-229, Vol. 10, No. 2.

Lee, J. H.; Allaire, P. E.; Tao, G. and Zhang, X. (2001). Integral Sliding-Mode Control of a Magnetically Suspended Balance Beam: Analisys, Simulation, and Experiment, *Transactions on Mechatronics*, pp. 338-346, Vol. 6, No. 3.

Ljung, L. (1987). *Systems Identification: Theory for the User*. Englewood Cliffs, Prentice-Hall.

Palazzolo, A. B. ; Jagannathan, S.; Kaskaf, A. F.; Monatgue, G. T. and Kiraly, L.J. (1993). Hybrid Active Vibration Control of Rotorbearing Systems Using Piezoelectric Actuators, *Journal of Vibration and Acoustics*, pp. 111-119, Vol. 115, No. 1.

Sagara, S. and Zhao, Z. Y. (1989). Recursive identification of transfer function matrix in continuous systems via linear integral filter, *International Journal of Control*, pp. 457-477, Vol. 50, No. 2.

Sagara, S. and Zhao, Z. Y. (1990). Numerical integration approach to on-line identification of continuous systems, *Automatica*, pp. 63-74. Vol. 26, No. 1, 1990

Sandler, B.Z. (1999). *Robotics: designing the mechanism for automated machinery*. San Diego, CA: Academic Press.

Sheu, G.J., Yang, S.M. and Yang, C.D. (1997). Design of Experiments for the Controller of Rotor Systems With a Magnetic Bearing, *Journal of Vibration and Acoustics*, pp. , Vol. 119, No. 2, 1997.

Sira, H., Márquez, R., Rivas, F. and Llanes, O. (2005) *Control de sistemas no lineales: Linealización aproximada, estendida, exacta*. Editorial pearson. Serie Automática, Robótica.

Soderstrom, T. and Stoica, P. (1989). *System Identification*. New York: Prentice-Hall.

Sun, L.; Krodkiewski, J. and Cen, Y. (1997). Control Law Synthesis for Self-Tuning Adaptive Control of Forced Vibration in Rotor Systems. *2nd International Symposium MV2 on Active Control in Mechanical Engineering*, S9-25-37. Lyon, France.

Thearle, E.L. (1932). A New Type of Dynamic Balancing Machine, *ASME Journal of Applied Mechanics*, pp. 131-141, No. 54.

Vance, J.M. (1988). *Rotordynamics of Turbomachinery*. New York: John Wiley and Sons Inc.

Zhou, S. and Shi, J. (2001). Active Balancing and Vibration Control of Rotating Machinery: a survey, *The Shock and Vibration Digest*, pp. 361-371, Vol. 33, No. 4, 2001.

Zhou, S. and Shi, J. (2002). Optimal one-plane active balancing of a rigid rotor during acceleration, *Journal of Sound and Vibration*, pp. 196-205, ISSN 0022-460X, Vol. 249, No. 1, 2002.

Chapter 10

Transverse Vibration Control for Cable Stayed Bridge Under Construction Using Active Mass Damper

Hao Chen, Zhi Sun and Limin Sun

Additional information is available at the end of the chapter

1. Introduction

Since the erection of the Stromaund Bridge of Sweden in 1956, cable-stayed bridge, as an efficient and economic bridge type to surmount a long-distance obstacle, has attracted more and more interests both from bridge engineering community and from the society and government. Nowadays, the cable stayed bridge is the most competitive type for the bridge with the span of 300-1000 meters. For a cable supported bridge, which is generally quite flexible and of low damping, its vibration under ambient excitation (such as the wind and ground motion excitation) and operational loading (such as the vehicle and train loads) is quite critical for its safety, serviceability and durability. Vibration control countermeasures, such as the installation of the energy dissipating devices, are thus required [1, 2]. Structural active control, which applies a counter-force induced by a control device to mitigate structural vibration, has been widely proposed for the vibration control of cable stayed bridges and proven to be efficient by many researchers [3-6].

Although the vibration response of a fully erected cable-stayed bridge should be controlled, a cable-stayed bridge under construction, which is of low damping and not as stable as the completed structure, is generally more vulnerable to dynamic loadings. During the construction stage, the cable pylons were generally erected firstly and the cable and main girders are then hang on the pylons symmetrically in a double-cantilever way. With the increase of the cantilever length, the bridge is more and more flexible. When the girder is on its longest double-cantilever state, the bridge is the most vulnerable to the external disturbance (such as the ambient wind fluctuation and ground motions). Moreover, if the cables, pylons and main girder of the bridge are all steel components and thus the damping of the bridge is very low, its vibration under ambient excitations will be quite large. The vibration reduction countermeasures are thus in great demand. Frederic conducted several mock-up tests representing a cable-stayed bridge during the construction stage [7]. Since the control objective was set to reduce the girder vertical vibration response or cable parametric vibration response, the active tendon was installed as the control device. While for a cable stayed bridge under uncontrollable ambient excitations, structure will vibrate not only in the vertical direction but also in the transverse direction. Moreover, since the bridges are generally designed to carry the vertical loads, the unexpected transverse loads, especially the transverse dynamic loads, will induce structural safety and durability problems. It is thus of crucial importance to install some vibration reduction devices to control bridge transverse vibration. For this type of structure vibration control problem, the active mass damper (AMD) or the active tuned mass damper (ATMD) will be a competent candidate [8, 9].

In this chapter, the general procedure and key issues on adopting an active control device, the active mass damper (AMD), for vibration control of cable stayed bridges under construction are presented. Taking a typical cable stayed bridge as the prototype structure; a lab-scale test structure was designed and fabricated firstly. A baseline FEM model was then setup and updated according to the modal frequencies measured from structural vibration test. A numerical study to simulate the bridge-AMD control system was conducted and an efficient LQG-based controller is designed. Based on that, an experimental implementation of AMD control of the transverse vibration of the bridge model was performed.

2. Model structure description and vibration test

The lab-scale bridge model studied in this chapter is designed according to a prototype cable-stayed bridge, the Third Nanjing Yangtze River Bridge located in Jiangsu Province of China. Since the prototype bridge is of all the characteristics of a modern cable-stayed bridge, the test model is assumed to be a good test bed to study the feasibility of active structural control applied to cable-stayed bridges under construction. The test model was designed and fabricated to simulate the longest double cantilever state during the construction stage of the prototype bridge. Since structural dynamic response control is the main focus of this chapter, the test model is preliminarily designed according to the dynamic scaling laws. However, concerning the restriction of test conditions, some modifications were made during the detailed design of the model bridge [10]. Fig. 1 shows the dimension of the designed model bridge. The bridge is composed of a 1.433 meters high cable pylon, a 3.08 meters long main girder, and six couples of stay cables. The cross section of the main girder is a rectangular of 16 mm wide and 10 mm high. At two ends of the main girder, the 3.6 kg and 3.8 kg weight AMD orbits were installed on the side span and main span respectively. The stay cables are made of steel wire with the diameter of 1 mm. At the upper end of each cable, an original 30 cm long spring was installed and adjusted to simulate the cable force. The cable forces were adjusted to provide supporting force to the main girder and to force it to match the designed layout of the test bridge. Table 1 shows the length, Young's Modulus and computed cable force for the cables. All of the components were made of steel. Since the model consists of only one cable pylon and no other piers, it is symmetric with respect to the cable pylon.

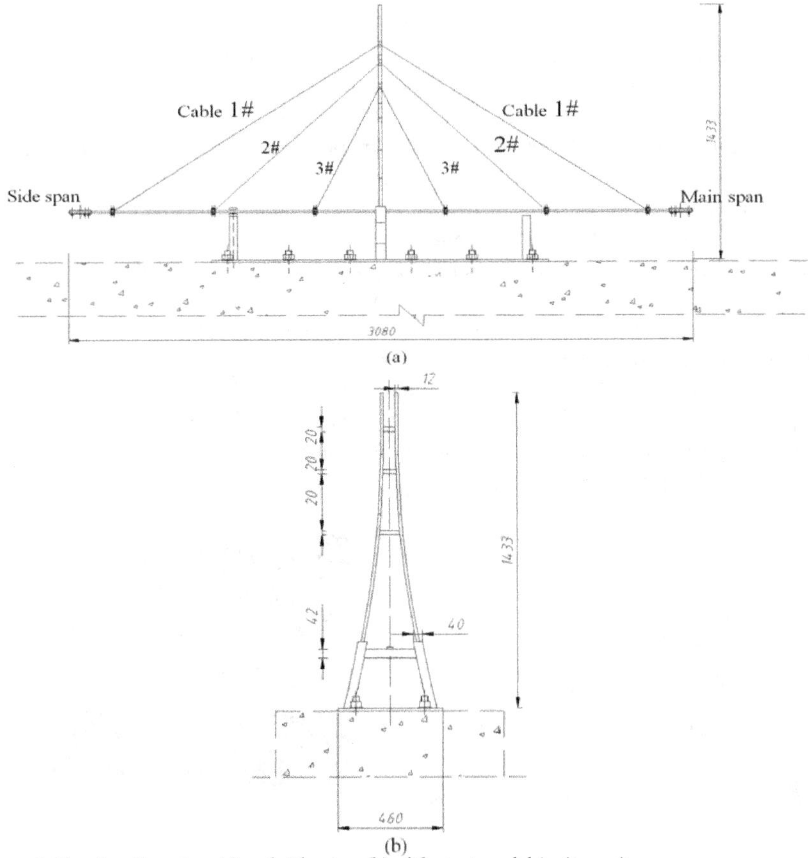

Figure 1. The elevation view (a) and side view (b) of the test model (unit: mm)

Cable No.	1#	2#	3#
Young's Modulus (Mpa)	451.7	327.0	216.0
Length with spring (m)	1.62	1.17	0.77
Spring Force (N)	32.0	13.9	7.2

Table 1. Cable parameters

Num.	1	2	3	4	5	6	7	8
Sensitivity (mv/g)	134.6	140.8	127.4	134.6	149.2	138.3	138.5	130.8
Location (m)	0.06	0.32	0.62	1.02	2.06	2.46	2.76	3.02

Table 2. Sensor sensitivity and location distances from the tip end of the side span

Vibration tests under forced excitations were conducted to identify the dynamic characteristics of the bridge model. Eight accelerometers (as described in Table 2) were distributed along the main girder both on the side span and the main span to collect structural acceleration responses at a sampling frequency of 50 Hz (as shown in Fig. 2a). A transverse impulse force was acted on the cantilever end of the side span to excite the structure. Concerning that the bridge model is symmetrical about the cable pylon and thus of repeated or close frequency modes, a modal identification algorithm of the capacity to identify the close modes, the wavelet based modal identification method developed by the research group, is used [11]. During the analysis, the mother wavelet function adopted is the complex Morlet function with the central frequency of 300 Hz and the scale increment is set to be 0.25 during the analysis. Fig. 2b shows the wavelet scalgram of a set of response measurement on the tip end of the side span. As shown in the figure, structural transverse vibration responses were dominated by two modes at the scale of 168.0 and 176.5, which correspond to the vibration modes of the natural frequencies of 1.701 Hz and 1.786 Hz, respectively. This figure also shows that the adopted modal identification method can separate these two close modes successfully. Structural modal parameters can then be estimated and the results are shown in Table 3.

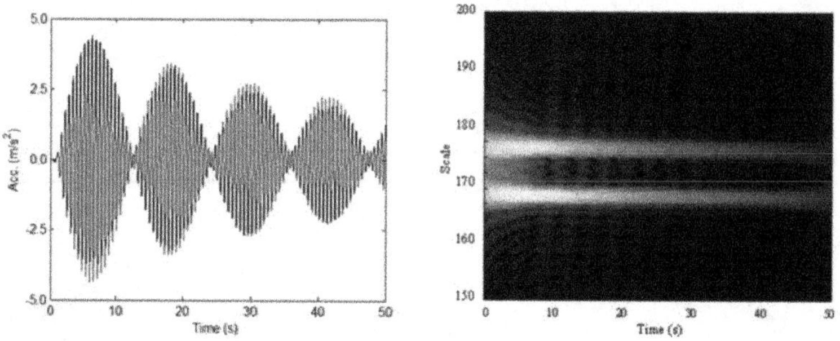

(a) (b)

Figure 2. Collected (a) free decay acceleration response and (b) its wavelet scalogram

3. Numerical modeling and model updating

For structural active control, a baseline numerical model is generally required for controller design. A FEM bridge model is thus setup in ANSYS according to the design diagram of the bridge model. The cable is modeled using a 3D uniaxial tension-only truss element. Equivalent modulus for the cables without spring are computed using Ernst formula and then series wound equivalent modulus for the cable with spring can be established. Other structural elements are all modeled using 3D elastic beam element. Cables are connected to the main girder using rigid beam element. Additional masses were modeled using isotropic mass element. The cable pylons and the main girder are linked by coupling the horizontal projective intersection points of the lowest transverse beam of the pylon with the main girder. The six DOFs at the feet of the pylon are fixed.

		The first transverse symmetrical bending (TSB) mode	The first transverse anti-symmetrical bending (TAB) mode
Frequency	Test	1.701 Hz	1.786 Hz
	Initial Model	1.848 Hz	1.856 Hz
	Updated Model	1.787 Hz	1.786 Hz
Mode Shape	Test	[1.00 0.86 0.70 0.32 0.00 0.30 0.70 0.90 0.99]	[1.00 0.84 0.72 0.28 0.00 -0.32 -0.70 -0.86 -0.99]
	Initial Model	[1.00 0.74 0.46 0.17 0.00 0.13 0.38 0.60 0.81]	[1.00 0.74 0.47 0.17 0.00 -0.14 -0.38 -0.60 -0.81]
	Updated Model	[1.00 0.74 0.46 0.17 0.00 0.17 0.46 0.74 1.00	[1.00 0.74 0.46 0.17 0.00 -0.17 -0.46 -0.74 -1.00]
MAC*	Initial Model	0.9620	0.9620
	Updated Model	0.9722	0.9718

* MAC (modal assurance criteria) is defined to be a correlation coefficient of two mode shape vectors.

Table 3. The computed modal parameters before and after model updating compared with the tested mode parameters

Taken eigenvalue analysis of the numerical model, structural natural frequencies and mode shape vectors can be computed. Table 4 shows the computed modal parameters of the transverse bending modes. As shown in the table, since the sum of the effective mass of the first transverse anti-symmetric bending mode (TAB) and the first transverse symmetric bending mode (TSB) is 80.1% of the sum of the effective mass of all transverse modes, these 2 transversal modes dominant the transverse vibration of the bridge. Fig. 3 shows the mode shape of these two transverse bending modes.

No.	Frequency (Hz)	Mode description	Effective mass (kg)
1	1.848	The first TAB mode	6.00
2	1.856	The first TSB mode	6.00
3	19.802	The second TAB mode	0.94
4	19.848	The second TSB mode	0.94

Table 4. The computed natural frequencies for the first 4 transverse modes of the main girder

Comparing the natural frequencies obtained from the eigenvalue analysis on the numerical model and the vibration test on the test bridge, some differences can be observed (as shown in Table 3). A model updating process is thus conducted to get an accurate baseline numerical model. The results of the sensitivity analysis show that the vertical and transverse vibration modes are sensitive to the tip mass magnitude, while the Young's module of the cable is critical for vertical bending modes. So these two parameters were updated: the tip masses of two spans were updated from 3.8kg and 3.6kg to 4.16kg, respectively, and the spring modulus was updated from 220N/m to 203N/m. Table 4 shows the natural frequencies and mode shapes of the updated model. As shown, the modal parameters of the first TSB and TAB modes have a good match with the modal test results.

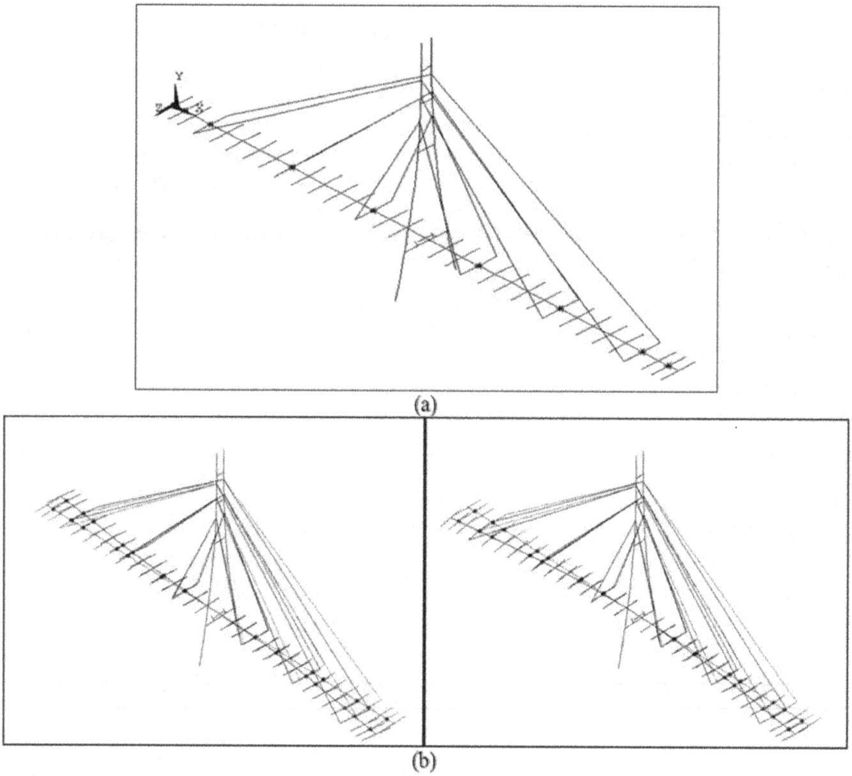

Figure 3. The baseline FEM model (a) setup in ANSYS and the computed mode shape (b) of the first transverse anti-symmetric and symmetric bending mode

4. Control system simulation

Based on the baseline numerical model updated according to dynamic measured structural modal parameters, a system simulation study is conducted. For a bridge-AMD system, its governing equation of motion is

$$M_s\ddot{X}(t)+C_s\dot{X}(t)+K_sX(t)=F_e(t)+D^TT^Tf_d(t) \tag{1}$$

$$m_d\ddot{x}_d(t)+f_d(t)=0 \tag{2}$$

$$f_d(t)=-c_d[\dot{x}_d(t)-TD\dot{X}]+b_du(t) \tag{3}$$

where, M_s, C_s, and K_s are the mass, damping and stiffness matrices of the bridge structure; $X(t)=\left[x_1\cdots x_i\cdots x_n\right]'$ is a n-dimensional vector, in which x_i is the displacement response vector of the ith DOF of the structure and n is the number of the DOF of the structure; $F_e(t)$

is the excitation force vector acting on the structure; $D=\begin{bmatrix}0 & I & 0\end{bmatrix}$ is the AMD location matrix, in which 0 and I are zero and identity matrix with appropriate sizes; $f_d(t)$ is the actuation force on the structure applied by the AMD ; T is the transfer matrix from the affiliated nodes of AMD to the principle nodes of the structure ; $m_d = diag\left[m_{d1} \cdots m_{di} \cdots m_{dl}\right]$ is the mass matrix of the AMDs, in which m_{di} is the mass of the ith AMD; $x_d(t)$ are absolute displacement of the AMD to the bridge structure, respectively; c_d is the viscous coefficient of the AMD; b_d is the force-voltage coefficient of the AMD and $u(t)$ is the control input voltage.

Since the numerical model of a complex structure is generally of a large amount of DOFs, for example in this study the FEM model obtained from the last section is of 1188 DOFs, this will induce great computation difficulty to design the controller according to this so called full order model. A reduced order model is thus required. In this study, the critical modal reduction method is adopted for this purpose because this method can greatly reduce structural DOFs and the reduced order model is of clear physical meaning [12, 13]. Concerning a structure whose vibration is dominated by the first r modes, its dynamic response can be approximately expressed as $X(t) \cong \sum_{i=1}^{r} \phi_i Y_i(t)$ and the full order model can thus be reduced to an r-order reduced order model. The governing equation of motion for this reduced order model is

$$\hat{M}_s \ddot{Y}(t) + (\hat{C}_s - c_d \phi^T D^T T^T T D \phi)\dot{Y}(t) + \hat{K}_s Y(t) + \phi^T D^T T^T c_d \dot{x}_d(t) = \hat{F}_e(t) + \phi^T D^T T^T b_d u(t) \quad (4)$$

$$m_d \ddot{x}_d(t) - c_d \dot{x}_d(t) + c_d T D \phi \dot{Y} = -b_d u(t) \quad (5)$$

where, $Y(t)$ is the modal response vector superposed by the modal responses of the rth principle modes; ϕ is the generalized mode shape matrix composed of the mode shape vectors of the first r principle modes; $\hat{M}_s = \phi^T M_s \phi$, $\hat{C}_s = \phi^T C_s \phi$ and $\hat{K}_s = \phi^T K_s \phi$ are structural generalized mass, damping and stiffness matrices, respectively; $\hat{F}_e = \phi^T F_e$ is the generalized excitation vectors. In this study, since the vibration responses of the first two transversal vibration modes are the most accountable for structural transverse vibration, structural modal mass, stiffness and damping matrices are computed using the modal parameters of these two modes. Correspondingly, the governing equation of motion for the reduced order bridge-AMD system can be derived.

For this reduced order bridge-AMD system, its state space equation is

$$\dot{x}_r = A_r x_r + B_r u + E_r w \quad (6)$$

$$z_r = C_r^z x_r + D_r^z u + E_r^z w \quad (7)$$

$$y_r = C_r^y x_r + D_r^y u + E_r^y w \quad (8)$$

where, x_r is an a dimensional state-space vector, $a = 2r + l$, r and l are the number of DOF for the reduced order structure and the number of the installed AMD, respectively; A_r is an $a \times a$ system matrix; u is the control input vector for the l AMDs; B_r is an $a \times l$ AMD location matrix; w is the generalized modal excitation vector; E_r is an $a \times r$ excitation matrix ; z_r is an l dimensional control output vector; y_r is a q dimensional observer output vector. The system matrices can be expressed as:

$$x = \begin{Bmatrix} Y \\ \dot{Y} \\ \dot{x}_d \end{Bmatrix};\ A_r = \begin{bmatrix} 0 & I & 0 \\ -\hat{M}_s^{-1}\hat{K}_s & -\hat{M}_s^{-1}\hat{C}^* & -\hat{M}_s^{-1}\phi^T D^T T^T c_d \\ 0 & m_d^{-1} c_d H & m_d^{-1} c_d \end{bmatrix};\ B_r = \begin{bmatrix} 0 \\ \hat{M}_s^{-1}\hat{H} b_d \\ -m_d^{-1} b_d \end{bmatrix};\ E_r = \begin{bmatrix} 0 \\ \hat{M}_s^{-1} \\ 0 \end{bmatrix};$$

$$C_r^z = \alpha \begin{bmatrix} I \\ \begin{bmatrix} 0 & I \end{bmatrix} A_r \end{bmatrix};\ D_r^z = \alpha \begin{bmatrix} I \\ \begin{bmatrix} 0 & I \end{bmatrix} B_r \end{bmatrix};\ E_r^z = \alpha \begin{bmatrix} 0 \\ \begin{bmatrix} 0 & I \end{bmatrix} E_r \end{bmatrix};$$

$$C_r^y = \beta \begin{bmatrix} I \\ \begin{bmatrix} 0 & I \end{bmatrix} A_r \end{bmatrix};\ D_r^y = \beta \begin{bmatrix} I \\ \begin{bmatrix} 0 & I \end{bmatrix} B_r \end{bmatrix};\ E_r^y = \beta \begin{bmatrix} 0 \\ \begin{bmatrix} 0 & I \end{bmatrix} E_r \end{bmatrix};$$

where $\hat{C}^* = \hat{C}_S - \hat{H} c_d \hat{H}^T$; $\hat{H} = \begin{bmatrix} -TD\phi & I \end{bmatrix}^T$; 0 and I are zero and identity matrix with appropriate sizes; α and β are appropriately selected weighting matrix to adjust the optimize objective of the controller, respectively. The control output vector z_r and the observer output vector y_r are the displacement, velocity and acceleration responses of the bridge structure or the AMDs.

The control output and observer output matrices C_r^z, D_r^z, C_r^y, and D_r^y are determined according to the sensor and actuator placement. The sensor placement should basically satisfy the following observability criteria $rank \begin{bmatrix} A_r - \lambda I \\ C_r^y \end{bmatrix} = a$, where λ is an arbitrary complex number [14]. For a system with n modes of repeated or close frequency, n sensors are required for full state response measurement. In this study, to provide redundant channels to collect structural response, eight accelerometers, the same as aforementioned in the modal test, are installed along the main girder during the control process. The actuators are placed on the girder by checking the following controllability criteria $a = rank\begin{bmatrix} A_r - \lambda I, B_r \end{bmatrix}$. In this study, different schemes for AMD placement will be discussed in detail in the following sections.

5. Controller design

The design of a controller is very important for the success of structural active vibration control. In this study, the LQG control algorithm is adopted since this control algorithm can offer excellent control performance and is of good robustness as shown in some preliminary study [10]. During the controller design process, the excitation is assumed to be a stationary white noise, and the following cost function is set as the control objective:

$$J = \lim_{t\to\infty} E\left[\int_0^t (z_r{}^T Q z_r + u^T R u)dt\right] \tag{9}$$

where z_r, system output variables, are set to be the transverse displacement or acceleration response at the tip ends; Q, a square matrix of the same order as z_r, equals to the multiplication of an identity matrix with a parameter q; u is a control force variable; R, the active force weight matrix, is set to be an identity matrix of the same order as the number of AMD applied. The design of the *LQG* controller is to adjust the weight parameter q via optimizing the performance of the system with compensator under the limitation of energy supply. The design of the controller relies on the full state feedback vector X_r. Since limited sensors are mounted on the structure, this full state vector cannot be directly measured but be estimated from the sensor measurements. When the excitation forces w and the measurement noise v are uncorrelated white noise process, the Kalman-Bucy filter is employed to get an estimation of the state vector X_r [15].

To obtain a good controller for experimental implementation, a series of numerical analysis with different value of weight parameter q are conducted. During the numerical analysis, structural modal damping coefficients are set to be the same as the real measured modal damping ratios. A scaled *El Centro* earthquake time history, whose dominant frequency band covers the first 2 transverse modal frequencies of the bridge, is adopted to excite the bridge in the numerical study. Two AMD placement strategies are employed for the comparison of optimal actuator location. These two strategies are the strategy of one AMD placed at the tip end of the main span (named S1) and the strategy of two AMDs placed at the tip ends of both spans (named S2). The AMDs are simulated to be the two electric servo-type AMD carts provided by Quanser Inc. with the following expression of the actuation force

$$f_d^1(t) = 8.246\dot{x}_d(t) + 1.42u(t) \tag{10}$$

$$f_d^2(t) = 12.576\dot{x}_d(t) + 1.73u(t) \tag{11}$$

To simulate a more practical control condition during experimental implementation, the following constraint condition is adopted: 1) The discrete digital computation is employed for the controller computation with the sampling frequency of 500Hz; 2) The precision of the A/D converter is set to be 12-bits and the range of the input voltage is set to be ± 5 V; 3) The measurement noise of a root mean square (RMS) value of 0.015 V is added into each channel of the acceleration responses, which corresponds to the 0.3% of voltage range of the A/D converter; 4) The maximum actuation voltage is set to be ± 5 V with the corresponding RMS voltage of 1.67 V and the maximum actuation displacement is set to be ± 0.08 m with the corresponding RMS displacement of 0.027 m.

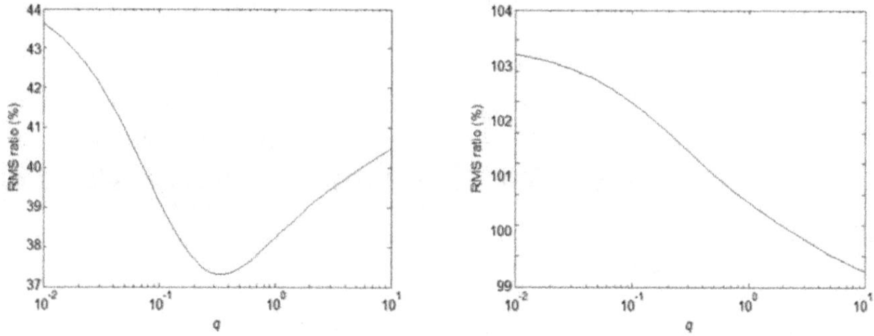

Figure 4. The ratio of the controlled acceleration RMS value to uncontrolled RMS value at the tip ends of the side span (a) and main span (b) with respect to the weighting parameter q for S1 control

For S1 control, a series of numerical studies simulating the control system with one AMD cart installed on the tip end of the main span of the bridge, whose actuation force is expressed as Eq. (10), are conducted when the weighting parameter q varies from 0.01 to 10. Fig. 4 shows the relative RMS ratio of the controlled acceleration response to the uncontrolled acceleration response at the tip end of both main span and side span with the varying of q. As shown in the figure, $q = 0.398$ are set for S1 to achieve an optimal control performances. This figure also tells that for S1 control, the tip acceleration response of the main span can be well controlled; however, a good control performance for the tip acceleration at the side span cannot be achieved by adjusting the weight parameter. Fig. 5 shows the controlled and uncontrolled tip acceleration response of the side span and the main span when q is set to be 0.398. This figure also tells that the well designed controller can greatly reduce the acceleration response of the tip end of the main span but cannot mitigate the vibration of the side span.

For S2 control, the control system with two AMD carts, whose actuation force expressions are shown as Eq. (10) and (11), installed on the tip ends of both spans of the bridge, is simulated. Numerical studies are conducted when the weighting parameter q varies from 0.01 to 100. Fig. 6 shows the relative RMS ratio of the controlled acceleration response to the uncontrolled acceleration response at the tip end of both spans with the varying of q. As shown in the figure, when $q = 9.1$, the control system provides the most optimal control performances. This figure also tells that for S2 control, the tip acceleration response of both the main span and the side span can be well reduced. Fig. 7 shows the controlled and uncontrolled tip acceleration time response of the side span and the main span when q is set to be 9.1. This figure also tells that the well designed controller can greatly reduce the acceleration response at the tip end of both spans.

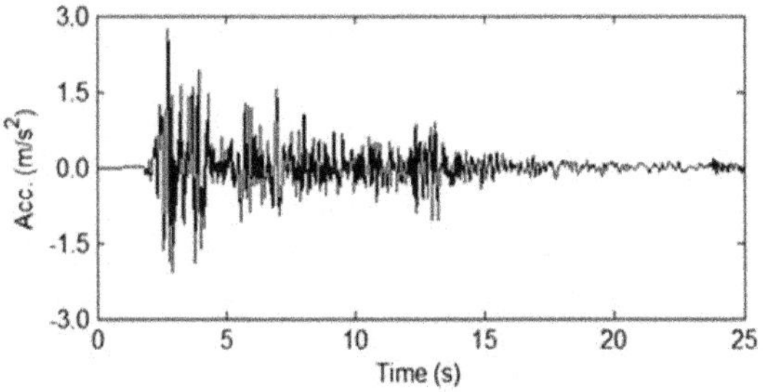

(a) *El Centro* seismic acceleration record

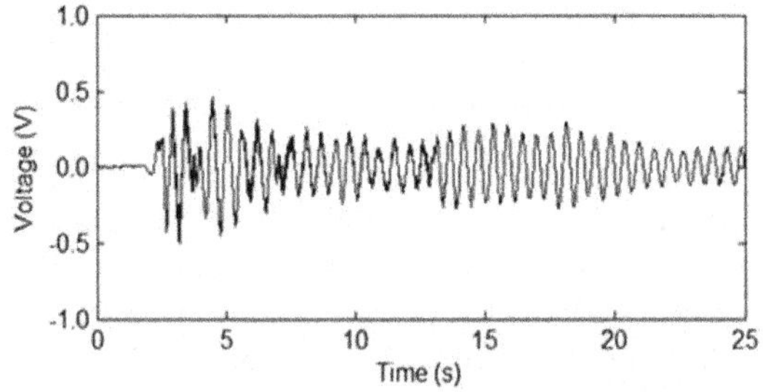

(b) Driving voltage of the AMD cart on the tip of the main span

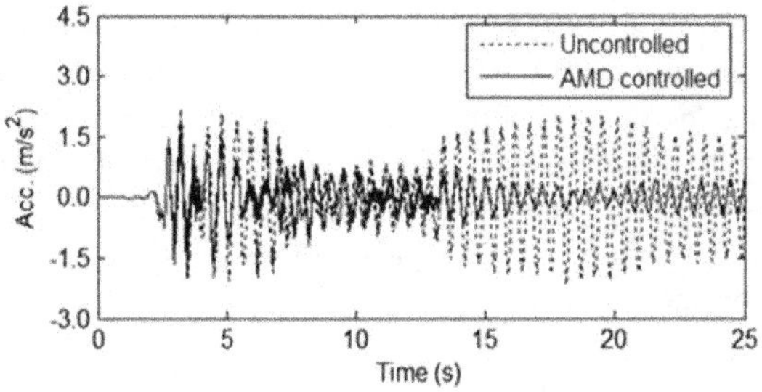

(c) Tip acceleration of the main span

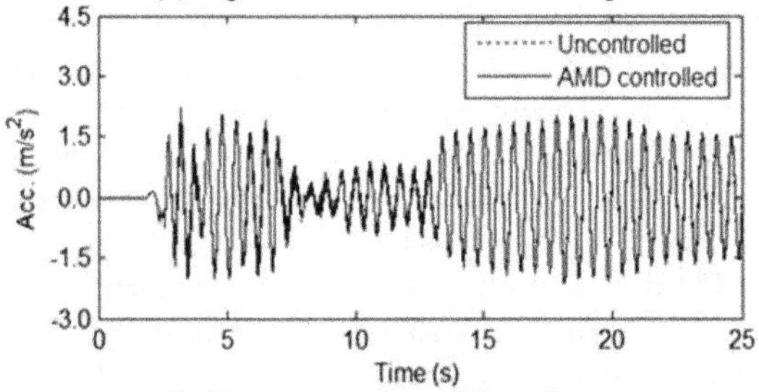

(d) Tip acceleration of the side span

Figure 5. Excitation (a), driving voltage (b), main span tip acceleration (c), and side span tip acceleration (d) time histories of the bridge under El Centro seismic excitation for S1 control

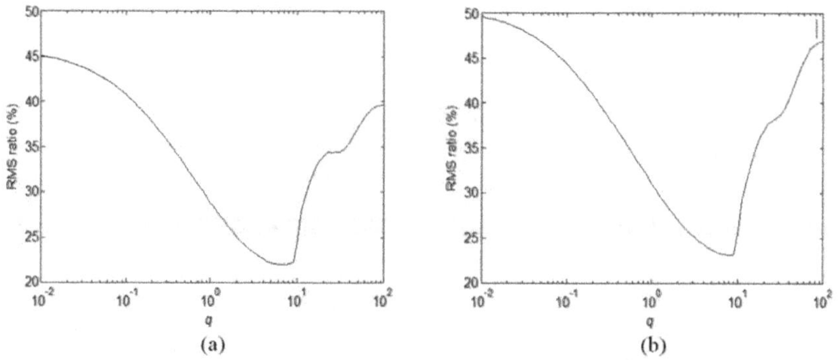

Figure 6. The ratio of the controlled acceleration RMS value to uncontrolled RMS value at the tip ends of the side span (a) and main span (b) with respect to the weighting parameter q for S2 control

The control performance comparison of these two AMD placing strategies tell that for the cable-stayed bridge studied, which is of two dominant transverse vibration modes with close frequencies, the single AMD control strategy (S1) can only reduce structural vibration response of the AMD-instrumented span, and the double AMD control strategy (S2) can achieve a good control performance for reducing structural response of both spans. These observations are verified by checking the controllability criteria. If the control system is of two close eigenvalues, at least two actuators are required to ensure the system is controllable. Moreover, since structural dominant transverse modes are anti-symmetric and symmetric bending modes, the shift of the AMD position along one span of the bridge will only proportionally vary the coefficients of B_r . If two AMDs are placed at one span of the bridge, the corresponding two columns of B_r are linear dependant. Therefore, the two AMDs must be placed at both the side span and the main span respectively to ensure the controllability of the bridge.

6. Experimental implementation

To verify the feasibility of the AMD control for transverse vibration reduction of cable-stayed bridge under construction, an experimental study on the fabricated test model is conducted in the Bridge Testing Laboratory of Tongji University. During the experiment, the S2 control strategy was adopted according to the conclusion obtained from the above numerical simulation study. Fig. 8 shows the layout of the experimental setup. As shown in the figure, the control system includes a data acquisition system, a central control computer, and two AMD carts. The data acquisition system consists of eight accelerometers, whose sensitivity is checked using dynamic calibration method; Dspace signal amplifier and filter; a general purpose data acquisition and control board MultiQ-3, which has 8 single ended analog inputs, 8 analog outputs, 16 bits of digital input, 16 bits of digital output, 3 programmable timers and up to 8 encoder inputs decoded in quadrature (option 2E to 8E). The central control computer is of 512 Mb memory and 1.0 GHz Intel Celeron processor. The

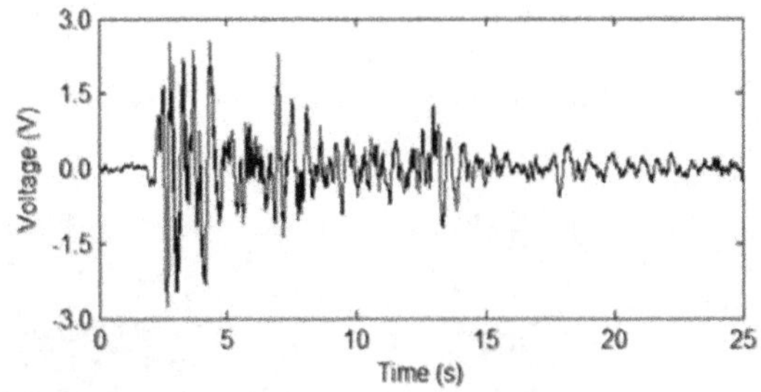

(a) Driving voltage of the AMD cart on the tip of the main span

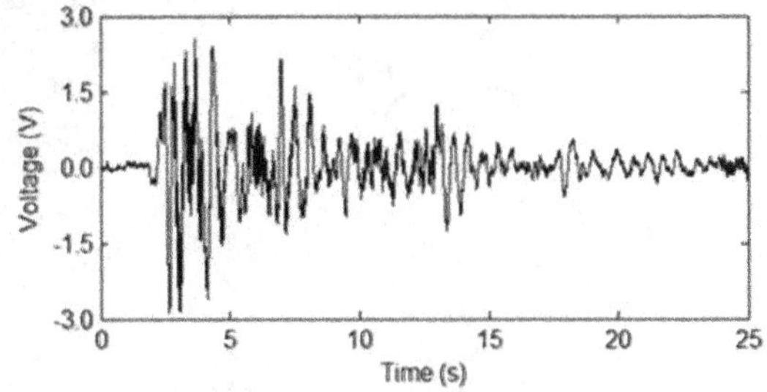

(b) Driving voltage of the AMD cart on the tip of the side span

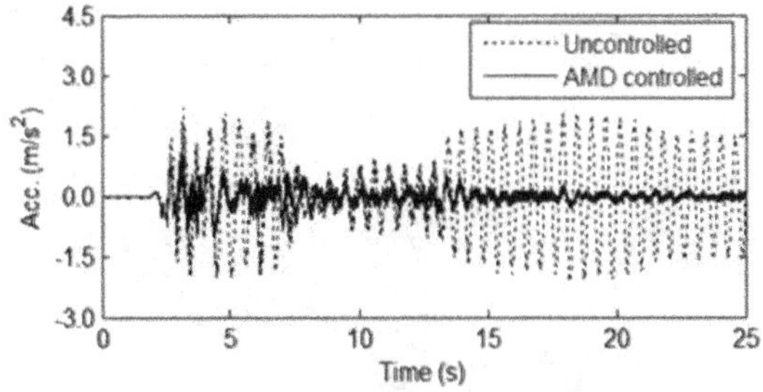

(c) Tip acceleration of the main span

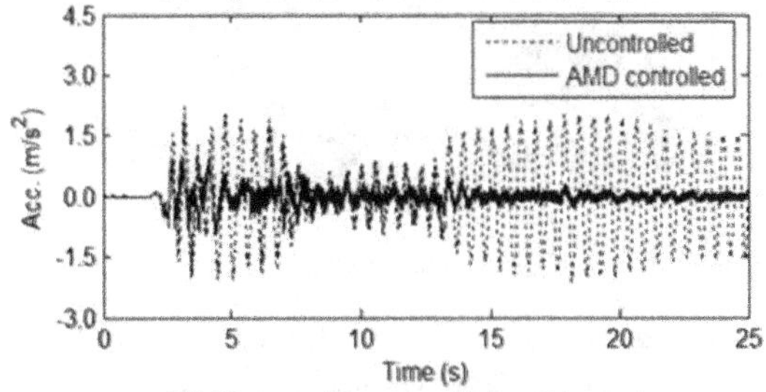

(d) Tip acceleration of the side span

Figure 7. Driving voltage of the main span (a) and side span (b), and tip acceleration time histories of main span (c) and side span (d) of the bridge under *El Centro* seismic excitation for S2 control

AMD carts (as shown in Fig. 8) are electric servo type [16]. They are 0.645 kg weight. Their track length is 32 cm and the max safe input voltage is ±5V. The system is designed under the constraint of avoiding the AMD to knock the baffle. If the AMD position exceeds the safe range of the orbit, the system will be shut down.

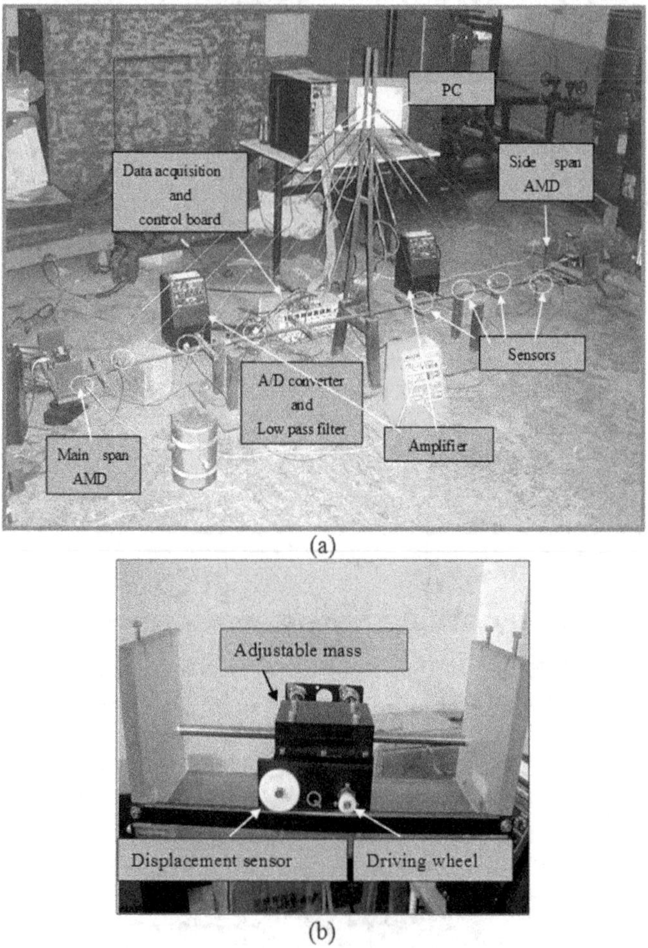

Figure 8. The experimental setup (a) and the AMD devices (b)

During the experiment, the tip ends of the main girder were pulled transversely using steel wires firstly to generate an initial displacement in this direction. The wires were then cut suddenly and the bridge started to vibrate due to this initial potential energy imported. Several seconds later, the power of the AMD control system was turned on and structural vibration response was recorded. Three case studies were conducted to check the performance of the AMD control system under different excitation schemes. For the first case, case C1, the tip ends of both the main span and the side span were pulled in the same direction. After the steel wires were cut, the transverse symmetric bending mode of the bridge was excited. For case C2, the tip ends of the two spans were pulled in the opposite direction to excite the transverse anti-symmetric bending mode of the bridge. Case C3 simulated bridge free vibration under an initial displacement of the main span. Both anti-symmetric and symmetric transverse bending modes of the bridge would thus be excited.

For test case C1, Fig. 9 shows the AMD driving voltage, recorded tip acceleration responses with or without AMDs, and their Fourier spectrums. As shown in these figures, when the power of the AMD carts was still turned off, structural acceleration responses were already reduced. When the power is turned on, the free decay ratios of structural responses were further increased. That verifies the performance of the active control system on structural response reduction. The comparison of the uncontrolled and AMD controlled Fourier spectrum magnitude of structural responses also verifies this statement. Moreover, as shown, when the AMD carts were installed, the peaks of Fourier spectrum were shifted to the left-hand-side, which meant that the natural frequencies of the system were decreased. Table 5 shows the peak and RMS acceleration response of the structure with and without AMD. As shown in the table, after the AMD was mounted, the peak and RMS accelerations recorded at the tip point of the side span were reduced 44.1 % and 82.1 %, respectively. For the tip point of the main span, the recorded peak and RMS accelerations were reduced 31.1 % and 81.4 %, respectively. For test cases C2 and C3, the peak and RMS acceleration responses of the structure with and without AMD were recorded. As shown in Table 5, the transverse vibration responses for these sensor-mounted points were also efficiently reduced: For case C2, structural peak and RMS acceleration responses were respectively reduced 28.3% and 65.4% for the tip point of the side span and 22.0% and 68.4% for the tip point of the main span; For case C3, structural peak and RMS acceleration responses were reduced 65.8% and 85.6% respectively for the tip point of the side span and 40.5% and 76.7% for the tip point of the main span. Moreover, concerning that the controller is designed via numerical studies on a reduced order model, the good control performances obtained on the bridge model tells that the control algorithm adopted in this study is of good robustness.

Case	Peak acc. of sensor 1		RMS acc. of sensor 1		Peak acc. of sensor 8		RMS acc. of sensor 8	
	No AMD	With AMD	No AMD	With AMD	No AMD	With AMD	No AMD	With AMD
C1	4.431	2.479	1.826	0.326	4.312	2.970	1.810	0.336
C2	4.198	3.012	1.341	0.464	3.563	2.779	1.327	0.419
C3	5.186	1.773	1.800	0.260	5.833	3.473	1.820	0.424

Table 5. Peak and RMS acceleration at the tip ends of the bridge with and without AMDs for the experimental cases

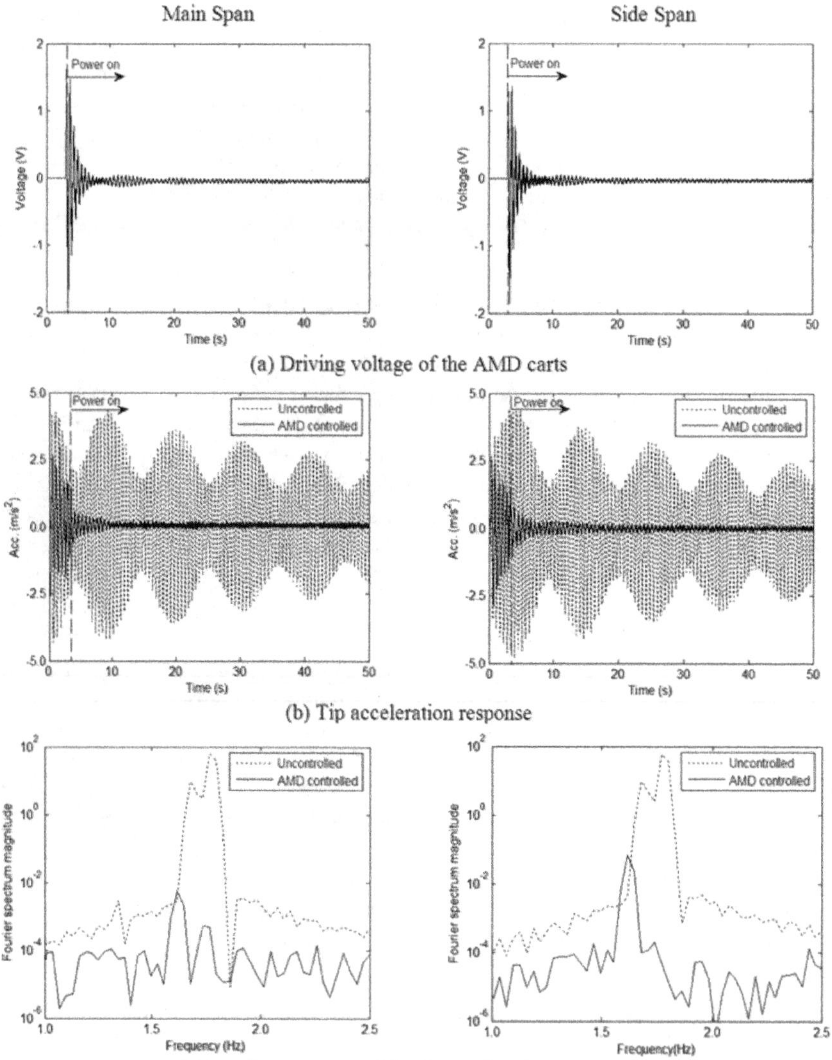

(a) Driving voltage of the AMD carts

(b) Tip acceleration response

(c) Fourier spectrum of the tip acceleration responses

Figure 9. The driving voltage (a) and tip acceleration (b) time histories and Fourier spectrum (c) of responses for control case C1

7. Concluding remarks

In this chapter, the active mass dampers are implemented for vibration control of a lab-scale cable stayed bridge in double cantilever construction state. The results of both numerical simulation and experimental study verified that the proposed AMD control technique is applicable and efficient for the control of transverse vibration of cable-stayed bridge under construction.

For the cable-stayed bridge studied in this chapter, structural vibration test showed that the bridge was of two dominant transverse bending modes with close frequencies. The numerical study verified that for the control of such a structure with repeated frequencies; at least two AMDs should be installed for a good control performance. Moreover, the placement of those two AMDs should be carefully studied.

For the control of a linear, time-invariant system, an accurate and complete system models are generally required. However, for the bridge-AMD system studied in this chapter, it is very difficult, if not impossible, to set up such a numerical model due to the complex layout of the bridge structure. Since structural vibration responses are generally governed by some dominant vibration modes and the objective of structural vibration control is to reduce but not to completely restrain structural vibration responses, this study verified that a reduced order model constructed from the critical modes is good enough for the controller design to achieve an excellent vibration reduction performance. Considering the differences between the numerical model and the real structure, the control algorithm adopted in this study must be robust to the vibration property change of the controlled structure. The results of experimental studies show that the vibration of the test model can be well controlled using the controller designed from the numerical studies. That means the *LQG* control algorithm is of good robustness for real application.

This study is an initial work on AMD control of transverse vibration of a cable stayed bridge under construction before it can be used for real applications. Although the experimental study verified the efficiency of the adopted AMD control for structural response reduction under given excitation, some primary issues for real application of the AMD control technique, such as how to deal with time delay, how to reduce the requirement on power supply of the control system, and et al., are not addressed in this study. Further laboratory studies or field applications on some real bridges will be conducted in the coming future to discuss these issues and make the technique more efficient and practical for real applications.

Author details

Hao Chen
Institute of Engineering Mechanics, China Earthquake Administration, Sanhe, Hebei, China

Zhi Sun and Limin Sun
State Key Laboratory for Disaster Reduction in Civil Engineering, Tongji University, Shanghai, China

Acknowledgement

This research is partially supported by the National High-tech R&D Program of China (863 Program) (Grant No. 2006AA11Z109), and Shanghai Rising Star Tracking Program (Grant No. 09QH1402300). These supports are greatly appreciated.

8. References

[1] Boonyapinyo, V., Aksorn, A., and Lukkunaprasit, P. (2007). Suppression of aerodynamic response of suspension bridges during erection and after completion by using tuned mass dampers. *Wind and Structures, An Int\'l Journal*, 10(1): 1-22.

[2] Ubertini, F. (2010). Prevention of suspension bridge flutter using multiple tuned mass dampers. *Wind and Structures, An Int\'l Journal*, 13(3): 235-256.

[3] Warnitchai, P., Fujino, Y., Pacheco, B.M. and Agret, R. (1993). An experimental study on active tendon control of cable-stayed bridges. *Earthquake Engineering and Structural Dynamics*, 22: 93-111.

[4] Achkire, Y. and Preumont, A. (1996). Active tendon control of cable-stayed bridges. *Earthquake Engineering and Structural Dynamics*, 25(6): 585-597.

[5] Dyke, S.J., Caicedo, J.M., Turan, G., Bergman, L.A., Hague, S. (2003). Phase I benchmark control problem for seismic response of cable-stayed bridges. *Journal of Structural Engineering, ASCE*, 129(7): 857–872.

[6] Caicedo, J.M., Dyke, S.J., Moon, S.J., Bergman, L.A., Turan, G., Hague, S. (2003). Phase II benchmark control problem for seismic response of cable-stayed bridges. *Journal of Structural Control*, 10: 137–168.

[7] Frederic, B. and Andre, P. (2001). Active tendon control of cable-stayed bridges: a large-scale demonstration. *Earthquake Engineering & Structural Dynamics*, 30: 961–979.

[8] Shelley, S.J., Lee, K.L., Aksel, T. and Aktan, A.E. (1995). Active control and forced vibration studies on highway bridges. *Journal of Structural Engineering, ASCE*, 121(9): 1306-1312.

[9] Korlin, R. and Starossek, U. (2007). Wind tunnel test of an active mass damper for bridge decks. *Journal of Wind Engineering and Industrial Aerodynamics*, 95: 267-277.

[10] Chen, H. (2007). Experimental study on active control of cable-stayed bridge under construction. Master Thesis. Tongji University.

[11] Sun, Z., Hou, W., Chang, C. C. (2009). Structural system identification under random excitation based on asymptotic wavelet analysis. *Engineering Mechanics*, 26(6), 199-204 (in Chinese).

[12] Schemmann, A.G. (1997). Modeling and active control of cable-stayed bridges subject to multiple-support excitation. PhD thesis. Standford University.

[13] Xu, Y. and Chen, J. (2008). Modal-based model reduction and vibration control for uncertain piezoelectric flexible structures. *Structural Engineering and Mechanics*, 29(5): 489-504.

[14] Laub, A.J. and Arnold, W.F. (1984). Controllability and observability criteria for multivariable linear second-order models, *IEEE Transactions on Automatic Control*, AC-29(2).

[15] Wu, J.C., Yang, J.N., and Schmitendorf W.E. (1998). Reduced-order H_∞ and *LQR* control for wind-excited tall buildings. *Engineering Structures*. 20(3): 222-236.

[16] Quanser Consulting Inc. (2002). Active Mass Damper: Two-Floor (AMD-2), User Manual.

Chapter 11

Dynamic Responses and Active Vibration Control of Beam Structures Under a Travelling Mass

Bong-Jo Ryu and Yong-Sik Kong

Additional information is available at the end of the chapter

1. Introduction

The dynamic deflection and vibration control of an elastic beam structure carrying moving masses or loads have long been an interesting subject to many researchers. This is one of the most important subjects in the areas of structural dynamics and vibration control. Bridges, railway bridges, cranes, cable ways, tunnels, and pipes are the typical structural examples of the structure to be designed to support moving masses and loads.

While the analytical studies on the dynamic behavior of a structure under moving masses and loads have been actively performed, a small number of experimental studies, especially for the vibration control of the beam structures carrying moving masses and loads, have been conducted. It is, therefore, strongly desired to conduct both analytical and experimental studies in parallel to develop the algorithm that controls effectively the vibration and the dynamic response of structures under moving masses and loads.

The dynamic responses and vibrations of structures under moving masses and loads were initially studied by (Stokes, 1849; Ayre et al., 1950) who tried to solve the problem of railway bridges. This type of study has been actively performed by employing the finite element method (Yoshida & Weaver, 1971). (Ryu, 1983) used the finite difference method to study the dynamic response of both the simply supported beam and the continuous beam model carrying a moving mass with constant velocity and acceleration. (Sadiku & Leipholz, 1987) utilized the Green's function to present the difference of the solutions for the moving mass problem without and with including the inertia effect of a mass. (Olsson, 1991) studied the dynamic response of a simply supported beam traversed by a moving object of the constant velocity without considering the inertia effect of moving mass. (Esmailzadeh & Ghorash, 1992) expanded Olsson's study by including the inertia effect of the moving mass. (Lin, 1997) suggested the effects of both centrifugal and Coliolis forces should be taken into account to obtain the dynamic deflection. (Wang & Chou, 1998) conducted nonlinear vibration of Timoshenko beam due to a moving force and the weight of beam. Recently, (Wu, 2005) studied dynamic analysis of an inclined beam due to moving loads.

Most studies on the dynamic response of a beam carrying moving mass or loads are analytical, but a small number of experimental investigations are recently conducted in parallel. The studies on the dynamic response of a beam caused by moving masses or loads have been also conducted domestically. For the studies on the vibration control of moving masses or loads, (Abdel-Rohman & Leipholz, 1980) applied the active vibration control method to control the beam vibration caused by moving masses. They applied the bending moment produced by tension and compression of an actuator to the beam when the vibration of a simply supported beam carrying a moving mass occurs. With the active control, the passive control approaches have been proposed in many engineering fields. (Kwon et al., 1998) presented an approach to reduce the deflection of a beam under a moving load by means of adjusting the parameters of a conceptually second order damped model attached to a flexible structures.

Recently, the piezoelectric material has been used for the active vibration control. (Ryou et al., 1997) studied the vibration control of a beam by employing the distributed piezoelectric film sensor and the piezoelectric ceramic actuator. They verified their sensor and actuator system by observing the piezoelectric film sensor blocked effectively the signal from the uncontrolled modes.

(Bailey & Hubbard Jr., 1985) conducted the active vibration control on a thin cantilever beam through the distributed piezoelectric-polymer and designed the controller using Lyapunov's second and direct method. (Kwak & Sciulli, 1996) performed experiments on the vibration suppression control of active structures through the positive position feedback(PPF) control using a piezoelectric sensor and a piezoelectric actuator based on the fuzzy logic. (Sung, 2002) presented the modeling and control with piezo-actuators for a simply supported beam under a moving mass. Recently, (Nikkhoo et al., 2007) investigated the efficiency of control algorithm in reducing the dynamic response of beams subjected to a moving mass. After that, (Prabakar et al., 2009) studied optimal semi-active control of the half car vehicle model with magneto-rheological damper. (Pisarski & Bajer, 2010) conducted semi-active control of strings supported by viscous damper under a travelling load.

In this chapter, firstly, dynamic response of a simply supported beam caused by a moving mass is investigated by numerical method and experiments. Secondly, the device of an electromagnetic actuator is designed by using a voice coil motor(VCM) and used for the fuzzy control in order to suppress the vibration of the beam generated by a moving mass. Governing equations for dynamic responses of a beam under a moving mass are derived by Galerkin's mode summation method, and the effect of forces (gravity force, Coliolis force, inertia force caused by the slope of the beam, transverse inertia force of the beam) due to the moving mass on the dynamic response of a beam is discussed. For the active control of dynamic deflection and vibration of a beam under the moving mass, the controller based on fuzzy logic is used and the experiments are conducted by VCM(voice coil motor) actuator to suppress the vibration of a beam.

2. Theoretical analyses

2.1. Governing equations of a simply supported beam traversed by a moving mass

The mathematical model of a beam traversed by a moving mass is shown in Fig. 1, where l is the length of a beam, v is the velocity of a moving mass, t is time, $w(z,t)$ is the transverse displacement, z and y are the axial and the transverse coordinates, respectively.

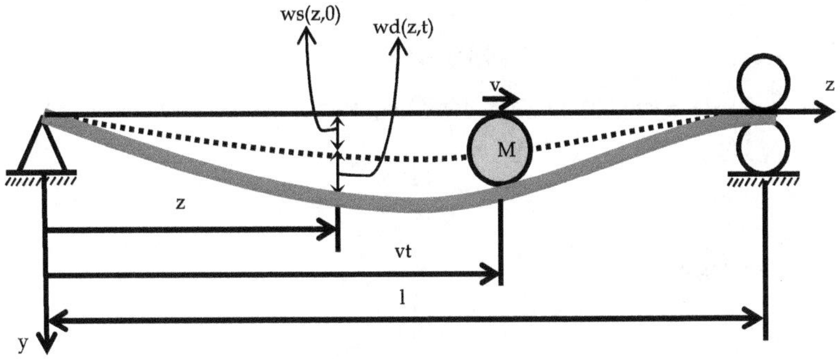

Figure 1. A mathematical model of a simply-supported beam subjected to a moving mass.

The governing equation of the system can be expressed as,

$$EI\frac{\partial^4 w(z,t)}{\partial z^4}+m\frac{\partial^2 w(z,t)}{\partial t^2}$$
$$= mg + Mg\delta(z-vt) - M\left[\frac{\partial^2 w(z,t)}{\partial z^2}v^2 + 2\frac{\partial^2 w(z,t)}{\partial z\partial t}v + \frac{\partial w(z,t)}{\partial z}a + \frac{\partial^2 w(z,t)}{\partial t^2}\right]\delta(z-vt) \tag{1}$$

Where E is Young's modulus of elasticity, I is the cross-sectional area moment of inertia, m is the mass per unit length of a beam, M is the mass of a moving mass, $\delta(z-vt)$ is Dirac delta function, and a is the acceleration of a moving mass. When the mass of a beam is not negligible, the static deflection should be considered. Therefore, the deflection of a beam can be expressed as the sum of the initial static deflection and the dynamic deflection.

$$w(z,t) = w_s(z) + w_d(z,t) \tag{2}$$

Employing Galerkin's mode summation method, the displacement of a beam, $w(\xi,t)$, may be assumed as

$$w_s(\xi) = \sum_{i=1}^{\infty} A_i \phi_i(\xi) \tag{3}$$

$$w_d(\xi,t) = \sum_{i=1}^{\infty} q_i(t)\phi_i(\xi) \tag{4}$$

$$w(\xi, t) = \sum_{i=1}^{\infty}\left[A_i + q_i(t) \right] \phi_i(\xi) \tag{5}$$

where the dimensionless displacement, $\xi = z / l$. The shape function $\phi_i(\xi)$ is the comparison function and has the general form of

$$\phi_i(\xi) = C_1 \sin \beta_i \xi + C_2 \cos \beta_i \xi + C_3 \sinh \beta_i \xi + C_4 \cosh \beta_i \xi \tag{6}$$

For the simply supported beam, the shape function, $\phi_i(\xi)$, may be written as

$$\phi_i(\xi) = \sin i\pi\xi \tag{7}$$

The eigenvalue $\beta_i = i\pi$ and the circular natural frequency w_i have the following relationship

$$w_i^2 = \frac{EI}{m}\left(\frac{\beta_i}{l}\right)^4 \tag{8}$$

From the following equation,

$$\frac{EI}{l^4}\frac{\delta^4 w_s(\xi)}{\delta\xi^4} = mg \tag{9}$$

The static deflection, $w_s(\xi)$, has the following solution form

$$w_s(\xi) = \frac{2mgl^4}{EI\pi^5}\sum_{n=1}^{\infty}\frac{1-(-1)^n}{n^5}\sin n\pi\xi \tag{10}$$

The maximum static deflection δ becomes

$$\delta = w_s\left(\frac{1}{2}\right) = \frac{4}{\pi^5}\frac{mgl^4}{EI}\sum_{n=1}^{\infty}\frac{1}{(2n-1)^5} = \frac{4mgl^4}{EI\pi^5} \tag{11}$$

Substituting the solution of Eq. (5) into Eq. (1) and performing inner product the shape function of $\phi_n(\xi)$ and rearranging provides the equation of motion as

$$\begin{aligned}
\ddot{q}_n(t) + \frac{EI}{m}\left(\frac{n\pi}{l}\right)^4 q_n(t) = {} & \frac{2gM}{ml}\sin n\pi v^* t - \frac{2M}{ml}\sum_{i=1}^{\infty}\left[\left\{\ddot{q}_i(t) - (i\pi v^*)^2 q_i(t)\right\}\sin i\pi v^* t\right. \\
& \left. + \left\{2i\pi v^* \dot{q}_i(t) + i\pi a^* q_i(t)\right\}\cos i\pi v^* t\right]\sin n\pi v^* t \\
& + \frac{4M}{ml}\frac{mgl^4}{EI\pi^5}\left[(\pi v^*)^2\sum_{i=1}^{\infty}\frac{1-(-1)^i}{i^3}\sin i\pi v^* t\right. \\
& \left. - \pi a^* \sum_{i=1}^{\infty}\frac{1-(-1)^i}{i^4}\cos i\pi v^* t\right]\sin n\pi v^* t
\end{aligned} \tag{12}$$

where, the dimensionless variables and parameter are defined as

$$\gamma = \frac{M}{ml},\ \delta = \frac{4mgl^4}{EI\pi^5},\ \tau = \frac{\omega_1}{\pi}t,\ v_0 = \frac{v}{v_{cr}} = \frac{\pi v^*}{\omega_1^2},\ a_o = \frac{\pi^2 a^*}{\omega_1^2},\ \varphi_n = \frac{q_n}{\delta} \tag{13}$$

Eq. (12) can be expressed in matrix form as

$$[M(\tau)]\left\{\ddot{\varphi}(\tau)\right\} + [C(\tau)] + \left\{\dot{\varphi}(\tau)\right\} + [K(\tau)]\{\varphi(\tau)\} = \{f(\tau)\} \tag{14}$$

where, the matrix components are

$$m_{ij}(\tau) = \delta_{ij} + 2\gamma \sin(i\pi v_o \tau)\sin(j\pi v_o \tau) \tag{15}$$

$$c_{ij}(\tau) = 4\gamma (i\pi v_o)\sin(i\pi v_o \tau)\cos(j\pi v_o \tau) \tag{16}$$

$$\begin{aligned} k_{ij}(\tau) = {} & n^4\pi^2\delta_{ij} - 2\mu(i\pi v_o)^2 \sin(i\pi v_o\tau)\sin(j\pi v_o\tau) \\ & + 2\mu(i\pi a_o)\sin(i\pi v_o\tau)\cos(j\pi v_o\tau) \end{aligned} \tag{17}$$

$$\begin{aligned} f_i(\tau) = {} & \gamma\pi^2\left[\frac{\pi}{2} + v_0^2 \sum_{n=1}^{\infty}\frac{1-(-1)^n}{n^3}\sin(n\pi v_o\tau)\right. \\ & \left. - a_o\sum_{n=1}^{\infty}\frac{1-(-1)^n}{n^4}\cos(n\pi v_o\tau)\right]\sin(i\pi v_o\tau) \end{aligned} \tag{18}$$

The response of Eq. (14) may be analyzed using Runge-Kutta integration method.

2.2. Designing fuzzy controller

A general linear control theory is not applicable for the study since it is the time variant system and has a large non-linearity, as presented in Eq. (14). The fuzzy controller is effectively applicable for the study since the controller is designed by considering the characteristics of the vibration produced only, without taking into account the dynamic characteristics of the system.

The fundamental structure of the fuzzy controller applied to the simply supported beam carrying a moving mass is shown in Fig. 2.

The system output measured from the laser displacement sensor is transferred to the fuzzy set element by fuzzification. The decision making rule was properly designed by the controller designer based on the dynamic characteristics of the system to be studied. It produced the control input by defuzzifying the fuzzified output calculated from the measured values of the system.

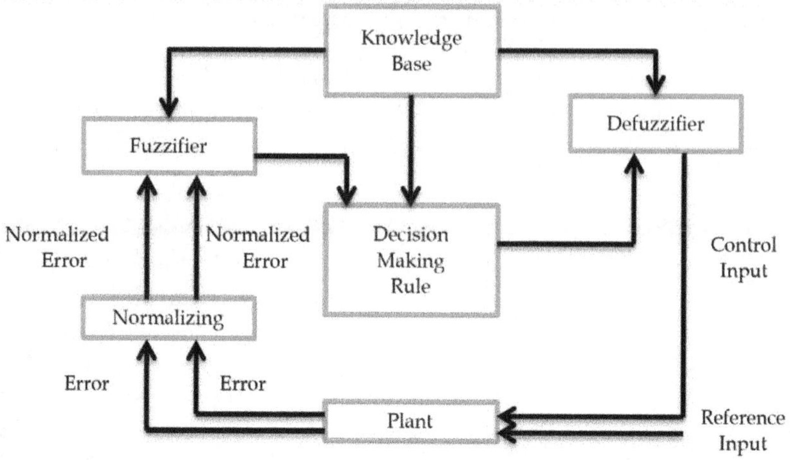

Figure 2. Flow-chart of fuzzy diagram.

The designer's experience, the professional's knowledge, and the dynamic characteristics of the system to be controlled are added by the designer in every process. The advantage of the fuzzy control theory using this type of design method is to design the controller easily when the characteristics or trends of the system are known or predictable to a certain extent, but the mathematical modelling for control is very difficult and complicated like the one for the study.

The system output obtained from the standardization process was fuzzified by formulating a fuzzy set. There are various ways to formulate a fuzzy set. A fuzzy set, which divides the range from -1 to 1 into 7 equal sections and has a triangular function as a membership function, was constructed and used for the present study as shown in Fig. 3.

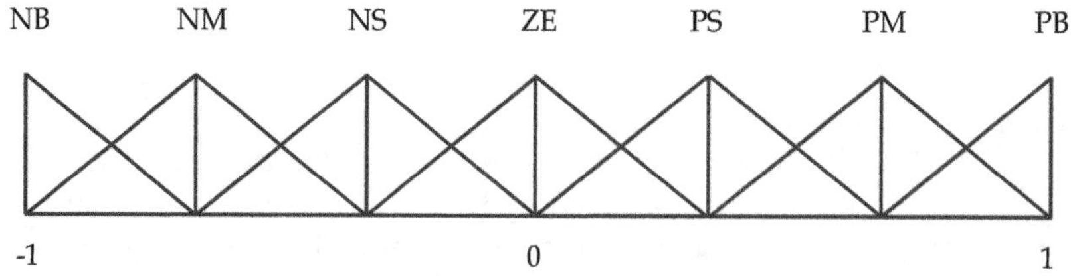

Figure 3. Fuzzy set with triangular membership function

It is desired to utilize the time response curve of a typical second-order system to a step input, as shown in Fig. 4, to determine the fuzzy rule for designing a fuzzy controller. At the point of "a", the control input should be PB(Positive Big) since the error is NB(Negative Big) and the time rate of the error is ZE(zero). At the point of "b", the control input is desired to be NS(Negative Small) if the error is ZE and the time rate of the error is PS(Positive Small). From the same logic, the control input at the point of "c" is needed to be PM(Positive Medium). Such a logical reasoning may be justified from the point of professional.

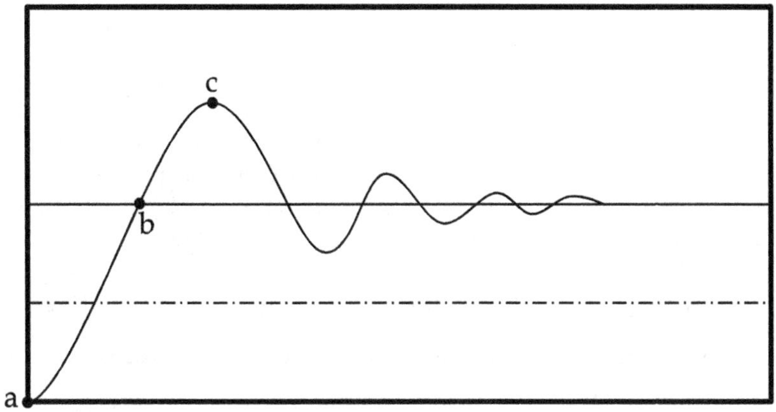

Figure 4. Typical response of controlled system.

μ_e \ $\dot{\mu}_e$	NB	NM	NS	ZE	PS	PM	PB
NB	PB	PB	PM	PM	PS	PS	ZE
NM	PB	PM	PM	PS	PS	ZE	NS
NS	PM	PM	PS	PS	ZE	NS	NS
ZE	PM	PS	PS	ZE	NS	NS	NM
PS	PS	PS	ZE	NS	NS	NM	NM
PM	PS	ZE	NS	NS	NM	NM	NB
PB	ZE	NS	NS	NM	NM	NB	NB

Table 1. Fuzzy rule with 7 elements

By expanding the logic stated above, the fuzzy rule may be constructed as shown in Table 1 for all cases of the fuzzy set consisted of seven elements.

where the errors e and $\dot{e}$ may be defined by the vibration displacement y, the vibration velocity $\dot{y}$, and the vibration displacement and the velocity to be controlled y_d and $\dot{y}_d$, respectively as follows

$$e = y - y_d \tag{19}$$

$$\dot{e} = \dot{y} - \dot{y}_d \tag{20}$$

The normalized error μ_e and $\dot{\mu}_e$ are defined by the initial target values y_o and $\dot{y}_o$ as

$$\mu_e = \frac{e}{y_o}, \quad \dot{\mu}_e = \frac{\dot{e}}{\dot{y}_o} \tag{21}$$

The actual control input was obtained by defuzzifying the fuzzified control input that was found based on the fuzzy rule from the normalized error and the time rate of the error obtained through the fuzzification process. For the study, the normalized control input was found by employing the min-max centroid method for the defuzzification.

3. Numerical analysis results of dynamic response to a moving mass

In order to produce numerical analysis results of dynamic response of a beam traversed by a moving mass, Runge-Kutta integration method was applied to Eq. (14). The dynamic deflection caused by a moving mass with constant velocity was investigated for the study, even though the dynamic deflection by a moving mass with constant acceleration can also be obtained from Eq. (14).

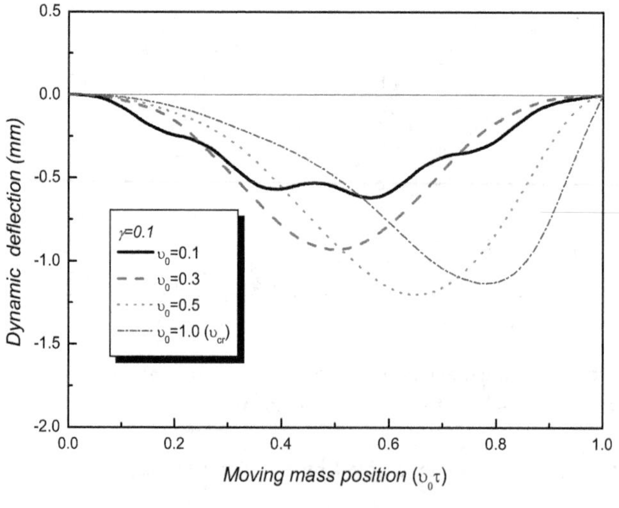

Figure 5. Dynamic deflection at the position of moving mass($v_0\tau$) for γ =0.1.

Figs. 5-7 show the dynamic responses at the dimensionless position $v_0\tau$ of the moving mass for the dimensionless velocity of the moving mass v_0=0.1, 0.3, 0.5, 1.0 when the mass ratio of the beam to the moving mass γ=0.1, 0.5, 1.0. Fig. 5 through Fig. 7 present the dynamic deflection only without including the static deflection. For γ=0.1, i.e. the mass of the beam is 1/10 of the mass of the moving mass, the first mode of an uniform beam was obtained as shown in Fig. 5, when the velocity of the moving mass is relatively low as v_0=0.1. For v_0 ≥0.5, the location of the maximum deflection shifts to the right end of the beam, where the moving mass arrives at. And also, the maximum dynamic deflection increases in general as the velocity of a moving mass increases. It is, however, observed that the maximum dynamic deflection decreases at the critical velocity v_{cr}. Fig. 6 of γ =0.5 through Fig. 7 of γ =1.0 show a similar trend to Fig. 5, but the difference from Fig. 5 is that the location of the maximum deflection shifts to the right end of the beam for all the velocity ranges from v_o =0.1 to the critical velocity as the velocity of a moving mass increases.

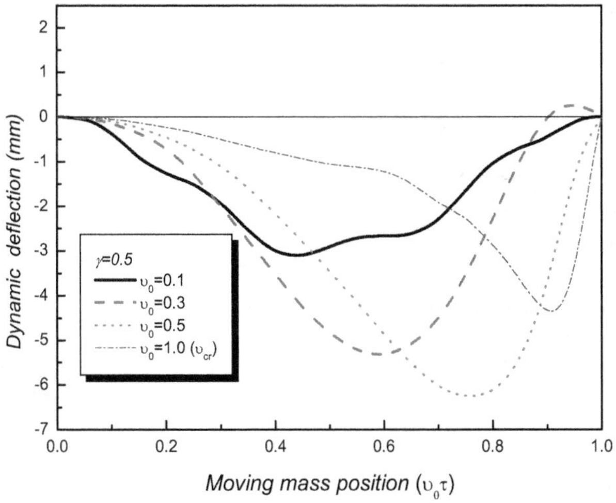

Figure 6. Dynamic deflection at the position of moving mass($v_o\tau$) for γ =0.5.

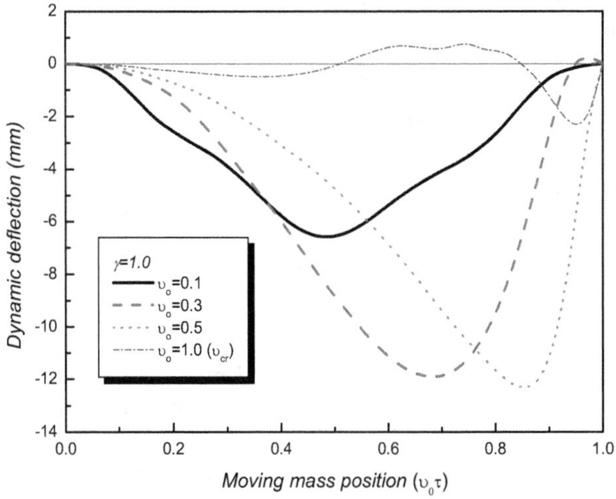

Figure 7. Dynamic deflection at the position of moving mass($v_o\tau$) for γ =1.0.

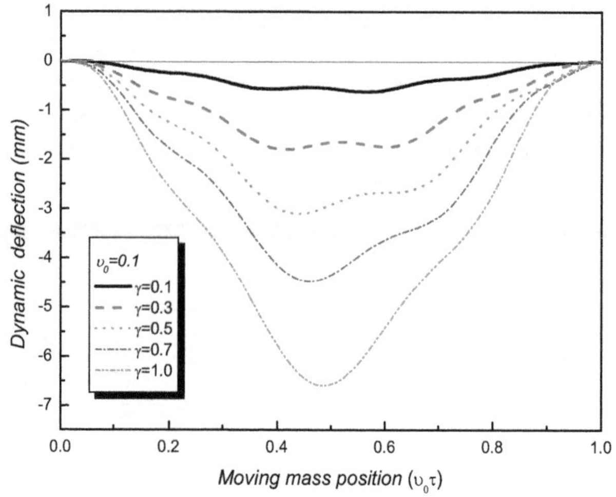

Figure 8. Dynamic deflection at the position of moving mass($v_0\tau$) for v_0=0.1

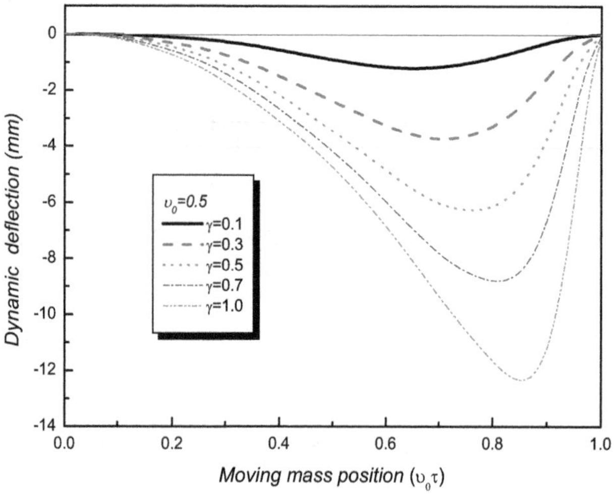

Figure 9. Dynamic deflection at the position of moving mass ($v_0\tau$) for v_0=0.5

Figs. 8-10 show the dynamic deflection at the dimensionless position of the moving mass for various mass ratios of γ when the dimensionless velocity of a moving mass v_0=0.3, 0.5, 1.0. When the velocity of the moving mass is as slow as v_0=0.1, the maximum dynamic deflection increases and its location slightly shifts to the right end as the mass ratio, γ, increases as shown in Fig. 8. And also, the moving mass position for the maximum deflection is near the middle point of the beam. For v_0=0.5 as shown in Fig. 9, the maximum dynamic deflection increases and the moving mass position for the maximum dynamic deflection shifts to the right end as the mass ratio, γ, increases. For the critical velocity v_0=1.0 as shown in Fig. 10, however, the maximum dynamic deflection decreases as the mass ratio, γ, increases for ≥0.7 while the moving mass position for the maximum dynamic deflection still shifts to the right end of the beam.

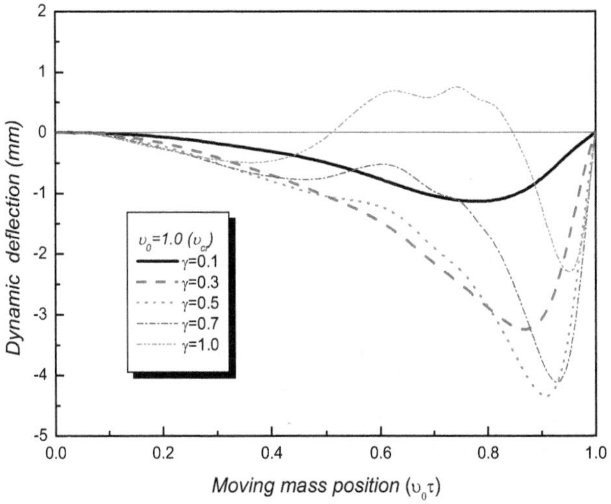

Figure 10. Dynamic deflection at the position of moving mass ($v_0\tau$) for v_0=1.0

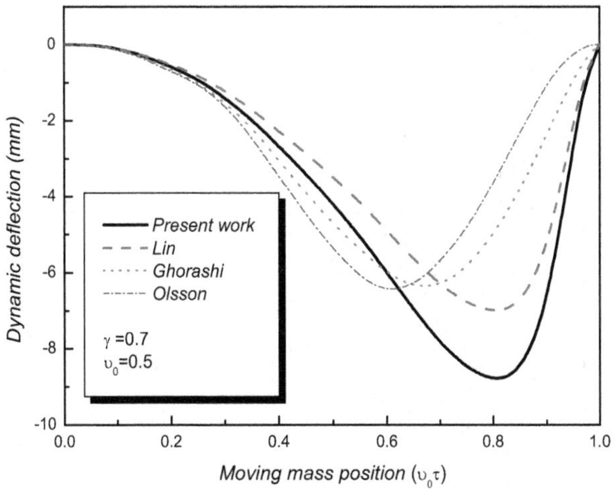

Figure 11. Comparison present results with other previous results for dynamic deflections (v_0=0.5)

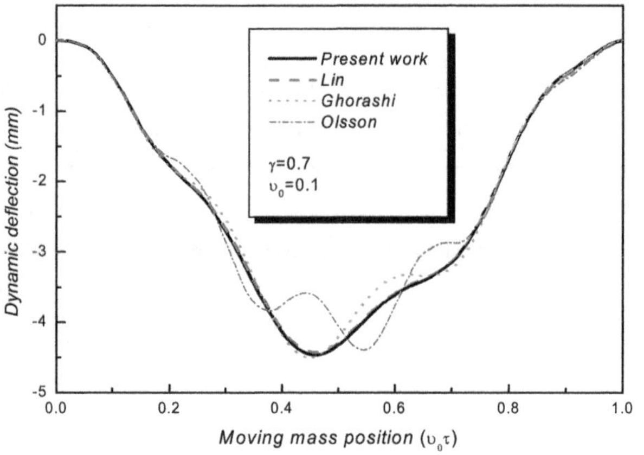

Figure 12. Comparison present results with other previous results for dynamic deflections (v_o =0.1)

Figs. 11 and 12 present the comparison of current results on the dynamic deflection and those from previous studies (references [6], [7], [8]) that do not consider as many effects of a moving mass as the present study does, for γ =0.7 and v_o=0.1, 0.5. As can be seen in these figures, the previous results are somewhat different from the present results. It is, therefore, believe that all the effects of a moving mass should be considered to predict more precisely the dynamic deflection of a beam traversed by a moving mass. This is confirmed by comparing with the experimental results presented in the Section 4.

4. Experimental apparatus and experiments

4.1. Experimental apparatus

An experimental apparatus was set up to control the dynamic deflection and the vibration of a simply supported uniform beam traversed by a moving mass as shown in Fig. 13.

The test beam has a groove along its length to help various sizes of a moving mass to run smoothly. The details of the test beam are shown in Table 2.

Material	Aluminum 6061
Modulus of Elasticity(Gpa)	7.07e+10
Density(kg/m^3)	2700
Mass(g)	283.0
Length (mm)	1000.0
Width (mm)	32.0
Thickness (mm)	4.0
Groove width (mm)	10.0
Groove depth (mm)	2.0

Table 2. Details of the test beam.

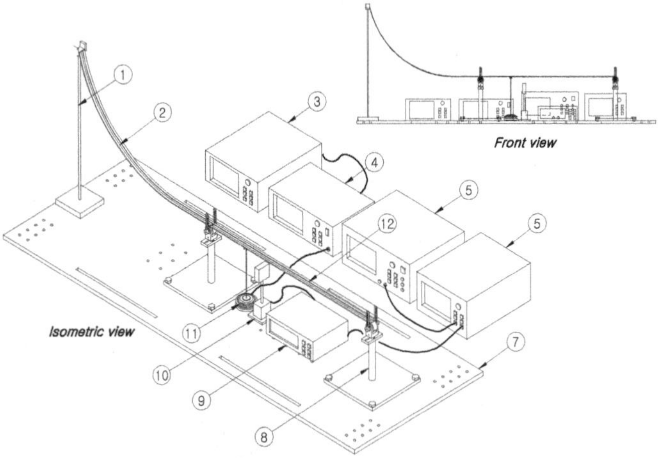

①Guide beam support pole ②Guide beam ③Power supply ④Amplifier ⑤FFT analyzer ⑥Digital oscilloscope ⑦Base ⑧Simply-supports ⑨Laser displacement meter ⑩Laser sensor ⑪Actuator ⑫Test beam

Figure 13. Experimental set up.

Steel balls were chosen as moving masses to reduce the friction with the test beam. The details of the moving masses are shown in Table 3

Type of moving mass	Mass (g)	Diameter (mm)	Materials
M_1	67.0	25.35	Steel
M_2	151.0	33.30	
M_3	228.0	38.05	

Table 3. Details of the moving masses.

Photo. 1 depicts the experimental set-up for the study. A non-contact laser displacement sensor, a sensor controller, and a digital memory oscilloscope are installed to measure the dynamic response of the test beam traversed by a moving mass.

Photo 1. Photograph of experimental set-up.

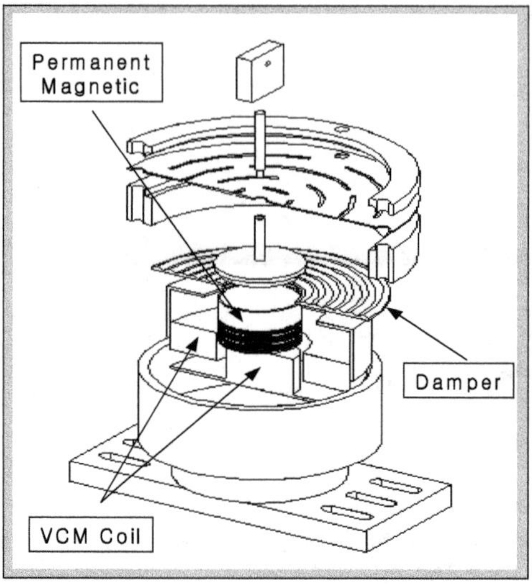

Figure 14. Schematic diagram of the front view of voice coil motor.

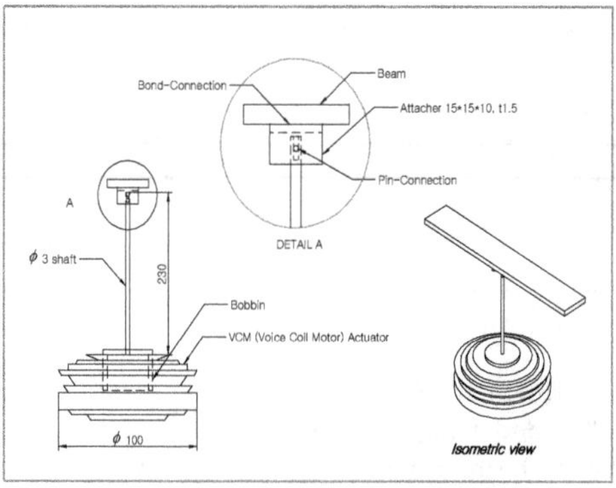

Figure 15. Details of a voice coil motor actuator.

Figs 14-15 show schematic diagram of the front view and details for the structure of a voice coil motor(VCM) actuator, respectively. The active actuator was reconstructed by using a commercial speaker. In the structure of the actuator, permanent magnet and voice coil wound in bobbin of the speaker were used, and an attacher and a shaft were produced in order to deliver control force from actuator to the beam. The actuator was built to suppress the dynamic deflection and the vibration of the test beam caused by a moving mass. The actuator is able to generate the control input by using a voice coil motor. The actuator is desired to be installed above and below the test beam to apply the force from the magnetic field to the beam. However, the control input is applied through a slender rod which connects the control actuator to the bottom of a beam, since the upper part of the beam should be free from any bar for a moving mass running. Photo. 2 depicts a test beam, moving masses and an actuator.

Photo 2. Photograph of a test beam, moving masses and an actuator.

4.2. Experiments

Experiments are conducted to measure the dynamic response first, and then to control the vibration of a beam traversed by a moving mass. Three different mass ratios and three different velocity ratios of a moving mass were chosen to investigate the dynamic response of the system. The moving mass was released freely at the designated location of the guide beam and traveled the horizontal part of the guide beam of 210 mm before entering and eventually passing through the test beam. The dynamic response was measured at 385 mm from the entrance of the test beam by a non-contact laser displacement meter. The output signal was amplified before being monitored on an oscilloscope.

In order to suppress the dynamic deflection and vibration of a beam traversed by a moving mass, the control input is supplied by an actuator as shown in Fig. 13. The control input in accordance with the fuzzy logic is applied to the system through a voice coil motor at the same time when the moving mass enters the test beam. The signal of the dynamic response was amplified by an amplifier and then displayed on an oscilloscope.

Such experiments on the dynamic deflection and the vibration control were performed in order for the selected masses and velocities of moving masses. The details of the moving mass used are shown in each figure.

4.3. Experimental results and discussion

4.3.1 Experimental results of the dynamic response and discussion

Figs. 16-18 present both the experimental results and the analytical results from the numerical simulation on the dynamic response of a simply supported beam traversed by a moving mass. Figs. 16-18 show both analytical and experimental results on the dynamic response for three different velocities and magnitudes of a moving mass (v =1.299(m/s), v =1.880(m/s), v =2.450(m/s), M =67(g), M =151(g), M =228(g)).

The analytical results agree well with the experimental results in both the magnitude and the shape of the dynamic response curve for two velocities(v =1.299(m/s), v =1.880(m/s))

of a moving mass. For v =2.450(m/s), however, analytical and experimental results are somewhat different as shown in Fig. 18. It is thought that the difference becomes greater as a higher velocity of a moving mass produces a stronger impact at the joint of the guide beam and the test beam. For the residual vibration region, two different initial conditions of which the moving mass leaves the beam are believed to make the difference between two results as explained previously.

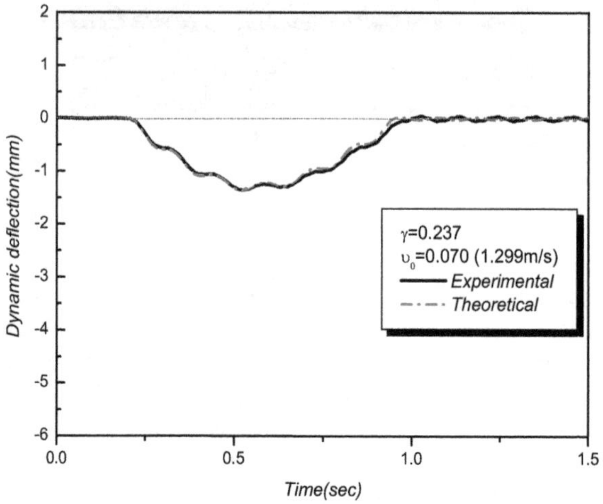

Figure 16. Comparison theoretical results with experimental ones for beam deflections (γ =0.237 and v_o =0.07)

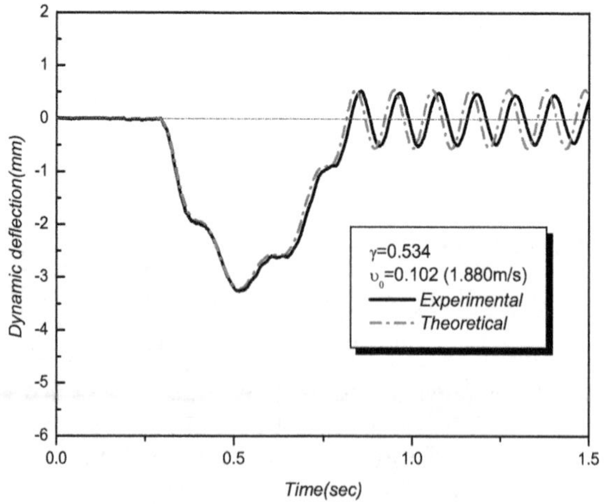

Figure 17. Comparison theoretical results with experimental ones for beam deflections (γ =0.534 and v_o =0.102)

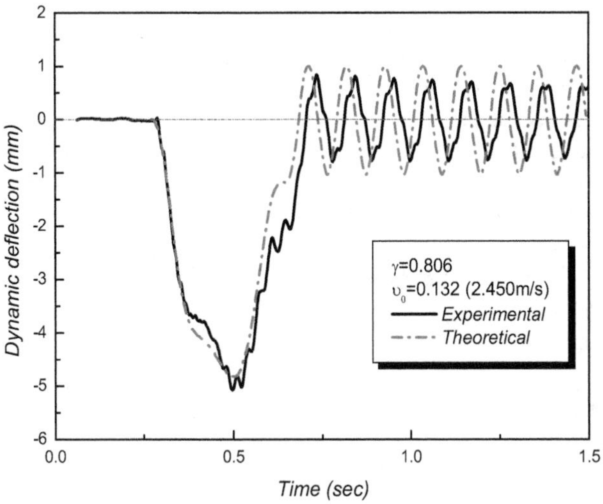

Figure 18. Comparison theoretical results with experimental ones for beam deflections (γ =0.806 and v_o =0.132)

4.3.2. Experimental results of vibration control of a beam

A control using a voice-coil motor was conducted to suppress the amplitude of dynamic response and vibration of a beam traversed by a moving mass. The Matlab simulation was performed to find the best location of the actuator for the control input.

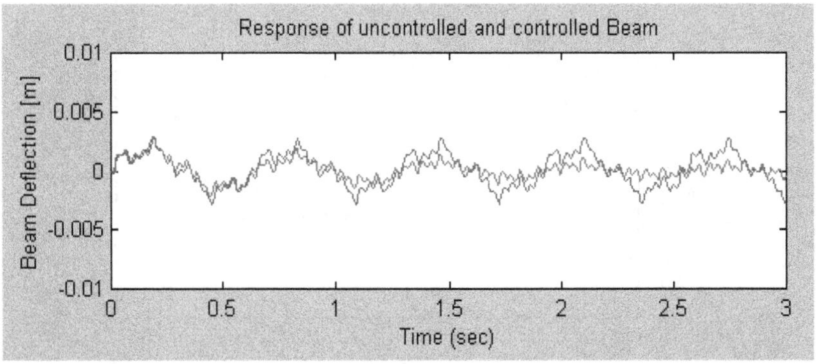

Figure 19. Response of uncontrolled and controlled beam for arbitrary disturbances (actuator position = l/10)

The results are presented in Figs. 19-20. The simulation was performed by investigating the response to the disturbance applied to a certain point for various locations of the actuator. Two typical simulation results are presented in Figs. 19-20. As shown in these figures, a better controlled result was obtained when the actuator is located at 3/10 of the beam length l. And also, the node of the third mode is believed to be a good place for locating the actuator since it is intended to control up to the second mode for the study. The position of 3/10 of the beam length l from the entrance of the test beam was finally selected for the location of the actuator. The experiments for control were conducted after locating the actuator at $3l/10$ based on the simulation results shown in Figs. 19 and 20.

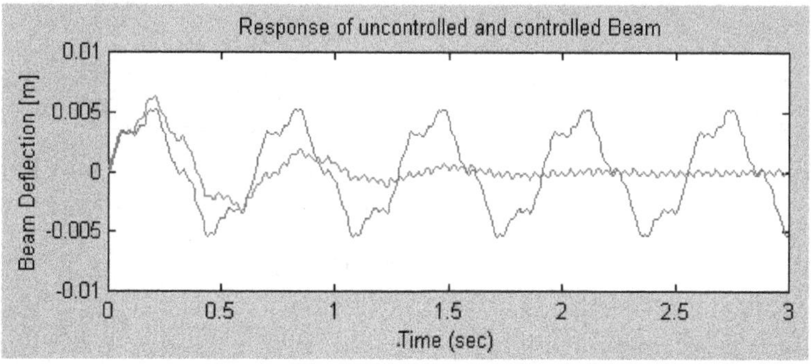

Figure 20. Response of uncontrolled and controlled beam for arbitrary disturbances (actuator position = $3l/10$)

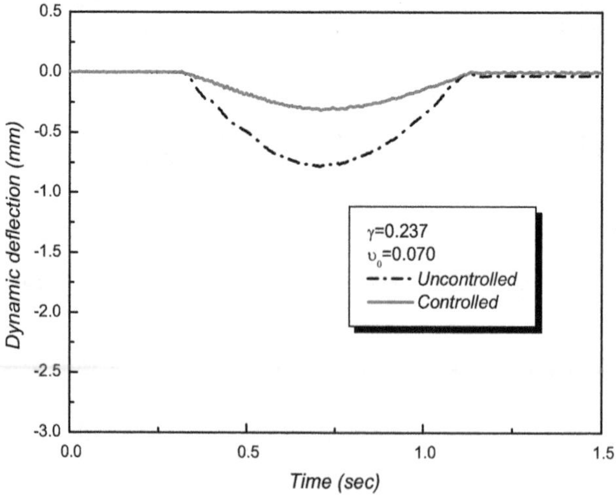

Figure 21. Dynamic deflections of uncontrolled and controlled beam for γ =0.237 and v_0 =0.07.

The experimental results under both uncontrolled and controlled conditions are presented in Figs. 21-23.

The experimental results of the dynamic deflection in 4.3.1 and the test results for the uncontrolled case in 4.3.2 are supposed to be identical for the same mass ratio γ and velocity ratio v_0. But they are different because the stiffness and the damping provided by the slender rod that connects the actuator and the beam change the dynamic characteristics of the system.

Therefore, it is necessary to include the slender rod to simulate the dynamic deflection curve for the uncontrolled cases shown in Figs. 21-23. It is, however, very difficult to develop the mathematical governing equation for such a system. Thus, the present study focuses on the control effect only by applying the control input from the measured dynamic responses of the system with the control actuator.

For three different velocities(v =1.299(m/s), v =1.880(m/s), v =2.450(m/s)), in general, the dynamic deflections are controlled to be less than 50% of the uncontrolled ones. And also, it does help to suppress the residual vibration occurred after the moving mass passes through the beam.

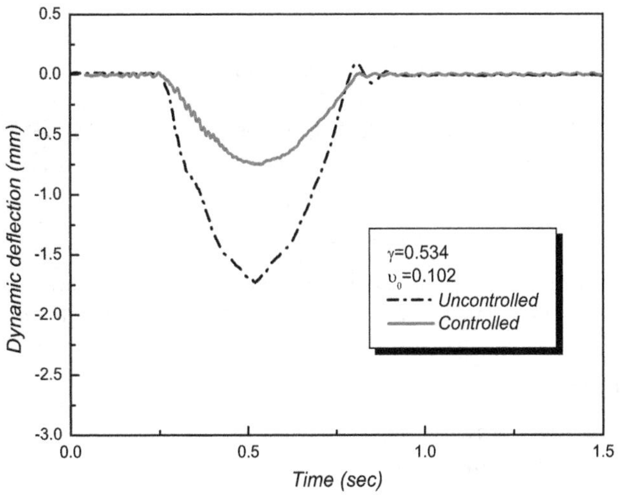

Figure 22. Dynamic deflections of uncontrolled and controlled beam for γ =0.534 and v_0 =0.102.

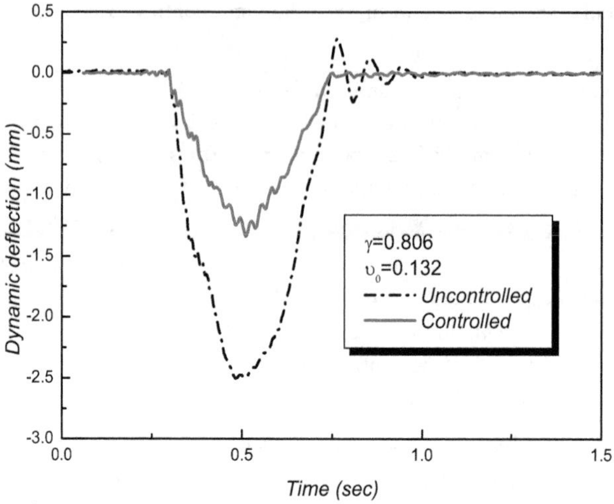

Figure 23. Dynamic deflections of uncontrolled and controlled beam for γ =0.806 and v_0 =0.132.

5. Conclusion

The following results were obtained from the fuzzy control studies on the dynamic response and the vibration of a simply supported beam traversed by a moving mass with a constant velocity. Firstly, the position of a moving mass at the maximum dynamic deflection moves to the right end of a beam as the mass ratio of a moving mass γ increases for a lower velocity ratio of a moving mass v_0 than the critical velocity. Secondly, the position of a moving mass at the maximum dynamic deflection moves to the right end of a beam as the velocity of a moving mass increases for a given mass ratio of a moving mass γ.

Thirdly, the experimental results of the dynamic deflection of a beam traversed by a moving mass agree well with the simulation results.

Fourthly, the dynamic deflection and the residual vibration of a beam traversed by a moving mass were successfully reduced more than 50 % through the fuzzy control.

Author details

Bong-Jo Ryu and Yong-Sik Kong
Hanbat National University / Hanwha L&C, South Korea

6. References

[1] Strokes, G. G. (1849). Discussion of a Differential Equation Relation to the Breaking of Rail Way Bridges. *Transactions of the Cambridge Philosophical Society*, Vol.85, pp.707-735.

[2] Ayre, R. S.; Ford, G. & Jacopsen, L. S. (1950). Transverse Vibration of a Two Span Beam under Action of Moving Constant Force. *Transactions of the ASME, Journal of Applied Mechanics*, Vol.17, pp.1-2.

[3] Yoshida, D. M. & Weaver, W. (1971). Finite Element Analysis of Beams and Plates with Moving Loads. *Publication of International Association for Bridge and Structural Engineering*, Vol. 31, No. 1, pp.179-195.

[4] Ryu, B. J. (1983). Dynamic Analysis of a Beam Subjected to a Concentrated Moving Mass, Master Thesis, Yonsei University, Seoul, Korea.

[5] Sadiku, S. & Leipholz, H. H. E. (1987). On the Dynamics of Elastic Systems with Moving Concentrated Masses. *Ingenieur-Archiv*, Vol. 57, pp. 223-242.

[6] Olsson, M. (1991). On the Fundamental Moving Load Problem. *Journal of Sound and Vibration*, Vol. 145, No. 2, pp. 299-307.

[7] Esmailzadeh, E. & Ghorashi, M. (1992). Vibration Analysis of Beams Traversed by Moving Masses, *Proceedings of the International Conference on Engineering Application of Mechanics*, Vol. 2, pp. 232-238, Tehran, Iran.

[8] Lin, Y. H. (1997). Comments on Vibration Analysis of Beams Traversed by Uniform Partially Distributed Moving Masses. *Journal of Sound and Vibration*, Vol. 199, No. 4, pp. 697-700.

[9] Wang, R. T. & Chou, T. H. (1998). Nonlinear Vibration of Timoshenko Beam due to a Moving Force and the Weight of Beam. *Journal of Sound and Vibration*, Vol. 218, No. 1, pp. 117-131.

[10] Wu, J. J. (2005). Dynamic Analysis of an Inclined Beam due to Moving Load. *Journal of Sound and Vibration*, Vol. 288, pp. 107-131.

[11] Abdel-Rohman, M. & Leipholz, H. H. E. (1980). Automatic Active Control of Structures, North-Holland Publishing Co. & Sm Publications.

[12] Kwon, H. C.; Kim, M. C. & Lee, I. W. (1998). Vibration Control of Bridges under Moving Loads. *Computers and Structures*, Vol. 66, pp. 473-480.

[13] Ryou, J. K.; Park, K. Y. & Kim, S. J. (1997). Vibration Control of Beam using Distributed PVDF Sensor and PZT Actuator. *Journal of Korean Society of Noise and Vibration Engineering*, Vol. 7, No. 6, pp. 967-974.

[14] Bailey, T. & Hubbard Jr., J. E. (1985). Distributed Piezoelectric-Polymer Active Vibration Control of a Cantilever Beam. *Journal of Guidance, Control, and Dynamics*, Vol. 8, No.5, pp.605-611.

[15] Kwak, M. K. & Sciulli, D. (1996). Fuzzy-Logic Based Vibration Suppression Control Experiments on Active Structures. *Journal of Sound and Vibration*, Vol. 191, No. 1, pp.15-28.

[16] Sung, Y. G. (2002). Modeling and Control with Piezo-actuators for a Simply Supported Beam under a Moving Mass. *Journal of Sound and Vibration*, Vol. 250, No. 4, pp.617-626.

[17] Nikkhoo, A.; Rofooei, F. R. & Shadnam, M. R. (2007). Dynamic Behavior and Modal Control of Beams under Moving Mass. *Journal of Sound and Vibration*, Vol. 306, pp. 712-724.

[18] Prabakar, R. S.; Sujatha, C. & Narayanan, S. (2009). Optimal Semi-active Preview Control Response of a Half Car Vehicle Model with Magneto-rheological Damper. *Journal of Sound and Vibration*, Vol. 326, pp. 400-420.

[19] Pisarski, D. & Bajer, C. I. (2010). Semi-active Control of 1D Continuum Vibrations under a Travelling Load. *Journal of Sound and Vibration*, Vol. 329, pp. 140-149.

Chapter 12

Optimal Locations of Dampers/Actuators in Vibration Control of a Truss-Cored Sandwich Plate

Kongming Guo and Jun Jiang

Additional information is available at the end of the chapter

1. Introduction

In the engineering application of many metal structural components, the damping property is an important factor to be considered. The approaches of adding damping can be mainly divided into two categories: passive vibration control and active vibration control. Passive vibration control with the use of damping materials is widely adopted in the engineering areas due to the simplicity for implementation and the high reliability, while active vibration control using various smart materials as an approach with great potentials has received a great deal of attention in recent years.

A traditional approach to passively suppress vibration of plate-like structure is using constrained layer damping [1] while active vibration control is often realized by bonding piezoelectric patch [2] to the surface of plate. The third approach, semi-active vibration control, adopts piezoelectric switching shunt techniques [3] and active constrained layer damping method [4]. However, the approach should also bond piezoelectric patch or constrained layer to the plate surface. These kinds of techniques are at the expense of adding considerable weight to the structure while the piezoelectric patch cannot generate large enough control force. Moreover, the surface of the plate will be changed. The Kagome sandwich structure [5] introduced in this chapter consists of a solid face sheet, a tetrahedral core and a planar Kagome truss as the back-plane. Because one face sheet of the Kagome structure has been replaced by a planar Kagome truss, the transverse displacement of its solid face sheet can be realized just by the in-plane tension-compression actuation forces if some specific rods in the planar Kagome truss are replaced by linear actuators. This character inspires the design of a new kind of sandwich plate, whose vibration control can be readily realized by replacing a very small portion of the rods only in the planar Kagome truss through cylindrical viscoelastic dampers to dissipate energy or through piezoelectric stack actuator to generate control forces to suppress the vibration due to the out-of-plane excitation.

It is well known that to keep the expense within the acceptable extent and meanwhile maintain the structure under control, the locations of small number of dampers/actuators used in passive/active vibration control will significantly influence the consequence of the control effect. Determining the optimal locations of dampers/actuators is a combinatory optimization problem since the possible location combinations are discrete. The approaches to achieve an optimal placement of the dampers/actuators can be classified into four categories. The first approach is using ad hoc iterative methods [6] to get the optimal combination, which often obtains the non-optimal solution. The second one solves the discrete problem in a continuous domain [7], which does not provide significant reduction in computational cost. The third method is directly solving the combinatory optimization by using various alternative search techniques like simulated annealing [8] and genetic algorithm [9]. The fourth approach adopts different effectiveness indices, like modal strain energy (MSE) [10] and eigensensitivity [11], to quantify the fitness of different locations. This kind of methods are very intuitive and can often get remarkable results even though they may not be the most optimal ones, while their computational costs are also very low. Moreover, the effectiveness index can help to generate a subset of location combinations that reduces the size of the problem to be dealt with.

This chapter contains three parts: In the first part the Kagome structure and its finite element model is introduced briefly. In the second part passive vibration control of the Kagome structure and optimization of locations of dampers are dealt with. This part is arranged as followings: (1) Building the finite element model of the viscoelastic damper. (2) To control vibration of single mode, fraction of axial MSE is chosen as an effectiveness index to decide the placement position of the dampers in the planar Kagome truss. (3) A practical optimization method of dampers in vibration control of broad-bandwidth vibration control is raised based on MSE. In the third part active vibration control of the Kagome structure and optimization of locations of actuators are discussed. Independent modal space control (IMSC) method is chosen aimed to control the vibration of the lower modes. The third part is arranged as follows: (1) The piezoelectric actuator is introduced. (2) The IMSC method is detailed and its stability is also analyzed. (3) The controllability of IMSC is discussed and an optimization method for the placement of actuators in the Kagome structure that uses both MSE and singular values is developed. (4) The validity of the optimization method is demonstrated through the eigenvalue analysis and the time-domain simulation.

1.1. Kagome sandwich structure

The Kagome based high authority shape morphing structure (or Kagome structure in short) [5] is a kind of sandwich truss-cored plate. In contrast to other sandwich truss-cored plates that have two solid face sheets, one face sheet of the Kagome structure is replaced by a planar Kagome truss while the other face sheet is still a solid one. The truss-core is a tetrahedral truss that lies in between the two faces (see figure 1). The ancient planar Kagome basket weave pattern truss is simultaneously static determinacy and stiff [12]. This feature of the planar Kagome truss enables its truss rods to be actuated in order to achieve arbitrary in-plane nodal displacements with minimal internal resistance.

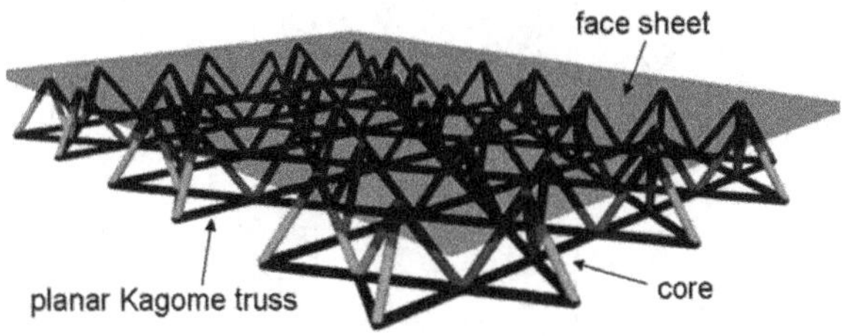

Figure 1. Schematic representation of the Kagome structure, the solid face sheet is shown in dark grey, the core in grey and the planar Kagome truss as one face plane in black.

The material and physical parameters of the Kagome structure studied in this paper are listed in table 1. There are total 1584 truss rods in the planer Kagome truss. Both the solid face sheet and the Kagome back plane of the structure are clamped.

Face sheet		Core truss and Kagome truss	
Material	Al alloy	Material	Stainless steel
Young's modulus	73.1GPa	Young's modulus	193GPa
Density	2700kg/m^3	Density	8030kg/m^3
Length	1.58m	Truss length	51mm
Width	1.50m	Section type	Circle
Depth	1.53mm	Radius	1.275mm

Table 1. Material parameters and size of face sheet and truss rods

MSC.PATRAN is used to build the finite element model of the Kagome structure which is shown in figure 2. The face sheet used is discretized using 1120 plate elements CQUAD4, while the 3167 truss-core and planar Kagome truss is modeled by simple beam elements CBAR. The first six mode shapes are shown in figure 3.

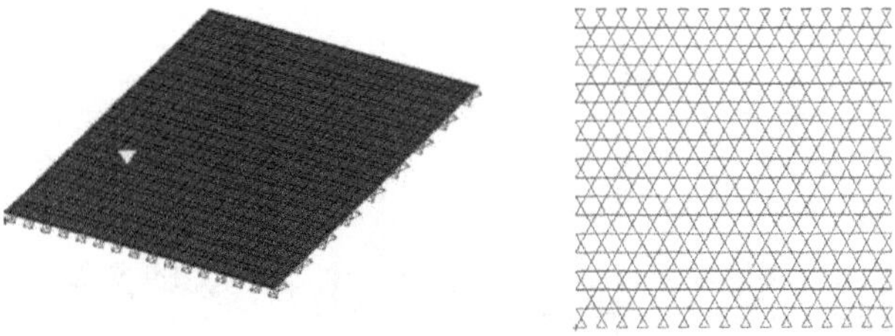

(a) (b)

Figure 2. (a) Finite element model of Kagome structure. The blue triangular represents the observe point of structure response below; (b) The back plane of the Kagome structure with a planar Kagome truss.

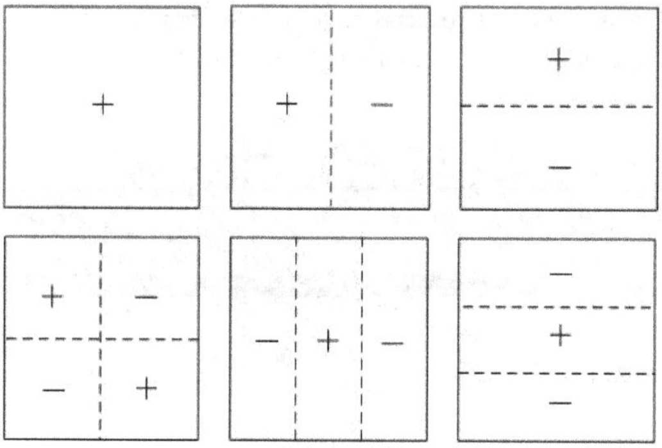

Figure 3. Modal shapes of mode 1-6. The symbol + and – represent the direction of the transverse vibration. The dashed line represents node line.

2. Passive control and damper placement optimization

2.1. Finite element model of viscoelastic damper

According to the feature of the planar Kagome truss, a kind of cylindrical sandwich shearing viscoelastic damper [13] with the same length as the truss members is adopted. This kind of damper is a bi-shearing sandwich structure composed of a core rod, a sleeve and a thin layer of viscoelastic material (see figure 4). When relative movement between the core rod and the sleeve exists, the viscoelastic material will undertake shearing deformation and dissipate energy. To ensure the load applied on the damper in the axial direction, the spherical hinges must be used in the connection between the damper and the truss members in order to avoid bending and torsion moments.

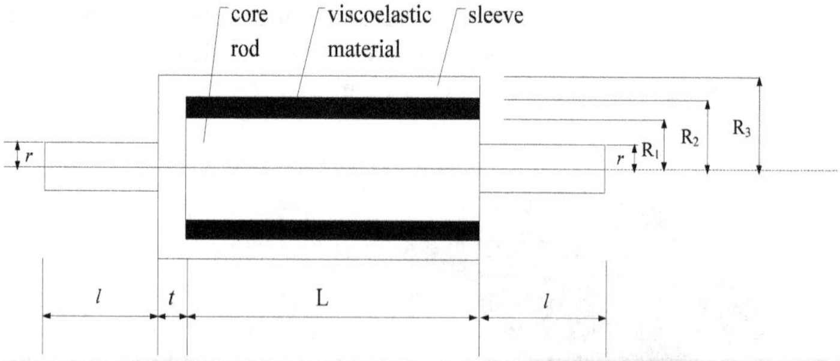

Figure 4. Sketch of the cylindrical sandwich shearing viscoelastic damper. The parameters are given as: L=29mm, t=2mm, l=10mm, r=3mm, R_1=8mm, R_2=10mm, R_3=12.4mm.

To set up a dynamic model of the cylindrical damper which is suitable for the incorporation into the finite element model of the Kagome structure, the theory of linear viscoelasticity is used and the constitutive relation for a viscoelastic material can be written as:

$$\sigma(t) = G(t)\varepsilon(0) + \int_0^t G(t-\tau)\frac{d\varepsilon(\tau)}{d\tau}d\tau \tag{1}$$

where σ is stress and ε represents strain. $G(t)$ is material relaxation function. This stress relaxation represents energy loss from the material. Takeing Laplace transform on (1) yields:

$$\sigma(s) = sG(s)\varepsilon(s) \tag{2}$$

where $sG(s)$ is called material modulus function.

Applying the Biot model [14], the modulus function of viscoelastic material can be written as a series of terms called mini-oscillator terms:

$$sG(s) = G^{\infty}\left(1 + \sum_{k=1}^{m} a_k \frac{s}{s+b_k}\right) \tag{3}$$

The factor G^{∞} represents the equilibrium value of the modulus, $\{a_k, b_k\}$ are positive constant determined by the shape of modulus function in the Laplace domain, $k=1,2,...,m$, while m is the total number of mini-oscillator terms. Figure 5 illustrates the mechanical analogy of Biot model.

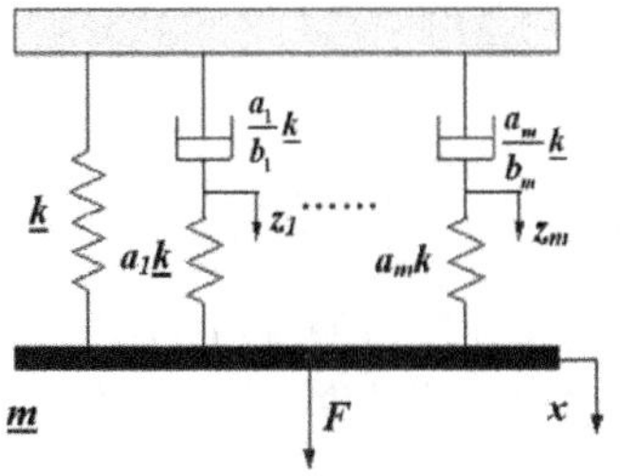

Figure 5. The mechanical analogy of Biot model

The finite element model of the dynamic equation of viscoelastic material by m mini-oscillator terms is written as:

$$[\tilde{M}]\{\ddot{q}\} + [\tilde{C}]\{\dot{q}\} + [\tilde{K}]\{q\} = \{\tilde{f}\} \tag{4}$$

where $q = \{\{x\} \ \{z_1\} \ \ldots \ \{z_m\}\}^T$ is the variable vector , among it $\{x\}$ represents the displacement vector of the damper, which is governed by the dynamical equation:

$$[M]\{\ddot{x}\} + [K]\left(G(t)\{x(0)\} + \int_0^t G(t-\tau)\{\dot{x}\}d\tau\right) = \{f\} \tag{5}$$

In (5) $[M]$ and $[K]$ are the mass and stiffness matrices of the damper. $\{z_1\},\ldots,\{z_m\}$ are the so-called dissipation coordinates. The symmetric coefficient matrices in (4) are given as:

$$\left[\tilde{M}\right]=\begin{bmatrix}[M] & [0] & \cdots & [0]\\ [0] & [0] & \cdots & [0]\\ \vdots & \vdots & \ddots & \vdots\\ [0] & [0] & \cdots & [0]\end{bmatrix}\quad \left[\tilde{C}\right]=\begin{bmatrix}[0] & [0] & \cdots & [0]\\ [0] & \frac{a_1\left[\bar{\Lambda}\right]}{b_1} & \cdots & [0]\\ \vdots & \vdots & \ddots & \vdots\\ [0] & [0] & \cdots & \frac{a_m\left[\bar{\Lambda}\right]}{b_m}\end{bmatrix}\quad \left[\tilde{K}\right]=\begin{bmatrix}[K](1+\sum_{k=1}^{m}a_k) & -a_1\left[\bar{R}\right] & \cdots & -a_m\left[\bar{R}\right]\\ -a_1\left[\bar{R}\right]^T & a_1\left[\bar{\Lambda}\right] & \cdots & [0]\\ \vdots & \vdots & \ddots & \vdots\\ -a_m\left[\bar{R}\right]^T & [0] & \cdots & a_m\left[\bar{\Lambda}\right]\end{bmatrix} \tag{6}$$

with

$$\left[\bar{K}\right]=G^{\infty}[K]\quad [K]=[R][\Lambda][R]^T\quad \left[\bar{\Lambda}\right]=G^{\infty}[\Lambda]\quad \left[\bar{R}\right]=G^{\infty}[R][\Lambda]\quad \{z\}=\left[\bar{R}\right]^T\{\hat{z}\} \tag{7}$$

where $[\Lambda]$ is a diagonal matrix of the nonzero eigenvalues of matrix $[K]$, and the corresponding normalized eigenvectors form the columns of matrix $[R]$.

The cylindrical damper shown in figure 4 can be regarded as a system with two degrees of freedom. The mass and stiffness matrices in (5) can be defined as:

$$[M]=\begin{bmatrix}m_c & \\ & m_s\end{bmatrix}\quad [K]=\frac{G^{\infty}\pi(R_1+R_2)L}{R_2-R_1}\begin{bmatrix}1 & -1\\ -1 & 1\end{bmatrix}=cG^{\infty}\begin{bmatrix}1 & -1\\ -1 & 1\end{bmatrix} \tag{8}$$

where m_c and m_s are respectively the mass of the core rod and the sleeve. Using equation (7), it can be derived:

$$\bar{\Lambda}=G^{\infty}\Lambda=2cG^{\infty}$$

$$\left[\bar{R}\right]=G^{\infty}[R]\Lambda=2cG^{\infty}\begin{Bmatrix}\sqrt{2}\\ -\sqrt{2}\end{Bmatrix} \tag{9}$$

So according to (6), the coefficient matrices are:

$$[M]=\begin{bmatrix}m_c & & & \\ & m_s & & \\ & & 0 & \\ & & & 0\end{bmatrix}\quad [C]=cG^{\infty}\begin{bmatrix}0 & & & \\ & 0 & & \\ & & \frac{2a_1}{b_1} & \\ & & & \frac{2a_2}{b_2}\end{bmatrix}\quad [K]=cG^{\infty}\begin{bmatrix}1+a_1+a_2 & -(1+a_1+a_2) & -a_1\sqrt{2} & -a_2\sqrt{2}\\ -(1+a_1+a_2) & 1+a_1+a_2 & a_1\sqrt{2} & a_2\sqrt{2}\\ -a_1\sqrt{2} & a_1\sqrt{2} & 2a_1 & \\ -a_2\sqrt{2} & a_2\sqrt{2} & & 2a_2\end{bmatrix} \tag{10}$$

We choose ZN-1 rubber as the viscoelastic material, and the parameters of Biot model are fitted as those in table 2:

Parameter	G^{∞}	a_1	a_2	b_1	b_2
Value	5.0013×10^5	2.8438	35.6028	830.1878	13758

Table 2. Fitting parameters of Biot model for ZN-1 at 30°C.

2.2. Vibration control of single mode

Since the rods in the planar Kagome truss are modeled by beam-like elements while the cylindrical dampers can be regarded as a kind of rod element that can not suffer bending and torsion moments, the fraction of axial modal strain energy should be used here. The total modal strain energy of the *i*th mode is written as:

$$E_i = \frac{1}{2}\{\phi\}_i^T [K]\{\phi\}_i \tag{11}$$

where $\{\phi\}_i$ is the mode shape of *i*th mode, and [*K*] is the global stiffness matrix of the structure.

The element *j*'s axial strain energy due to mode *i* is denoted by:

$$E_{ij}^a = \frac{EA}{2L^3}\left(\Delta x_j^1 \Delta\phi_{ij}^1 + \Delta x_j^2 \Delta\phi_{ij}^2 + \Delta x_j^3 \Delta\phi_{ij}^3\right)^2 \tag{12}$$

Here *L* is the length of the truss element. The quantities Δx_j^k, *k=1,2,3* are the components of node coordinate differences of element *j*. $\Delta\phi_{ij}^k$, *k=1,2,3* are the component of node differences in the mode shapes $\{\phi\}_i$ of element *j*. The fraction of axial modal strain energy (FAMSE in short) of element *j* in mode *i* is then define as:

$$\delta_{ij} = \frac{E_{ij}^a}{E_i} \tag{13}$$

This quantity will be used as the effective index to determine which rods are to be replaced by the dampers. Assuming that the damping of the structure is of Rayleigh type and the modal damping ratios of both the first and second modes are 1%, then damping ratios of other modes can be calculated. For modes 1 to 3, we choose 8 rods in the planar Kagome truss, which takes only 0.51% of the total rods in the planar Kagome truss, with significant FAMSE to be replaced by dampers. For the first mode and the second mode, the locations of dampers are the same. The placements of the dampers for every mode are shown in figure 6. It can be seen that the optimal positions of the dampers are all in the constrained boundaries.

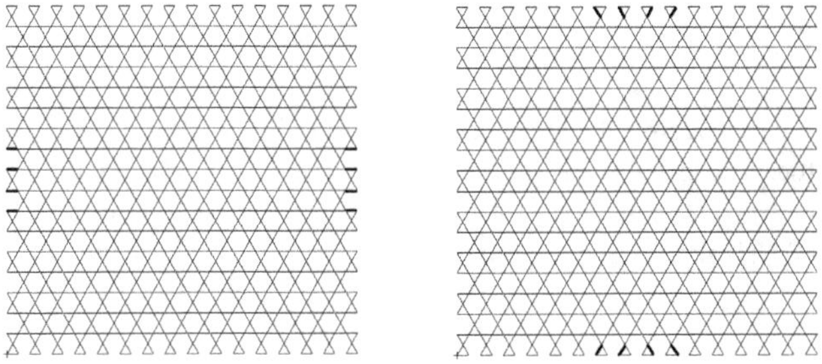

(a) (b)

Figure 6. The location of the dampers in the design for different controlled modes. The heavy solid line segments represent the dampers (a) for modes 1 and 2; (b) for mode 3.

The results of complex eigenvalue analysis are listed in table 3 for the vibration control on mode 1 and 2 and in table 4 for the vibration control on mode 3. It is noticed that, because the stiffness of damper is lower than the Kagome truss, the raise of the damping ratio is at the cost of losing stiffness of the structure.

Mode #	Damping ratio	Increment of damping ratio	Frequency(Hz)	Decrement of frequency(%)
1	0.0300	0.0200	112	5.08
2	0.0311	0.0211	208	4.59
3	0.0118	0.0016	231	0.43

Table 3. The result of complex modal analysis under the dampers placement design for mode 1&2.

Mode #	Damping ratio	Increment of damping ratio	Frequency(Hz)	Decrement of frequency(%)
1	0.0243	0.0143	113	4.24
2	0.0104	0.0004	217	0.46
3	0.0309	0.0207	222	4.31

Table 4. The result of complex modal analysis under the dampers placement design for mode 3.

2.3. Broad-bandwidth vibration control

In practice, the structure often suffers excitation with broad-bandwidth frequency, so a control method which can suppress the vibration of several modes is introduced below. Here FAMSE is continuously used as an index to optimize the location of the dampers. The target is to search the truss elements with remarkably large FAMSE among all the target modes. According to the excitation frequency band in the case studied here, the target modes are chosen to be the first six modes. In this paper the number of dampers used is 20, which takes only 1.26% of total rods in the planar Kagome truss.

The total FAMSE in mode i of the N=20 chosen elements (rods) is defined as:

$$\delta_i = \sum_j^{20} \delta_{ij},\ i = 1,\ldots 6 \tag{14}$$

Let e=minδ_i (i varies from 1 to 6) indicate the minimum of total FAMSE from mode 1 to mode 6 for the 20 chosen elements. The target of the optimization is to maximize e by selecting different combination of 20 elements. To simplify the optimization procedure, total FAMSE of all the six modes in element j is defined as:

$$\delta_j = \sum_i^{6} \delta_{ij}. \tag{15}$$

In the optimization procedure, 100 elements that have the top largest δ_j are first determined from 1584 truss rods. Then, 20 elements will be selected from the 100 elements. Assuming the placement of the dampers is symmetrical, and the computation of the optimization search can get further reduction. As above, the optimal positions of the dampers are in the constrained boundaries (see figure 7(a)). For the purpose of comparison, the dampers are also placed in a relatively uniform style as shown by figure 7(b).

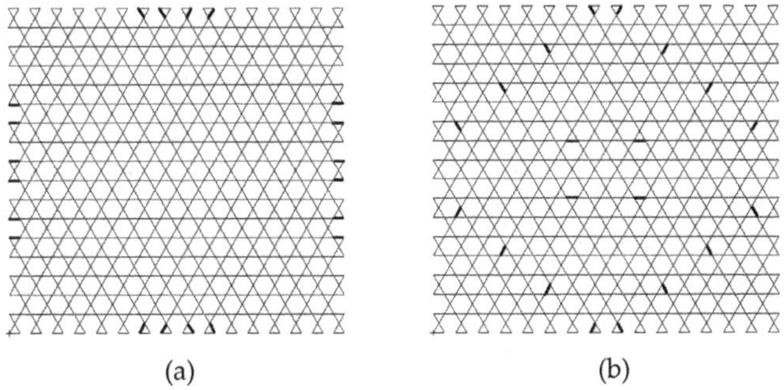

Figure 7. (a) The position of the dampers using FAMSE method; (b) The positions of the dampers in a relatively uniform style. The heavy solid line segments represent the dampers.

Mode#	Damping ratio	Increment of damping ratio	Frequency(Hz)	Decrement of frequency(%)
1	0.0460	0.0360	106	10.17
2	0.0329	0.0229	207	5.05
3	0.0382	0.0281	218	6.03
4	0.0260	0.0141	310	2.52
5	0.0365	0.0237	343	3.65
6	0.0321	0.0186	376	3.34

(a)

Mode#	Damping ratio	Increment of damping ratio	Frequency(Hz)	Decrement of frequency(%)
1	0.0243	0.0143	104	11.86
2	0.0201	0.0101	201	7.80
3	0.0226	0.0124	211	9.05
4	0.0235	0.0116	285	10.38
5	0.0258	0.0130	323	9.27
6	0.0248	0.0113	358	7.97

(b)

Table 5. The results of complex modal analysis with the dampers.(a) placement of dampers using FAMSE method; (b) Placement of dampers in a relatively uniform style.

The results of the damping ratios and the natural frequencies of the Kagome structure with dampers through complex eigenvalue analysis are shown table 5. It can be seen that the damping ratios of mode 1 to 6 have been considerably increased. The optimized placement of the dampers using FAMSE method achieves better results in comparison with those by a relatively uniform placement of the dampers (compare data in table 5(a) and 5(b)). Beside the smaller increment of damping ratios, the stiffness of the Kagome structure is significantly reduced as shown by the sharp decrease of natural frequencies of all six modes.

To validate this passive vibration control method, responses of the Kagome structure during the excitation of vertically downward distributed white noise load on the solid face sheet are computed. The noise is with finite bandwidth of 2000Hz. The single-side power spectrum density is $10N^2/Hz$.

The lateral responses of the Kagome structure, observed at the point shown in figure 2(a), in the cases without dampers, with the dampers in placement as shown in figure 7(a) and in figure 7(b) are drawn in figure 8. It is found that the response amplitudes are considerably reduced with the installation of the dampers in the structure. The response amplitudes in the case with optimized placement of the dampers are even smaller than those in the case of the relatively uniform placement of the dampers.

From the spectra of the responses shown in figure 8(b), the effectiveness of the passive vibration control method proposed in this paper can be easily seen. The advantage of the optimal placement of the dampers over the relatively uniform placement of the dampers can also be observed. Furthermore, due to decrease of stiffness, the shift of the resonance peaks is detectable.

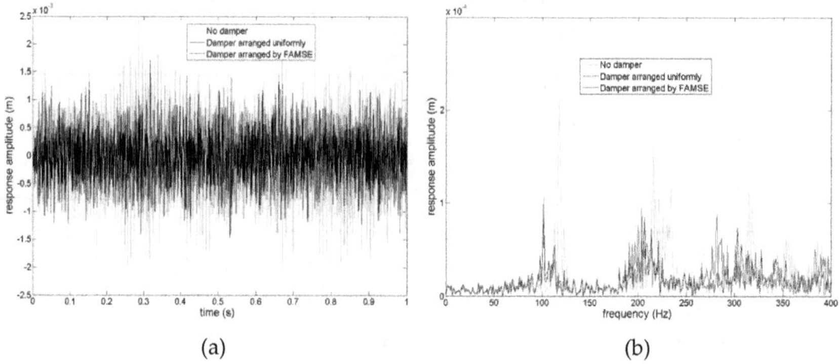

Figure 8. (a) The time history of response under the excitation of white noise; (b) Spectra of responses.

3. Active vibration control

3.1. Piezoelectric actuator

To realize active control, a rod-like piezoelectric actuators are adopted (see figure 9) to replace small number of rods in the planar Kagome truss of the Kagome structure. The actuator is mainly composed of a piezoelectric stack, outer casing, preloading spring and two link rods. A sphere joint is used to prevent the piezoelectric stack from bending and torsion moments.

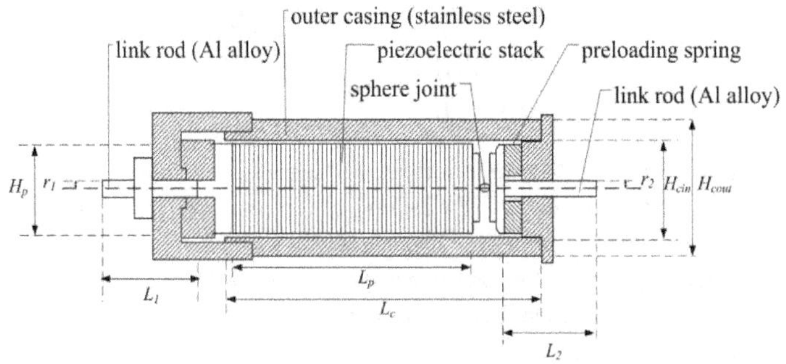

Figure 9. Sketch of the piezoelectric actuator. The parameters are given as: L_1=L_2=8mm, L_p=30mm, L_c=40mm, H_p=10mm, r_1=r_2=1.3mm, H_{cin}=12mm, H_{cout}=15mm.

After combining the finite element model of the two link rods with piezoelectric stack and condensing the internal degrees of freedom in two connected points, the finite element model of the actuator in its local coordinate can be written as:

$$\left[M^{\mathrm{e}}\right]\left\{\ddot{q}^{\mathrm{e}}\right\}+\left[K^{\mathrm{e}}\right]\left\{q^{\mathrm{e}}\right\}=\left\{f^{\mathrm{e}}\right\}-\left[K_{\mathrm{c}}^{\mathrm{e}}\right]\phi \tag{16}$$

where $[M^e]$ is the mass matrix, $[K^e]$ the stiffness matrix, $[K_c^e]$ the coupling matrix connecting the mechanical variables $\{q^e\}$and the electrical variables ϕ. The matrices are in the following forms [15]:

$$\left[M^{\mathrm{e}}\right]=\begin{bmatrix}\frac{1}{2}m_{\mathrm{p}}+m_1 & \\ & \frac{1}{2}m_{\mathrm{p}}+m_3+m_{\mathrm{t}}\end{bmatrix} \quad \left[K^{\mathrm{e}}\right]=\frac{k_1k_2k_3}{k_1k_2+k_2k_3+k_1k_3}\begin{bmatrix}1 & -1\\ -1 & 1\end{bmatrix} \quad \left[K_{\mathrm{c}}^{\mathrm{e}}\right]=nd_{33}k_2\begin{bmatrix}-1\\ 1\end{bmatrix} \tag{17}$$

Here m_p ,m_1, m_3 are masses of piezoelectric stack and two link rods respectively, and m_t is the total mass of outer casing and spring. The masses of other components are ignored. k_1 and k_3 are the stiffness of the two link rods while k_2 is the equivalent stiffness of the piezoelectric stack, the outer casing and the spring. n is the number of disks in the stack and piezoelectric strain constant d_{33} means induced strain in axial direction of the stack per unit electric field applied in axial direction. Under the designed size, the actuator can match the truss rod both in length and in axial stiffness. Choose actuate voltage V= $-\phi$, the control force of the actuator is given by:

$$u^{\mathrm{e}}=nd_{33}k_2V \tag{18}$$

3.2. Independent modal space control

After incorporating the actuator model into the Kagome structure, the global dynamic equations of the actuated Kagome structure without excitation could be written as:

$$\begin{aligned} &[M]\{\ddot{q}\}+[D]\{\dot{q}\}+[K]\{q\}=[B_s]\{u\} \\ &\{y_d\}=[C_d]\{q\} \\ &\{y_v\}=[C_v]\{\dot{q}\} \end{aligned} \tag{19}$$

where $\{q\}$, $\{\dot{q}\}$ and $\{\ddot{q}\}$ are the N-length vectors of displacement, velocity and acceleration, respectively , with N being the degree of freedoms of the structure. $[M]$, $[D]$ and $[K]$ are $N \times N$ mass, damping and stiffness matrices, respectively. $[B_s]$ is $N \times s$ spatial coupling matrix relative to the s-length physical control force vector $\{u\}$. The r-length output vectors $\{y_d\}$and $\{y_v\}$ are related to the displacement and velocity vectors through the matrices $[C_d]$ and $[C_v]$, respectively. s is the number of actuators while r is number of sensors. In IMSC method (see below), r and s equal the number of modes to be controlled.

In structure vibration, the lower modes often possess the most energy and play more critical roles. Meirovitch [16] developed an independent modal space control (IMSC) method aimed to control the vibration of the lower modes. In IMSC, the dynamic equation of structure is decoupled using modal analysis and then modal control forces are determined by various control laws. The design of control law and the placement of actuators are two independent steps in IMSC.

In IMSC, the modes are divided into two parts: controlled modes and residual modes. In what follows, the subscripts c and r will refer to the controlled and the residual modes respectively.

Assume that $[\psi]$ is the modal matrix composed of the controlled modes $[\psi_c]$ and residual modes $[\psi_r]$. Let:

$$\{q\}=[\psi]\{q_m\}=[\psi_c]\{q_{mc}\}+[\psi_r]\{q_{mr}\} \tag{20}$$

where $\{q_m\}$ is the vector of modal displacement consists of the controlled part $\{q_{mc}\}$ and the residual part $\{q_{mr}\}$. Substitute (20) into (19), and left multiple the first equation of (19) by $[\psi]^T$ to yield:

$$\begin{aligned} &\begin{Bmatrix} \ddot{q}_{mc} \\ \ddot{q}_{mr} \end{Bmatrix}+[D_d]\begin{Bmatrix} \dot{q}_{mc} \\ \dot{q}_{mr} \end{Bmatrix}+[\Lambda]\begin{Bmatrix} q_{mc} \\ q_{mr} \end{Bmatrix}=\begin{bmatrix} [B_{mc}] \\ [B_{mr}] \end{bmatrix}\{u\} \\ &\{y_d\}=[C_{dmc}]\{q_{mc}\}+[C_{dmr}]\{q_{mr}\} \\ &\{y_v\}=[C_{vmc}]\{\dot{q}_{mc}\}+[C_{vmr}]\{\dot{q}_{mr}\} \end{aligned} \tag{21}$$

where $[D_d]=diag(2\xi_i\omega_i)$, $[\Lambda]=diag(\omega_i^2)$, $[B_{mc}]=[\psi_c]^T\times[B_s]$, $[B_{mr}]=[\psi_r]^T\times[B_s]$, $[C_{dmc}]=[C_d]\times[\psi_c]$, $[C_{dmr}]=[C_d]\times[\psi_r]$, $[C_{vmc}]=[C_v]\times[\psi_c]$, $[C_{vmr}]=[C_v]\times[\psi_r]$, ξ_i and ω_i are the damping ratio and the circle frequency of mode i. Define the modal control forces of the controlled modes as:

$$\{f_c\}=[B_{mc}]\{u\} \tag{22}$$

and the equation of the controlled modes can be written as:

$$\{\ddot{q}_{mc}\} + [D_{dc}]\{\dot{q}_{mc}\} + [\Lambda_c]\{q_{mc}\} = \{f_c\} \tag{23}$$

where $[\Lambda_c]$ and $[D_{dc}]$ are diagonal matrices like $[\Lambda]$ and $[D_d]$ but only contain the elements corresponding to the controlled modes.

In IMSC, the modal control force vector $\{f_c\}$ is first determined then physical control force vector $\{u\}$ will be calculated from $\{f_c\}$. Equation (23) consists of n independent equations for each of the controlled mode. For the ith mode the equation is given as:

$$\ddot{q}_i + 2\xi_i\omega_i\dot{q}_i + \omega_i^2 q_i = f_i \tag{24}$$

when modal displacement and velocity are obtained, the control force will be taken in the form:

$$f_i = -[G_i]\{q_i \quad \dot{q}_i\}^{\mathrm{T}} \tag{25}$$

The modal velocity feedback control will be adopted:

$$[G_i] = [0 \quad 2\omega_i(\xi_{ic} - \xi_i)] = [0 \quad g_{vi}] \tag{26}$$

while ξ_{ic} is the designed damping ratio of mode i. If modal control force $\{f_c\}$ is determined, the physical control force vector $\{u\}$ can be calculated by using:

$$\{u\} = [B_{mc}]^{-1}\{f_c\} \tag{27}$$

In general, the modal states in (25) are not directly available. So an observer is needed to reconstruct the modal states from the physical signals measured by sensors. To do so, the part of equation (21) which governs the controlled modes is reformulated in state-space form as:

$$\begin{aligned}\{\dot{x}_c\} &= [A_c]\{x_c\} + [B_c]\{u\} \\ \{y_c\} &= [C_c]\{x_c\}\end{aligned} \tag{28}$$

where $[A_c]$ is $2n\times2n$ system matrix, $[B_c]$ is $2n\times n$ input matrix and $[C_c]$ is $n\times2n$ output matrix in the following forms:

$$\{x_c\} = \{\{q_{mc}\} \quad \{\dot{q}_{mc}\}\}^{\mathrm{T}} \quad [A_c] = \begin{bmatrix} 0 & [I] \\ -[\Lambda_c] & -[D_{dc}] \end{bmatrix} \quad [B_c] = \begin{bmatrix} [0] \\ [B_{mc}] \end{bmatrix} \quad [C_c] = \begin{bmatrix} [C_{dmc}] & 0 \\ 0 & [C_{dvc}] \end{bmatrix} \tag{29}$$

with $[I]$ being the identity matrix. The modal control force vector in state space can be written as:

$$\{F_c\} = -[G]\{x_c\} \tag{30}$$

where

$$[G]=\begin{bmatrix} 0 & 0 & \cdots & 0 & g_{v1} & 0 & \cdots & 0 \\ 0 & 0 & \cdots & 0 & 0 & g_{v2} & \cdots & 0 \\ \vdots & \vdots & \ddots & \vdots & \vdots & \vdots & \ddots & \vdots \\ 0 & 0 & \cdots & 0 & 0 & 0 & \cdots & g_{vn} \end{bmatrix} \tag{31}$$

Since modal states are not directly available, the estimated modal states $\{\hat{x}_c\}$ will be used instead, so $\{F_c\}$ is rewritten as:

$$\{F_c\}=-[G]\{\hat{x}_c\} \tag{32}$$

To get the estimated modal states from the measured signals, an observer, e.g., Kalman filter, is adopted in this paper to identify the modal displacements and velocities. The Kalman filter dynamics can be described as:

$$\{\dot{\hat{x}}_c\}=[A_c]\{\hat{x}_c\}+[B_c]\{u\}+[H_c](\{y_c\}-[C_c]\{\hat{x}_c\}) \tag{33}$$

The observer gain matrix $[H_c]$ may be determined by solving the following Riccati equation:

$$[P][A_c]^{\mathrm{T}}+[A_c][P]-[P][C_c]^{\mathrm{T}}[V]^{-1}[C_c][P]+[W]=0 \tag{34}$$

where $[P]$ is the solution, while $[W]$ and $[V]$ are, respectively, the covariance intensity matrices of process and measurement noises. The observer gain $[H_c]$ is obtained from:

$$[H_c]=[P][C_c]^{\mathrm{T}}[V]^{-1} \tag{35}$$

Since IMSC uses the reduced model, the situation may occur that the damping ratios of the controlled modes increase while the damping ratios of the residual modes may decrease (spillover). When the damping ratio of any residual modes drops below zero, the controlled system will become instable. To analyze the stability of the controlled system, the governing equation of the whole system with control will be developed. To get the equation of the whole system, the equations of residual modes in state-space is obtained as that in (28). Now the equations of the system with the observer for ISMC can be written as:

$$\begin{aligned} &\{\dot{x}_c\}=[A_c]\{x_c\}+[B_c]\{u\} \\ &\{\dot{x}_r\}=[A_r]\{x_r\}+[B_r]\{u\} \\ &\{y_c\}=[C_c]\{x_c\} \\ &\{\dot{\hat{x}}_c\}=[A_c]\{\hat{x}_c\}+[B_c]\{u\}+[H_c](\{y_c\}-[C_c]\{\hat{x}_c\}) \end{aligned} \tag{36}$$

To analyze the stability of the controlled system, the observer error vector, $\{e_c\}=\{x_c\}-\{\hat{x}_c\}$, is defined, and the governing equation is rewritten in the matrix form as:

$$\begin{pmatrix}\{\dot{x}_c\}\\ \{\dot{x}_r\}\\ \{\dot{e}_c\}\end{pmatrix}=\begin{pmatrix}[A_c]-[B_c^*][G] & 0 & [B_c^*][G]\\ -[B_r^*][G] & [A_r] & [B_r^*][G]\\ 0 & -[H_c][C_r] & [A_c]-[H_c][C_c]\end{pmatrix}\begin{pmatrix}\{x_c\}\\ \{x_r\}\\ \{e_c\}\end{pmatrix} \tag{37}$$

where

$$[B_c^*]=\begin{bmatrix}0\\ [I]\end{bmatrix}\quad [B_r^*]=\begin{bmatrix}0\\ [B_{mr}][B_{mc}]^{-1}\end{bmatrix} \tag{38}$$

The controlled system is stable when all the eigenvalues of system matrix have negative real parts. The term $[H_c][C_r]$ contaminates the sensor output by the residual modes, known as observation spillover. Meanwhile, the residual modes are excited via the term $[B^*_r][G]$ called control spillover. Spillover can make residual modes unstable, that is, the eigenvalues corresponding to these modes will have positive real parts. Since the control spillover solely has no effect on the eigenvalues of the controlled system, spillover can be avoided when the observation spillover is suppressed. To alleviate the observation spillover, equations (34) and (35) will be rewritten in the form [17]:

$$\begin{aligned}&[P][A_c]^T+[A_c][P]-[P][C_c]^T[\tilde{V}]^{-1}[C_c][P]+[W]=0\\ &[H_c]=[P][C_c]^T[\tilde{V}]^{-1}\end{aligned} \tag{39}$$

where:

$$[\tilde{V}]=[V]+[C_r][V_1][C_r]^T \tag{40}$$

$[V_1]$ is a weighting matrix which allows to desensitize the estimated states to the residual modes. The block diagram of the controlled system is shown in figure 10:

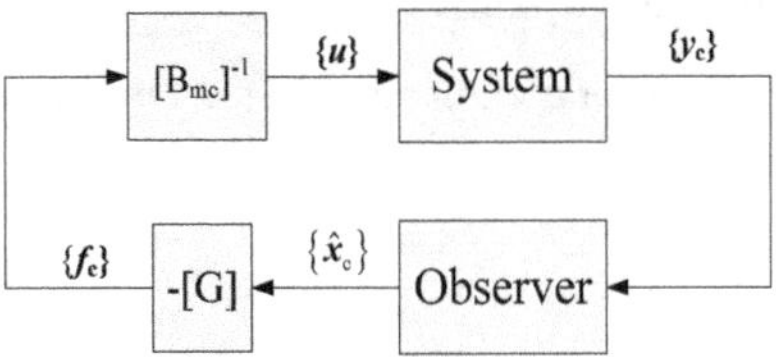

Figure 10. The block diagram of close-loop system.

3.3. Actuator placement optimization in ISMC method

For the placement of actuators in IMSC, Baruh [18] pointed out that the energy going into the controlled modes is independent of the actuator locations but the energy pumped into the uncontrolled modes depends on the actuator locations. Lindberg Jr [19] indicated that improperly chosen locations can make the actuator forces grow to infinity, that is, the system became uncontrollable. So the singular values of input matrix can be used to optimize the actuator locations in order to reduce the control effort and suppress spillover. Lammering [20] used the trace of the input matrix as the objective function to minimize the control energy and spillover, while electric potentials (something like MSE) was introduced as an effectiveness index to get a subset of suitable locations.

3.3.1. *Objective function for optimization*

From (26) it is known that the system is controllable if and only if $[B_{mc}]$ is of full rank. Otherwise, the modal control forces can not be realized physically. It is also known that the goal of actuator optimization is not just simply to meet the condition of controllability. The extent of controllability that can provide a quantitative measurement on the physical actuator force vector needed for a given modal control force vector is also of great interest [21]. For a modal control force vector designed to achieve the expected control effect, it is highly expected that through actuator placement optimization, smaller physical control forces will be required to realize a more controllable system. Of course, the physical control forces will grow to infinity if the system is uncontrollable.

Lammering [20] proposed an optimization method for the actuator placement by using the trace of matrix $\left(\left[B_{\mathrm{mc}}\right]^{\mathrm{T}}\right)^{-1}\left[B_{\mathrm{mc}}\right]^{-1}$ as the criterion in order to make the total control effort, i.e., the square of the norm of the physical control force $\{u\}^{\mathrm{T}}\{u\}$ to become minimum. The corresponding objective function is thus given by:

$$tr\left(\left(\left[B_{\mathrm{mc}}\right]^{\mathrm{T}}\right)^{-1}\left[B_{\mathrm{mc}}\right]^{-1}\right) \rightarrow \min \tag{41}$$

Different from the above method, the minimization of the maximal physical control force will be the objective in the present paper. Meanwhile, the singular values of the input matrix $[B_{mc}]$ are used to measure the controllability.

The singular value decomposition of input matrix $[B_{mc}]$ is done as the following:

$$\left[B_{\mathrm{mc}}\right]_{n\times n} = \left[U\right]_{n\times n}\left[S\right]_{n\times n}\left[Q\right]_{n\times n} \tag{42}$$

with $[U]^{\mathrm{T}}[U]= [Q]^{\mathrm{T}}[Q]=[I]$, $[S]=\begin{bmatrix}[\Sigma] & 0\\ 0 & 0\end{bmatrix}$ and $[\Sigma]=\mathrm{diag}[\sigma_1, \sigma_2,\ldots, \sigma_m]$, where σ_i is the ith singular value of $[S]$. Here assume the system is controllable, so $[B_{mc}]$ and $[S]$ are of full rank, and $[S] = [\Sigma]$.

To illustrate the relationship among modal control forces, singular values and physical control forces, we introduce a new modal coordinate $\{\eta\}$ through the coordinate transformation [21]:

$$\{q\} = [\psi][U]\{\eta\} = [\psi^*]\{\eta\} \tag{43}$$

where $[\psi^*]$ is also a modal matrix of the system which obviously satisfies the orthogonality condition with respect to the mass matrix $[M]$. Obviously $\{\eta\}$ can also be divided into the controlled modes $\{\eta_c\}$ and the residual modes $\{\eta_r\}$. Since the dynamical equation is decoupled in the modal coordinates, only the controlled part will be considered below. By substituting (43) into (19) a new form of equation governing the controlled modes can be obtained as:

$$\{\ddot{\eta}_c\} + [D_{dc}]\{\dot{\eta}_c\} + [\Lambda_c]\{\eta_c\} = [S]\{u^*\} \tag{44}$$

with $\{u^*\} = [Q]^T\{u\}$. Because $[Q]^T[Q]=[I]$, $\{u^*\}$and $\{u\}$ can be considered to be equivalence. Denote the right part of equation (44) as:

$$\{f_c^*\} = [S]\{u^*\} \tag{45}$$

Of course $\{f_c^*\}$ is still the modal control force. From (45) the follow relation between the modal control force $\{f_c^*\}$ and physical control force $\{u^*\}$ can be derived as:

$$\{u^*\} = [S]^{-1}\{f_c^*\} = diag[1/\sigma_1 \quad 1/\sigma_2 \quad \cdots \quad 1/\sigma_n]\{f_c^*\} \tag{46}$$

The relationship between the ith element of $\{u^*\}$ and $\{f_c^*\}$ is:

$$u_i^* = f_{ci}^* / \sigma_i \tag{47}$$

Here it is supposed that the elements in $\{f_c^*\}$ are within the same fixed range, which is supposed from $-b$ to $+b$. The maximal value u^*_{max} in $\{u^*\}$, on the other hand, will be:

$$|u_{max}^*| = b/\sigma_{min} \tag{48}$$

In this way, the amplitude of maximal element in $\{u^*\}$ depends on the minimal singular value σ_{min} of $[B_{mc}]$. If the minimal singular value of $[B_{mc}]$ is larger, the maximal physical control force will be smaller. So the optimization objective function becomes:

$$\sigma([B_{mc}])\Big|_{min} \rightarrow \max \tag{49}$$

3.3.2. *Selection of candidate locations though modal strain energy*

The computational cost of optimization may become too high to be affordable if all the combinations are calculated in order to find the most optimal combination that meets criterion (49). However, the computational cost of optimization can be significantly reduced when a proper subset of combinations may be first defined. In this part, FAMSE will also be used as the index to determine the subset. And it will be shown below that MSE solely can not properly determine the locations of actuators in the present case.

To investigate the suitability of FAMSE in IMSC method, the case to control one mode vibration is first studied. According to IMSC method, only one actuator is needed in the case, the input matrix [$B_{\rm mc}$] now has only one element denoted by:

$$B_{\rm mc} = \{\psi_i\}^{\rm T}[B_{\rm s}] = \frac{\Delta x_j^1 \Delta\phi_{ij}^1 + \Delta x_j^2 \Delta\phi_{ij}^2 + \Delta x_j^3 \Delta\phi_{ij}^3}{L} \tag{50}$$

while the axial MSE of the element j for the controlled mode i is given by:

$$E_{ij}^{\rm a} = \frac{EA}{2L^3}\left(\Delta x_j^1 \Delta\phi_{ij}^1 + \Delta x_j^2 \Delta\phi_{ij}^2 + \Delta x_j^3 \Delta\phi_{ij}^3\right)^2 = \frac{EA}{2L}\left(B_{\rm mc}\right)^2 \tag{51}$$

Assume $B_{\rm mc}$ is positive, the relationship between $B_{\rm mc}$ and axial MSE is:

$$B_{\rm mc} = \sqrt{\frac{2L}{EA}E_{ij}^{\rm a}} \tag{52}$$

So according to the magnitude of the value E_{ij}^{a}, the most proper location of the actuator can be chosen, that is, the axial MSE can work as an index for the actuator placement in the case.

Let us now examine the case of vibration control on multiple modes. Assume that the number of modes to be controlled is n by using the same number of actuators. The input matrix [$B_{\rm mc}$] is in the form:

$$[B_{\rm mc}] = \sqrt{\frac{2L}{EA}}\begin{bmatrix} \pm\sqrt{E_{11}^{\rm a}} & \pm\sqrt{E_{12}^{\rm a}} & \cdots & \pm\sqrt{E_{1n}^{\rm a}} \\ \pm\sqrt{E_{21}^{\rm a}} & \pm\sqrt{E_{22}^{\rm a}} & \cdots & \pm\sqrt{E_{2n}^{\rm a}} \\ \vdots & \vdots & \ddots & \vdots \\ \pm\sqrt{E_{n1}^{\rm a}} & \pm\sqrt{E_{n2}^{\rm a}} & \cdots & \pm\sqrt{E_{nn}^{\rm a}} \end{bmatrix} \tag{53}$$

where the signs of the elements in each row of the input matrix are determined according to the corresponding mode shape.

Below, through two simple examples, it will be demonstrated that if the locations of actuators are optimized just in accordance to the magnitudes of the values of axial MSE in the Kagome structure, the controllability of the system, or the condition of the full rank of the corresponding input matrix [$B_{\rm mc}$], can not be guaranteed. Let us investigate the control of the vibration accounting for the second and the third modes of the Kagome structure with two actuators by using IMSC. In figure 11, two kinds of placement of actuators are shown. It is noted that even though the axial MSE of the elements in the positions of the actuators are of the equal values for each mode due to the symmetry of the structure, the rank of the input matrix [$B_{\rm mc}$] differs greatly in the two cases.

In case (a) of figure 11, the input matrix has the form:

$$[B_{\mathrm{mc}}]=\sqrt{\frac{2L}{EA}}\begin{pmatrix}-\sqrt{E_{21}^{\mathrm{a}}} & \sqrt{E_{22}^{\mathrm{a}}}\\ -\sqrt{E_{31}^{\mathrm{a}}} & \sqrt{E_{32}^{\mathrm{a}}}\end{pmatrix}=\begin{pmatrix}-0.0134 & 0.0134\\ -0.0137 & 0.0137\end{pmatrix} \tag{54}$$

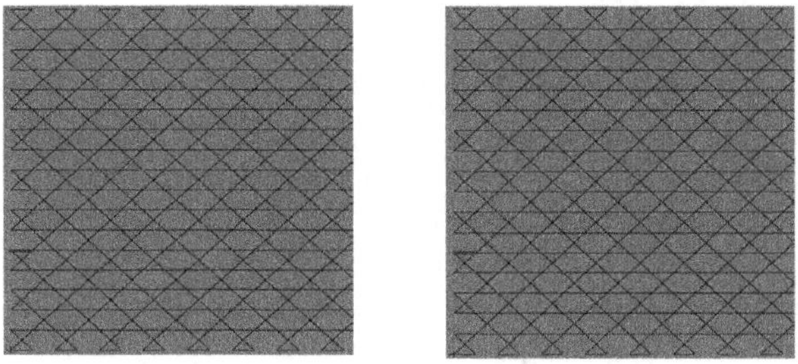

(a) (b)

Figure 11. Different locations of the actuators on the planar Kagome used to control mode 2 and 3. The heavy red line segments represent the actuators.

It is obvious that the matrix $[B_{mc}]$ is singular and the system becomes uncontrollable. Similar phenomenon was also found in [20]. On the other hand, the input matrix for case (b) in figure 11 is in the form:

$$[B_{\mathrm{mc}}]=\sqrt{\frac{2L}{EA}}\begin{pmatrix}-\sqrt{E_{21}^{\mathrm{a}}} & \sqrt{E_{22}^{\mathrm{a}}}\\ -\sqrt{E_{31}^{\mathrm{a}}} & -\sqrt{E_{32}^{\mathrm{a}}}\end{pmatrix}=\begin{pmatrix}-0.0134 & 0.0134\\ -0.0137 & -0.0137\end{pmatrix} \tag{55}$$

which is in full rank.

By examining the mode shapes of mode 2 and 3 as given in figure 3, it is seen that in case (a), the two actuators are in the locations of anti-phase for both of the two modes, while in case (b), the two actuators are in the positions of anti-phase for mode 2 but in the positions of in-phase for mode 3. So the consequence of the actuator placement on the effect of control is also close related to the mode shapes of the modes to be controlled beside the values of axial MSE.

3.3.3. A two-step optimization method

Based on above analysis, a two-step optimization method for the placement of actuators in the Kagome structure is proposed as follows:

Step 1. Denote the collection of N_i elements which contains the elements in the planar Kagome with the largest axial MSE of mode i by $\{U_i\}=\{ele_{1i},ele_{2i},\ldots,ele_{Ni}\}$. Since one actuator is used by IMSC method, an element is chosen from the subset $\{U_i\}$ for the control of mode i. When the total number of modes to be controlled is n, the number of candidate sets of actuator locations is $(N_i)^n$. Due to the existence of repeated elements in different $\{U_i\}$, the actual number is always less than $(N_i)^n$.

Step 2. Calculate the singular-values of the input matrix $[B_{mc}]$ of all the combinations, and the optimal combination should be the one that meet the objective function (49).

3.4. Numerical simulations

In this chapter the vibration control on the Kagome structure covering the bandwidth of the first six modes will be implemented. According to IMSC method, six actuators and six sensors are needed. Here, the accelerometers are used as sensors to measure the z-direction (out-of-plane) acceleration of the Kagome planar truss and the information of displacement and velocity can be derived through integration. The placement of sensors is determined by choosing the nodes where the peak of the mode shape of mode i, i=1~6, locates. The positions of sensors are shown in figure 12.

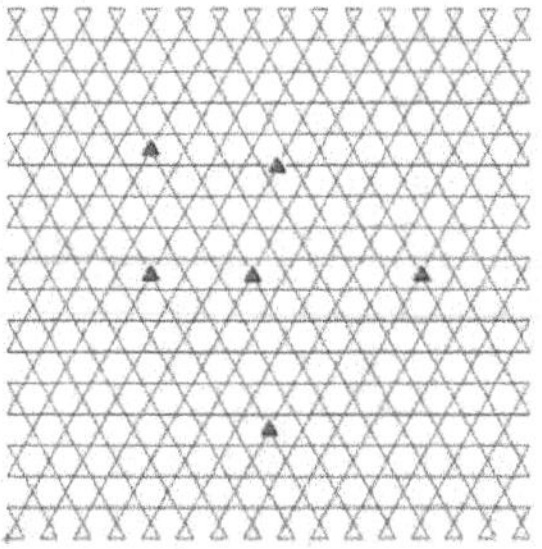

Figure 12. Locations of accelerometers in the planar Kagome (red triangles).

To obtain the optimal actuator locations, 12 elements in the planar Kagome of the Kagome structure which have the largest axial MSE in mode i are selected to form the subset $\{U_i\}$ (i=1~6). By using the two-step optimization method above, the optimized locations of actuators are obtained as shown in figure 13(a). For the purpose of comparison, the non-optimal locations of actuators chosen just in accordance with the same MSE and the optimized locations of actuators selected following the criterion in [20] from the same subsets are shown in figure 13(b) and 13(c) respectively. The corresponding singular-values of the input matrices for the three cases are listed below as: $[S]_a$=diag[0.0615,0.0448, 0.0334,0.0282,0.0218,0.0201],$[S]_b$=diag[0.0617,0.0480,0.0337,0.0287,0.0199,0.0107],$[S]_c$=diag[0.0570,0.0448,0.0339,0.0315,0.0297,0.0185].

It could be easily seen that the minimal singular-value in case (a) has the largest value, and that in case (c) has the second largest value. As demonstrated below, the minimal value of the singular values reflects the capability of control of the design system. To evaluate the results of the three types of actuator locations in figure 9, the damping ratios are calculated under the condition that with the same feedback gain matrices, [G] from (26), to achieve the desired damping ratios 0.35% for all six modes, while the original damping ratios of modes 1-6 is assume 1%. Because of coupling of the residual modes, the actual eigenvalues solved from equation (37) will be shifted away from the desired damping ratio for some controlled modes. Through eigenvalue analysis, the modal damping ratios of mode 1~6 under the three types of actuator locations are listed in table 3. It can be seen from table 3 that almost all the calculated damping ratios are 3.5%, while mode 4 is slightly smaller. Meanwhile, the damping ratios of all the residual modes are positive (not listed here), so the system is stable.

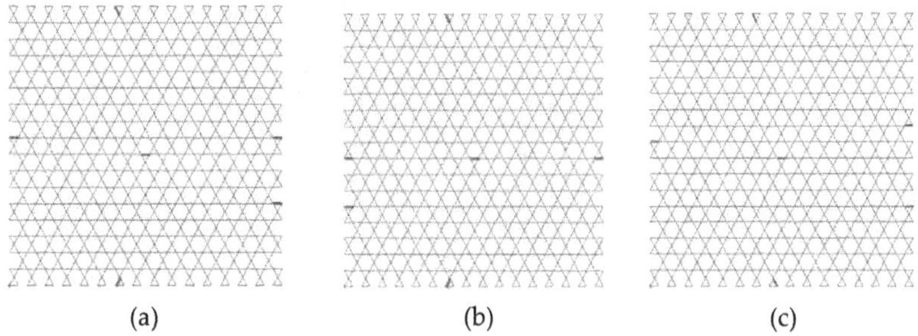

Figure 13. Locations of actuators in the planar Kagome. (a) Optimized by criterion (49). (b) A non-optimal just by using MSE. (c) Optimized by the criterion in [20].

To validate the efficiency of the present optimization method on the actuator placement in reducing physical control forces, the vibration control of the Kagome structure under the excitation of a vertically distributed noise load on the solid face sheet is simulated. The white noise is in a finite bandwidth of 1000Hz. The single-side power spectrum density is $0.3N^2/Hz$.

mode	damping ratio in case(a)	damping ratio in case(b)	damping ratio in case(c)
1	3.5%	3.5%	3.5%
2	3.5%	3.5%	3.5%
3	3.5%	3.5%	3.5%
4	3.45%	3.41%	3.22%
5	3.49%	3.50%	3.49%
6	3.49%	3.49%	3.49%

Table 6. The close-loop modal damping ratios of mode 1~6

The lateral responses of the Kagome structure, observed also at the triangular point shown in figure 2(a), are computed for four cases: without control and with control in the three types of actuator placements as shown in figure 14. It is found from figure 14(a) that the response amplitudes are considerably reduced through the control by using IMSC method. From the spectra of the responses shown in figure 14(b), the effectiveness of the control method can be more easily seen. The control effect under the three types of actuator locations is almost identical. This is also in agreement with what was demonstrated in table 6 and the finding in [18].

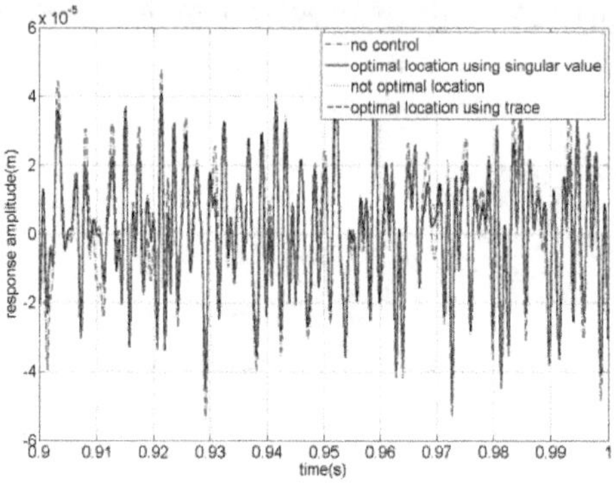

(a)

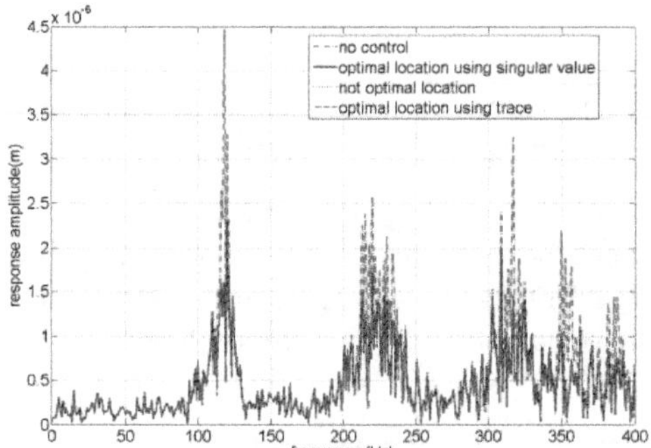

(b)

Figure 14. Comparison between uncontrolled and controlled responses under the excitation of white noise. (a) The time history of responses; (b) The spectra of responses.

From the feature of IMSC method, it is known that the main consequence of the types of actuator locations is to cause different requirement on the physical control forces provided by the actuators. To examine the achievement of the present method for the optimal placement of actuators, the root mean square (RMS) values and the maximal instantaneous values of the physical control forces of the six actuators for the three different cases of actuator locations shown in figure 13 are listed in table 7. The physical control forces of the actuator at the same position (the uppermost one) in the three cases are drawn in figure 15.It is easily seen that the actuator location (c) in figure 13 has a smallest RMS value of the physical forces, that is, an overall least control effort is required. The present optimal method, the actuator location (a) in figure 13, requires a least maximal instantaneous physical control force along with a quite comparable RMS value as that in the actuator location (c). The significant reduction of the maximal actuator force through the optimal placement of actuators has the advantage that the requirement on the capability of actuators is low. From an overall point of view, the present optimal method for the actuator placement with objective function (49) achieves a better result.

case	RMS value (N)	Maximal instantaneous value (N)
(a)	64.17	283.13
(b)	91.04	424.04
(c)	60.39	319.03

Table 7. RMS and maximal instantaneous values of physical forces

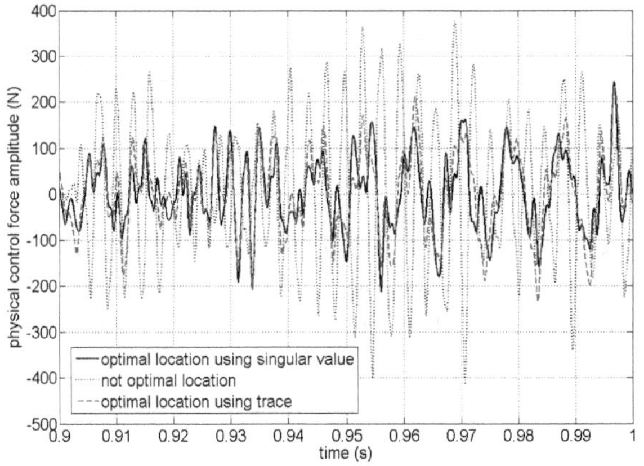

Figure 15. Comparison of the control force in the same actuator.

5. Conclusions

In this chapter, optimization of dampers/actuators placements in the passive/active vibration control of Kagome sandwich structure is studied. In the passive vibration control of Kagome sandwich structure, an optimal placement method of viscoelastic damper for both single mode control and broad-bandwidth control is raised based on FAMSE of the planar Kagome truss. At first, the damper placements for vibration control of mode 1 to mode 3 are carried out and only 0.51% rods in the planar Kagome truss are replaced by dampers in each case. Through complex modal analysis, it is demonstrated that the damping ratios have all gained an increment over 2% which shows the validity and effectiveness of the present method. For a broad-bandwidth vibration control, a method based on FAMSE is introduced to suppress the vibration in a bandwidth of the first six modes. Only 1.26% rods in the planar Kagome truss are replaced with dampers in this case. The increments of modal damping ratios are over 2% for the modes except for mode 4 and 6, whose increments are 1.41% and 1.86% respectively. Through time-domain simulations under a broad-bandwidth excitation, the reduction of the vibration responses and the spectrum peaks can be seen obviously. It is noticed that when rods in the planar Kagome truss are replaced by the dampers with low-stiffness, the increase of damping factors can be achieved at the cost of losing stiffness of the structure. It is important to get a balance between the increase of damping and the loss of stiffness of the structure.

In the active vibration control of Kagome sandwich structure, a two-step method for the optimal placement of actuators using both modal strain energy and singular value of input matrix is devised. The validity of this method is examined and verified through the eigenvalue analysis and the time-domain simulations. In this part, the stability of system under IMSC with an observer is analyzed and the method to suppress spillover is introduced. It is demonstrated that for the given modal control forces, the maximal physical control force is related to the minimal singular value of the input matrix. Through the optimal placement of actuators, an input matrix with a relatively larger minimal singular value should be realized so that a relatively smaller maximal physical control force is required to achieve the given modal forces. It is also found that MSE, which is widely used to choose the proper locations of dampers/actuators, can not be solely used to get the optimal locations of actuators in IMSC. Due to the influence of the symmetrical features in the Kagome structure and its mode shapes, an inappropriate placement of actuators just according to the magnitudes of MSE may result in the uncontrollability of the structure.

Author details

Kongming Guo and Jun Jiang*

State Key Laboratory for Strength and Vibration, Xi'an Jiaotong University, Xi'an, China

Acknowledgement

The work is supported by the National Fundamental Research Project (973) under the grant No. 2006CB601206.

6. References

[1] Torvik P J, Strickla D Z (1972) Damping additions for plates using constrained viscoelastic layers, Journal of the Acoustical Society of America, 51:985-991.

* Corresponding Author

[2] Ip K H ,Tse P C (2001) Optimal configuration of a piezoelectric patch for vibration control of isotropic rectangular plates, Smart materials and structures. 10: 395-403
[3] Corr L R, Clark W W (2002) Comparison of low-frequency piezoelectric switching shunt techniques for Structural damping, Smart materials and structures. 11:370-376.
[4] Baz A, Ro J (1996) Vibration control of plates with active constrained layer damping, Smart materials and structures. 5:272-280.
[5] dos Santos e Lucato S L, Wang J, Maxwell P, McMeeking R M , Evans A G (2004) Design and demonstration of a high authority shape morphing structure, Int. J. Solids Struct., 41: 3521–3543.
[6] Haftka R T, Adelman H M (1985) Selection of Actuator Locations for Static Shape Control of Large Space Structures by Heuristic Integer Programming, Computers and Structures , 20: 575–582.
[7] Burdisso R A ,Haftka R T (1990) Statistical analysis of static shape control in space structures AIAA Journal, 28:1504-1508.
[8] Chen G S, Bruno R J , Salama M (1991) Optimal placement of active/passive members in truss structures using simulated annealing ,AIAA Journal 29: 1327-1234.
[9] Rao S S , Pan T S , Venkayya V B (1991) Optimal Placement of Actuators in Actively Controlled Structures Using Genetic Algorithms, AIAA Journal, 29: 942–943
[10] Preumont A, Dufour J P , Malekian C (1992) Active damping by a local force feedback with piezoelectric actuators Journal of Guidance, Control and Dynamics 15:390-395
[11] Bilbao A, Avilés R, Aguirrebeitia J , Bustos I F (2009) Eigensensitivity-based optimal damper location in variable geometry trusses AIAA Journal, 47 : 576-590.
[12] Hutchinson R G , Wicks N, Evans A G , Fleck N A, Hutchinson J W (2003) Kagome plate structures for actuation, Int. J. Solids Struct , 40 :6969–6980.
[13] Guo X, Jiang J (2001) Passive vibration control of truss-cored sandwich plate with planar Kagome truss as one face plane, Science China-Technological Sciences , 54: 1113-1120.
[14] Zhang J, Zheng G T (2007) The Biot Model and Its Application in Viscoelastic Composite Structures, J of Vibration and Acoustics, 129:533-540.
[15] Xu B, Jiang J S (2004) Integrated optimization of structure and control for piezoelectric intelligent trusses with uncertain placement of actuators and sensors, Computational Mechanics 33: 406-412.
[16] Meirovitch L , Baruh H (1981) Optimal control of damped flexible gyroscopic systems, Journal of Guidance and Control, 4:157–163.
[17] Sesak J R, Likins P , Coradetti T (1979) Flexible Spacecraft Control by Model Error Sensitivity Suppression , The Journal of the Astronautical Sciences, 27: 131-156.
[18] Baruh H, Meirovitch.L (1981)On the placement of actuators in the control of distributed-parameter systems AIAA paper 81-0638: 611-619.
[19] Lindberg Jr R E, Longman R W (1984) On the number and placement of actuators for independent modal space control, Journal of Guidance and Control 7: 215-221

[20] Lammering R, Jia J H , Rogers C A (1994) Optimal placement of piezoelectric actuators in adaptive truss structures J.Sound Vib. 171: 67-85
[21] Liu Z S, Wang D J , Hu H C , Yu M (1994) Measures of modal controllability and observability in vibration control of flexible structures Journal of Guidance, Control and Dynamics 17: 1377-1380.

Permissions

The contributors of this book come from diverse backgrounds, making this book a truly international effort. This book will bring forth new frontiers with its revolutionizing research information and detailed analysis of the nascent developments around the world.

We would like to thank Mauricio Zapateiro De la Hoz and Francesc Pozo, for lending their expertise to make the book truly unique. They have played a crucial role in the development of this book. Without their invaluable contribution this book wouldn't have been possible. They have made vital efforts to compile up to date information on the varied aspects of this subject to make this book a valuable addition to the collection of many professionals and students.

This book was conceptualized with the vision of imparting up-to-date information and advanced data in this field. To ensure the same, a matchless editorial board was set up. Every individual on the board went through rigorous rounds of assessment to prove their worth. After which they invested a large part of their time researching and compiling the most relevant data for our readers. Conferences and sessions were held from time to time between the editorial board and the contributing authors to present the data in the most comprehensible form. The editorial team has worked tirelessly to provide valuable and valid information to help people across the globe.

Every chapter published in this book has been scrutinized by our experts. Their significance has been extensively debated. The topics covered herein carry significant findings which will fuel the growth of the discipline. They may even be implemented as practical applications or may be referred to as a beginning point for another development. Chapters in this book were first published by InTech; hereby published with permission under the Creative Commons Attribution License or equivalent.

The editorial board has been involved in producing this book since its inception. They have spent rigorous hours researching and exploring the diverse topics which have resulted in the successful publishing of this book. They have passed on their knowledge of decades through this book. To expedite this challenging task, the publisher supported the team at every step. A small team of assistant editors was also appointed to further simplify the editing procedure and attain best results for the readers.

Our editorial team has been hand-picked from every corner of the world. Their multi-ethnicity adds dynamic inputs to the discussions which result in innovative outcomes. These outcomes are then further discussed with the researchers and contributors who give their valuable feedback and opinion regarding the same. The feedback is then collaborated with the researches and they are edited in a comprehensive manner to aid the understanding of the subject.

Apart from the editorial board, the designing team has also invested a significant amount of their time in understanding the subject and creating the most relevant covers. They scrutinized every image to scout for the most suitable representation of the subject and create an appropriate cover for the book.

The publishing team has been involved in this book since its early stages. They were actively engaged in every process, be it collecting the data, connecting with the contributors or procuring relevant information. The team has been an ardent support to the editorial, designing and production team. Their endless efforts to recruit the best for this project, has resulted in the accomplishment of this book. They are a veteran in the field of academics and their pool of knowledge is as vast as their experience in printing. Their expertise and guidance has proved useful at every step. Their uncompromising quality standards have made this book an exceptional effort. Their encouragement from time to time has been an inspiration for everyone.

The publisher and the editorial board hope that this book will prove to be a valuable piece of knowledge for researchers, students, practitioners and scholars across the globe.

List of Contributors

Samuel da Silva, Vicente Lopes Junior and Michael J. Brennan
UNESP - Univ Estadual Paulista, Faculdade de Engenharia de Ilha Solteira, Departamento de Engenharia Mecânica, Av. Brasil 56, Centro, Ilha Solteira, SP, Brasil

Xingiian Jing
Department of Mechanical Engineering, Hong Kong Polytechnic University, Hung Hom, Kowloon, Hong Kong

Pablo Ballesteros, Christian Bohn, Wiebke Heins and Xinyu Shu
Institute of Electrical Information Technology, Clausthal University of Technology, Clausthal-Zellerfeld, Germany

Pablo Ballesteros, Xinyu Shu,Wiebke Heins and Christian Bohn
Institute of Electrical Information Technology, Clausthal University of Technology, Clausthal-Zellerfeld, Germany

Grzegorz Tora
Institute of Machine Design, Faculty of Mechanical Engineering, Cracow University of Technology, Cracow, Poland

Hamid Reza Karimi
Department of Engineering, Faculty of Engineering and Science, University of Agder, Grimstad, Norway

Mauricio Zapateiro and Francesc Pozo
Department of Applied Mathematics III, Universitat Politècnica de Catalunya, Barcelona, Spain

Ningsu Luo
Institute of Informatics and Applications, University of Girona, Girona, Spain

Tore Bakka and Hamid Reza Karimi
University of Agder, Norway

Andrés Blanco-Ortega, Jorge Colín-Ocampo, Marco Oliver-Salazar, Gerardo Silva-Navarro and Gerardo Ve-la-Valdés
Centro Nacional de Investigación y Desarrollo Tecnológico, CENIDET, México
Centro de Investigación y de Estudios Avanzados del IPN, CINVESTAV, México

Hao Chen
Institute of Engineering Mechanics, China Earthquake Administration, Sanhe, Hebei, China

Zhi Sun and Limin Sun
State Key Laboratory for Disaster Reduction in Civil Engineering, Tongji University, Shanghai, China

Bong-Jo Ryu and Yong-Sik Kong
Hanbat National University / Hanwha L&C, South Korea

Kongming Guo and Jun Jiang
State Key Laboratory for Strength and Vibration, Xi'an Jiaotong University, Xi'an, China

Printed in the USA
CPSIA information can be obtained
at www.ICGtesting.com
JSHW011457221024
72173JS00005B/1108

9 781632 405159